Western Wetland Flora

An Introduction to the Wetland and Aquatic Plants
of the Western United States

Mitella pentandra
FIVE-POINT BISHOP'S-CAP

STEVE W. CHADDE

Portions of this book were first published in 1991 as *Western Wetland Flora: Field Office Guide to Plant Species,* prepared for the USDA Soil Conservation Service (now Natural Resources Conservation Service) by Biotic Consulatnts, Inc. Robert Mohlenbrock prepared the descriptions, Mark Mohlenbrock prepared the illustrations. This new volume updates plant nomenclature to reflect recent taxonomic changes, replaces English measurement units with metric units, incorporates new distribution maps, and includes a number of additional species, especially recent invaders of western wetland and aquatic habitats.

WESTERN WETLAND FLORA
An Introduction to the Wetland and Aquatic Plants of the Western United States

Steve W. Chadde

Printed in the United States of America

ISBN: 9781951682378

Grateful acknowledgment is given to the Biota of North America Program (*www.bonap.org*) for permission to use their data to generate the distribution maps.

The author can be reached via email: *steve@chadde.net*

Ver. 2.0 (04/25/2020)

Introduction 11
Key to the Groups 13
Species Descriptions 16

GROUP 1. AQUATIC HERBS

GROUP 2. FERNS AND FERN-RELATIVES

Contents

GROUP 3. POACEAE (GRASS FAMILY)

GROUP 4. CYPERACEAE (SEDGE FAMILY)

Contents

GROUP 5. OTHER MONOCOTS (NON-CYPERACEAE)

Contents

GROUP 7. DICOT HERBS WITH MOST OR ALL OF THE LEAVES SIMPLE, AND OPPOSITE OR WHORLED

GROUP 8. DICOT HERBS WITH ALL THE LEAVES SIMPLE AND BASAL AND/OR ALTERNATE

Contents

GROUP 9. TREES, SHRUBS, WOODY VINES

Parnassia palustris - NORTHERN GRASS-OF-PARNASSUS

WETLANDS in the western United States are found in a wide range of settings, and their plant composition is determined by many factors including elevation, soils, climate, topography, water chemistry, etc. The resulting flora can be quite complex, and is made up of many, at first glance, similar-appearing species. The goal of *Western Wetland Flora* is to provide an introduction, in non-technical terms, to the aquatic and wetland plants of this region (generally, lands west of 100 degrees longitude), and to allow those with little or no botanical background to identify the plants found in lakes, ponds, rivers, marshes, swamps, bogs, etc. The *Flora* describes over 300 vascular plant species, both native and introduced, and a number of rapidly spreading invasive wetland plants, such as Kariba-weed (*Salvinia molesta*). Each species is illustrated with a line-drawing, a detailed description, and a county-level distribution map (counties having a verified occurrence of that plant). Together, these provide useful tools for identifying unknown plant specimens.

To aid identification, plants are placed into one of nine groups (such as aquatic plants, ferns, grasses, sedges, etc.; see **Key, page 13**). Within each group, plants are listed in alphabetical order by their scientific name. Users of the *Flora* should first use this Key, then turn to the group to identify an unknown plant. If the plant family is known, see page 331 for an index of all species described for a particular plant family. As the area treated in this work is vast, and as wetlands support a diverse flora, many additional plants not covered here are present in the region's wetland and aquatic environments. However, unknown plants may at least be identified to the correct family or genus by finding a similar-looking plant in this guide. Then, a more technical reference can be consulted.

Familiarity with the Group Key and characteristics of major plant families will greatly facilitate identification. Group Keys begin by separating seed-producing plants (Angiosperms) from those reproducing by spores (Ferns and Fern Allies). Angiosperms are further divided into Dicots and Monocots. Dicots include many familiar trees, shrubs, and "wildflowers", such as those of the Aster Family; Monocots include the grass, sedge, rush, and orchid families, among others. Woody plants (wetland trees, shrubs and woody vines) form another group.

Nomenclature of plant families, genera, and species generally follows that of the published volumes of *The Flora of North America* series (1993+), the *Synthesis of the North American Flora* (*www.bonap.org*), and *The Plant List*, a collaboration between the Royal Botanic Gardens (Kew), and the Missouri Botanical Garden (*www.theplantlist.org*).

A standard format is used to describe each species: scientific name, common name, plant family, general flowering period, habitat where most typically occurs, habit (annual, perennial, herb, grass, etc.; also whether native to the region or introduced), stem and leaf characteristics, features of the flowers and fruit, and wetland status (next page). Common synonyms (other formerly accepted scientific names) are listed. A regional map shows the verified presence or absence of each species within the western United States.

Wetland indicator status ratings are widely used in wetland delineation studies, and also provide information on each species' typical soil moisture regime. Five categories are defined: OBL, FACW, FAC, FACU and UPL (see the home of the National Wetland Plant List, *www.wetland-plants.usace.army.mil* for more information):

- **OBL** (OBLIGATE WETLAND): Plants that almost always occur in wetlands (i.e. almost always in standing water or seasonally saturated soils.
- **FACW** (FACULTATIVE WETLAND): Plants that usually occur in wetlands, but may occur in non-wetlands.
- **FAC** (FACULTATIVE): Plants that occur in wetlands and non-wetland habitats.
- **FACU** (FACULTATIVE UPLAND): Plants that usually occur in non-wetlands but may occur in wetlands.
- **UPL** (OBLIGATE UPLAND): Plants that almost never occur in wetlands (or in standing water or saturated soils).

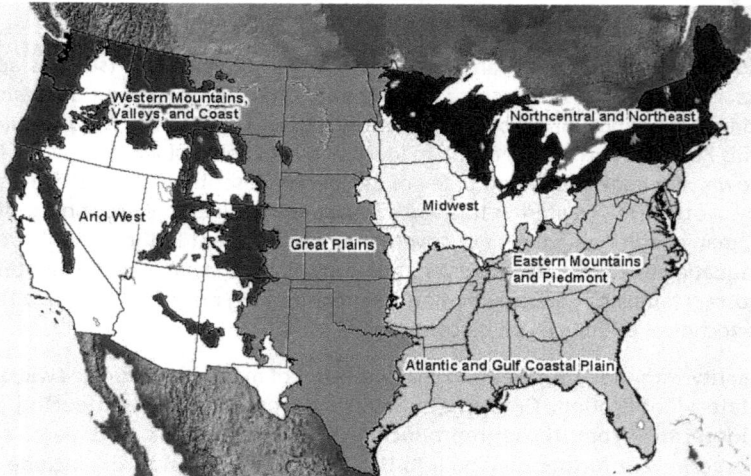

The western United States contain portions of three regions:
- **AW** Arid West
- **GP** Great Plains
- **WMV** Western Mountains, Valleys, and Coast

1 Ferns or related plants, reproducing by spores, not seeds **GROUP 2.**
.... **FERNS AND FERN-RELATIVES**, p. 307

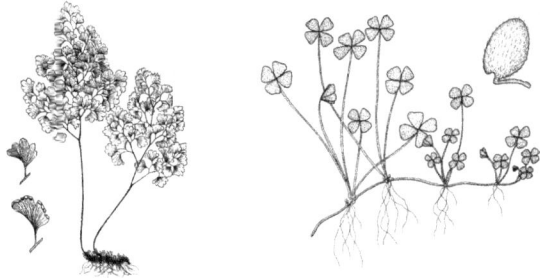

1 Plants reproducing by flowers and seeds **2**

2 Plants woody, either trees, shrubs, or vines .... **GROUP 9.**
.... **TREES, SHRUBS, WOODY VINES**, p. 307

2 Plants herbaceous, stems not woody **3**

3 Plants true aquatics; leaves submersed or floating on water surface
.... **GROUP 1. AQUATIC HERBS**, p. 16

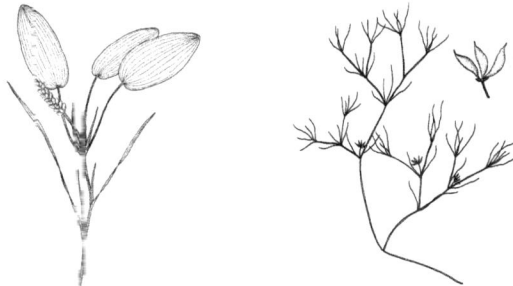

3 Plants not true aquatics; leaves not all submersed or floating on ..
water surface **4**

4 Monocots; leaves usually with parallel veins; petals and sepals, if
present, usually in 3's or multiples of 3's **5**

4 Dicots; leaves usually with net veins; petals and sepals usually in 4's or 5's or multiples of 4's or 5's **7**

5 Flowers with sepals and petals (green or brown or black in *Juncus*), the flowers not arranged in spikelets ...... **GROUP 5. MONOCOTS** .... **(OTHER THAN GRASSES OR SEDGES), p. 134**

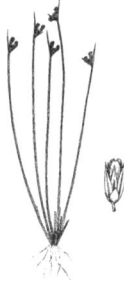

5 Flowers without sepals and petals; each flower subtended by one or more scales and borne in spikelets **6**

6 At least one or more scales at the base of each spikelet not subtend ing a flower; stems never triangular .... **GROUP 3. POACEAE** .... **(GRASS FAMILY), p. 63**

6 All the scales of a spikelet subtending a flower; stems sometimes triangular .... **GROUP 4. CYPERACEAE (SEDGE FAMILY), p. 101**

7 At least some of the leaves compound; that is, divided into distinct leaflets . . . . **GROUP 6, p. 165**

7 None of the leaves compound, although they may be deeply divided **8**

8 Most or all of the leaves opposite or whorled . . . . **GROUP 7, p. 197**

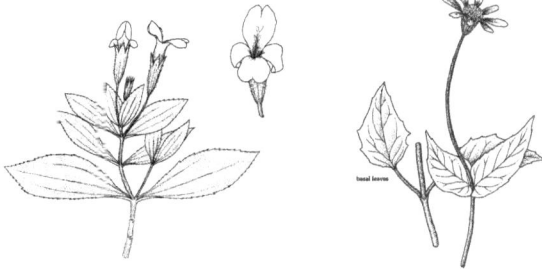

8 Most or all of the leaves alternate and/or basal . . **GROUP 8, p. 235**

gelatinous coating

> **FIELD NOTES** Leaves elliptic to oval with the leaf stalk attached to the center of the blade underside. Lower surface of the leaves and the stem conspicuously covered by a gelatinous material.

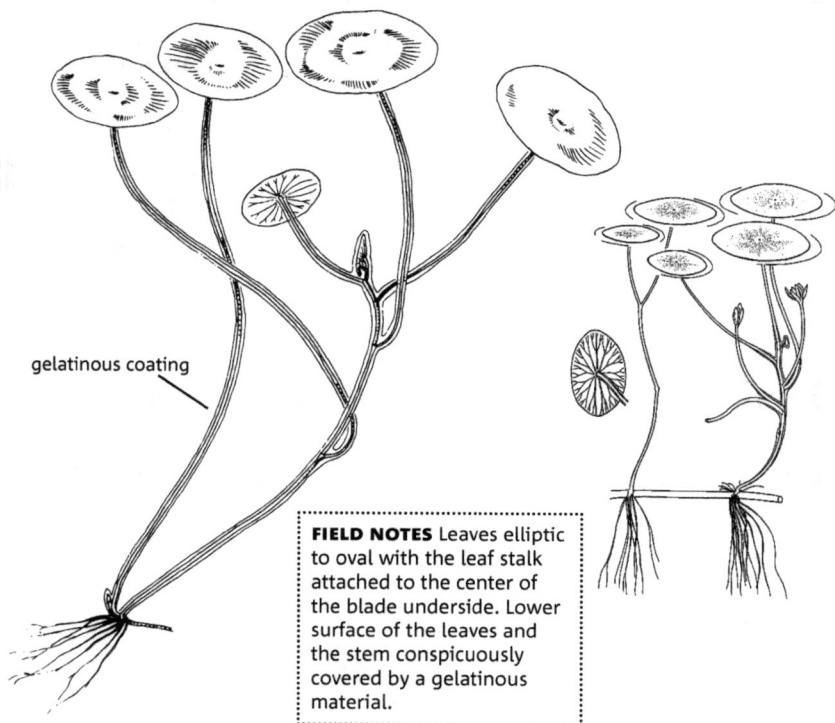

Brasenia schreberi J. F. Gmel.
WATERSHIELD

FAMILY Cabombaceae (Watershield)

FLOWERING June–September

HABITAT Ponds, lakes, streams.

HABIT Native, aquatic perennial herb, with slender, creeping rhizomes.

STEMS Spreading, much-branched, to 2 m long, covered with a gelatinous material.

LEAVES Alternate, floating, elliptic to oval, to 13 cm long, smooth, without teeth but sometimes with a slightly wavy edge, the lower surface covered by a gelatinous material; leaf stalks often longer than the blades, attached to the middle of the blade underside.

FLOWERS Borne singly from the axils of the leaves, dull red to purple, to 40 mm long, borne on stout stalks as long as the leaf stalks. **Sepals** 3-4, free from each other, green or red, lanceolate, to 20 mm long. **Petals** 3–4, free from each other, dull red to purple, lanceolate, to 20 mm long. **Stamens** 12–18, of 2 different lengths. **Pistils**: Several, free from each other, the ovaries superior.

FRUIT Club-shaped, leathery, to 13 mm long, several in a cluster, each containing 1-2 seeds, each fruit tipped by the persistent style.

NOTE The seeds are eaten by waterfowl.

WETLAND STATUS AW OBL | GP OBL | WMV OBL

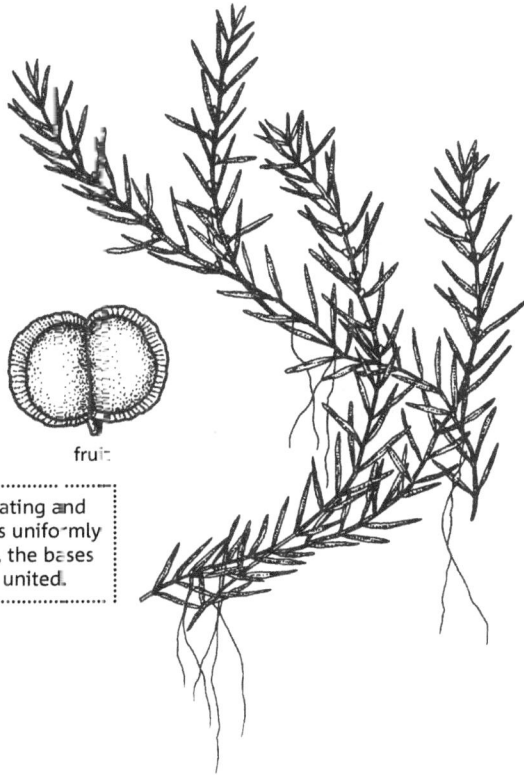

fruit

> **FIELD NOTES** Floating and submersed leaves uniformly linear-lanceolate, the bases of the leaves not united.

Callitriche hermaphroditica L.
AUTUMNAL WATER-STARWORT

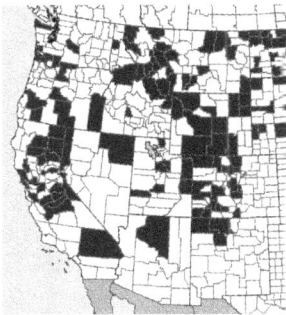

FAMILY Plantaginaceae (Plantain)

FLOWERING June–August

HABITAT In streams and edges of ponds, usually in shallow water.

HABIT Native perennial aquatic herb with slender rhizomes.

STEMS Slender, mat-forming, to 40 cm long, rooting at the nodes, smooth.

LEAVES Opposite, simple, the submersed and floating leaves uniformly linear-lanceolate, to 13 mm long, to 2 mm wide, pointed or rounded at the tip, smooth, without teeth, the bases not united.

FLOWERS 1–3 in the axils of the leaves, the male and female flowers borne separately but on the same plant. **Sepals** absent. **Petals** absent. **Stamens** 1. **Pistils**: Ovary superior; styles 2.

FRUIT Round in outline, 4-lobed, winged, notched at the tip, to 2 mm long, smooth, with the recurved style persistent.

NOTE This species, when growing in mats, provides cover for fish. Formerly placed in Callitrichaceae (Water-starwort Family).

WETLAND STATUS AW OBL | GP OBL | WMV OBL

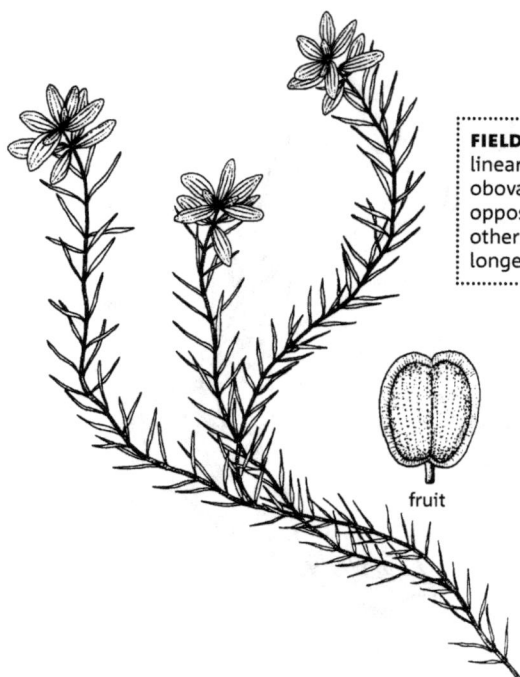

> **FIELD NOTES** Submersed leaves linear-lanceolate; floating leaves obovate to spatulate, with the opposite leaves attached to each other at their base. Fruit slightly longer than wide.

fruit

Callitriche palustris L.
SPINY WATER-STARWORT

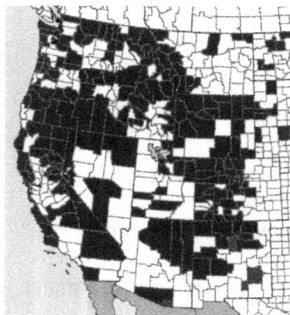

FAMILY Plantaginaceae (Plantain)

FLOWERING May–August

HABITAT In streams, roadside ditches, and edges of ponds and lakes, usually in shallow water.

HABIT Native perennial aquatic herb with slender rhizomes.

STEMS Slender, sometimes mat-forming, to 30 cm long, rooting at the nodes, smooth.

LEAVES Opposite, simple, smooth, without teeth, united at the base, the submersed leaves linear to linear-lanceolate, to 13 mm long, to 8 mm wide, with 1 vein, the floating leaves obovate to spatulate, to 13 mm long, to 8 mm wide, with 3 veins.

FLOWERS 1–3 in the axils of the leaves, the male and female flowers borne separately but on the same plant. **Sepals** absent. **Petals** absent. **Stamens** 1. **Pistils:** Ovary superior; styles 2.

FRUIT Ovoid, 4-lobed, winged, notched at the tip, to 2 mm long, a little longer than wide, smooth, with a recurved persistent style.

NOTE European Water-Starwort (*Callitriche stagnalis* Scop.) is an introduced, potentially invasive species found mostly in Washington, Oregon and northern California. Its fruit is more nearly round and slightly large than the otherwise similar *C. palustris*. Genus formerly placed in Callitrichaceae (Water-starwort Family).

SYNONYMS *Callitriche verna* L.

WETLAND STATUS AW OBL | GP OBL | WMV OBL

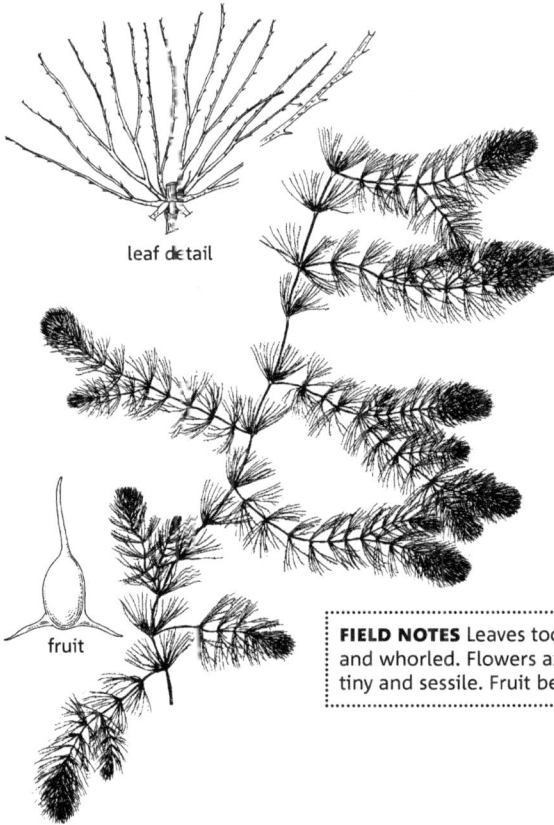

leaf detail

fruit

> **FIELD NOTES** Leaves toothed
> and whorled. Flowers axillary,
> tiny and sessile. Fruit beaked.

Ceratophyllum demersum L.
COMMON HORNWORT

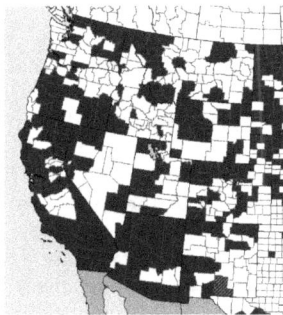

FAMILY Ceratophyllaceae (Hornwort)
FLOWERING July–September
HABITAT Lakes, ponds, slow streams.
HABIT Native perennial aquatic herb.
STEMS Much branched, elongated, smooth, varying in
length according to the depth of the water.
LEAVES Whorled, sessile, up to 40 mm long, smooth,
divided usually into 3-forked, nearly thread-like
segments, the segments flattened and toothed.
FLOWERS Solitary in the axils of the leaves, the male
and female borne separately, each subtended by an 8-
to 12-cleft bract. **Sepals** absent. **Petals** absent. **Stamens**
10–20. **Pistils:** one, the ovary superior.
FRUIT Achenes ellipsoid, flattened, smooth, wingless, the body 4–6 mm long, with 2
basal spurs up to 4 mm long and a beak 3–4 mm long.
NOTE The length and texture of the stems, the degree of toothing on the leaves, and
the characters of the fruit are all variable. The fruits are eaten by ducks.
WETLAND STATUS AW OBL | GP OBL | WMV OBL

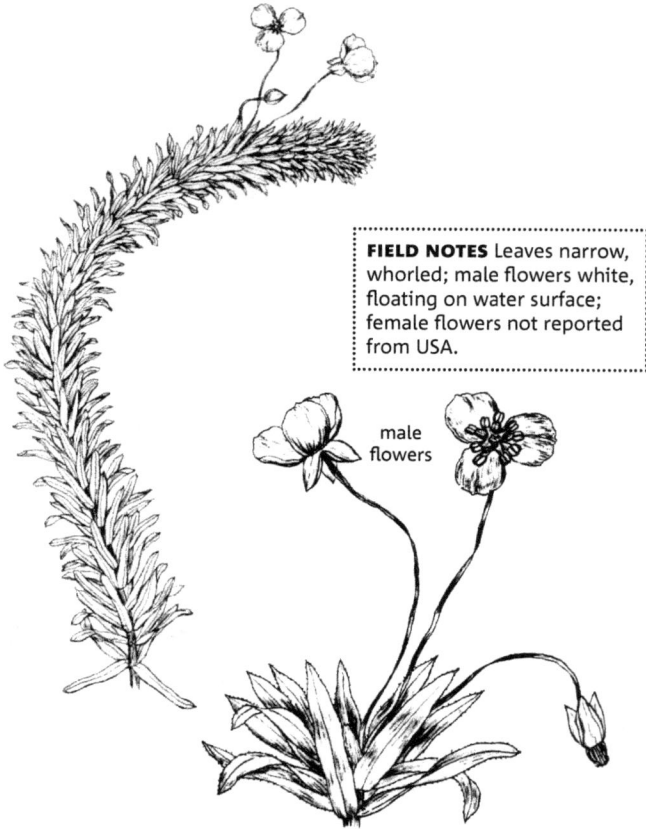

> **FIELD NOTES** Leaves narrow, whorled; male flowers white, floating on water surface; female flowers not reported from USA.

male flowers

Egeria densa Planch.
BRAZILIAN-WATERWEED

FAMILY Hydrocharitaceae (Tape-grass)

FLOWERING July–September.

HABITAT Occasionally established in ponds, pools, quiet streams and sloughs.

HABIT Introduced aquatic herb; plants submersed.

LEAVES in whorls of 4–6, the principal leaves 2–3.3 cm long, linear-lanceolate, tapering gradually to a finely toothed point; the lower leaves remote, the upper ones crowded.

FLOWERS at anthesis 15–20 mm wide; the staminate spathes with 2 or more exserted flowers with relatively showy white **petals** 9–11 mm long and 6–9 mm wide; stamens 9; the pistillate flowers not known in United States.

NOTE Staminate plants commonly cultivated in aquaria. Native of Argentina. Similar in appearance to *Elodea canadensis* but plants generally larger, and leaves in whorls of 4–6 versus whorls of 3 in *Elodea canadensis*.

SYNONYMS *Elodea densa* (Planch.) Casp.

WETLAND STATUS AW OBL | GP OBL | WMV OBL

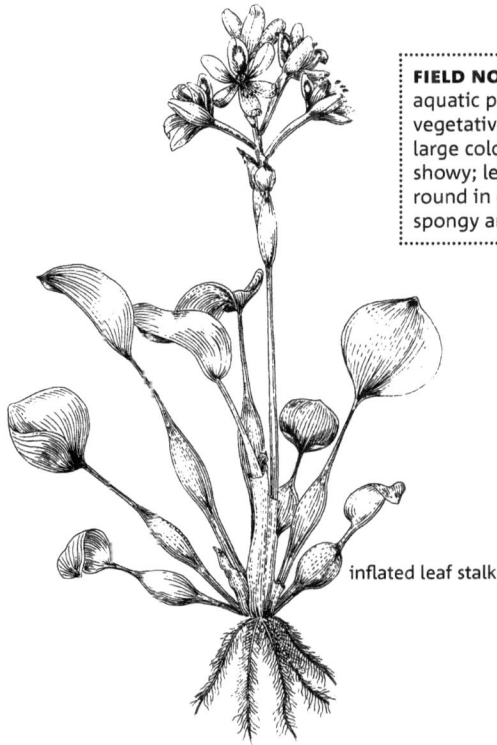

FIELD NOTES Floating aquatic plant, spreads vegetatively and may form large colonies; flowers showy; leaves often nearly round in outline; leaf stalk spongy and often inflated.

inflated leaf stalk

Eichhornia crassipes (Mart.) Solms
COMMON WATER-HYACINTH

FAMILY Pontederiaceae (Pickerelweed)

FLOWERING July-October (longer in warmer regions)

HABITAT Ditches, quiet streams, rivers and waterways, lakes, and ponds; usually very prolific and troublesome, often clogging and obstructing waterways.

HABIT Introduced, perennial aquatic herb.

STEMS stout, erect, often connected by stolons, rooting at the nodes, with long black pendant roots.

LEAVES forming a basal rosette, leaf stalks spongy and-often greatly inflated; leaf blades ovate, round or heart-shaped, 4–12 cm wide.

FLOWERS in a contracted panicle, 4–15 cm long, with several flowers. **Perianth** funnel-shaped; lilac, bluish-purple, or white, the upper lobe bearing a violet blotch with a yellow center. **Stamens** 6, unequal in length; **ovary** 3-chambered. Stalk of the inflorescence soon becoming goose-necked, forcing the dead flowers under the water.

FRUIT a many-seeded capsule.

NOTE Widely naturalized worldwide in warm regions (native to South America). Plants spread rapidly via vegetative means; considered world's most troublesome aquatic weed, and first reported in the USA in 1884.

WETLAND STATUS AW OBL | GP OBL | WMV OBL

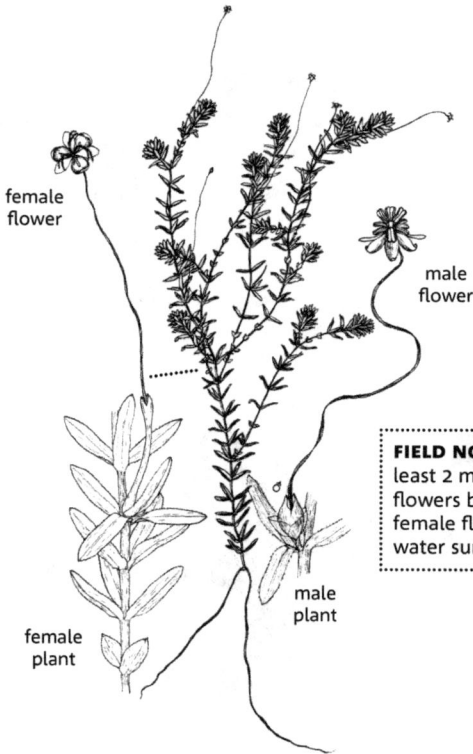

female
flower

male
flower

FIELD NOTES Most leaves at least 2 mm inch wide. Male flowers borne on stalks; female flowers reaching water surface.

male
plant

female
plant

Elodea canadensis Michx.
CANADIAN WATERWEED

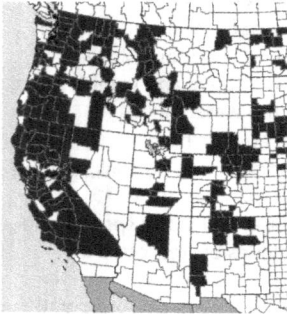

FAMILY Hydrocharitaceae (Tape-grass)
FLOWERING July–September
HABITAT Quiet water.
HABIT Native, aquatic perennial herb with thread-like stolons and either male or female flowers.
STEMS Slender, smooth, with equal branching.
LEAVES Upper and middle leaves in whorls of 3, the lower opposite, those of the male plants linear or linear-lanceolate, pointed at the tip, to 13 mm long, to 4 mm wide, those of the female plants broadly lanceolate, rounded or slightly pointed at the tip, to 15 mm long, to 6 mm wide.

FLOWERS Male and female flowers borne on separate plants from the axils of the leaves; **male flowers:** sepals 3, green, to 6 mm long; petals 3, white, to 6 mm long; stamens 9; **female flowers:** sepals 3, united at base to form a tube, green, to 2 mm long; petals 3, united at base to form a tube, white, to 3 mm long; ovary inferior; **stigmas** 3.
FRUIT Ovoid, smooth, to 6 mm long.
NOTE The tube of the female flowers becomes elongated so that the lobes of the sepals and petals reach the surface of the water. Reproduction mostly by detaching plant fragments, and also by winter buds ("turions"). *Elodea* is popular for use in aquaria.
WETLAND STATUS AW OBL | GP OBL | WMV OBL

water surface

female flower

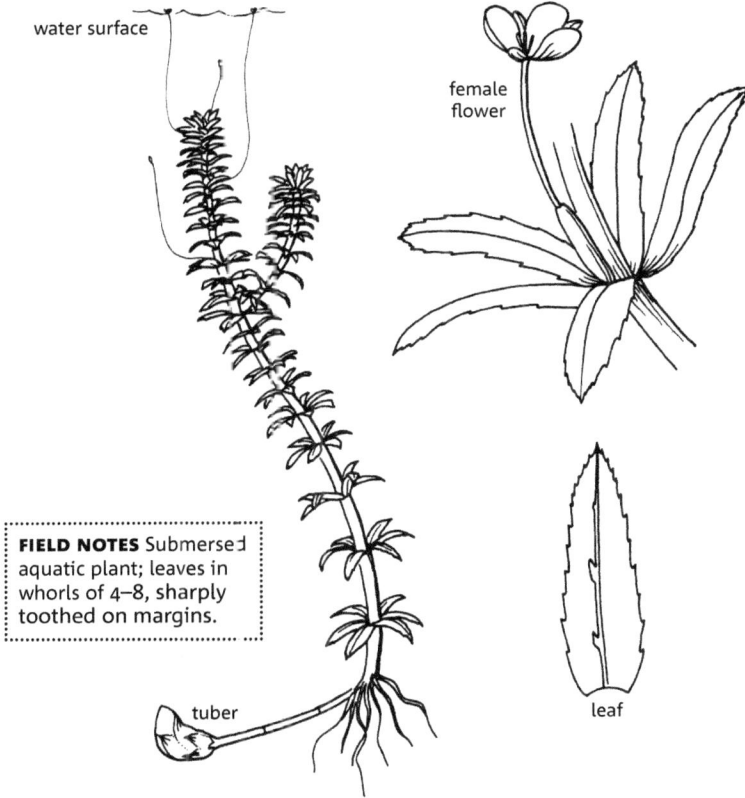

FIELD NOTES Submersed aquatic plant; leaves in whorls of 4–8, sharply toothed on margins.

tuber

leaf

Hydrilla verticillata (L. f.) Royle
WATER-THYME, FLORIDA ELODEA

FAMILY Hydrocharitaceae (Tape-grass)

FLOWERING June–August

HABITAT In canals, ponds, lakes and streams. It is strictly a submersed plant and cannot withstand extensive drying.

HABIT Introduced, submersed perennial herb, rooted to the bottom, with long, branching stems to 8 m long. The stems may break loose and form floating mats.

LEAVES Generally 4–8 per node, stalkless, 1–2 cm long, to 2 mm wide, oblong, coarsely sharp-toothed, tip generally sharp-pointed; usually with several prickles along midrib of leaf underside.

FLOWERS arise singularly from the spathe, and are found at or near the water surface, and from near the growing tip; flowers inconspicuous, measuring only 4–5 mm across the tip of a threadlike pedicel. **Male plants** rare; thus, seed formation is poor if it occurs at all. **Female flowers** with 6 **petals**, to 2 mm long; **stamens** 3, **stigmas** 3.

FRUIT Broadly ovoid, 5–6 mm long.

NOTE Native to Eurasia and potentially invasive.

WETLAND STATUS AW OBL | GP OBL | WMV OBL

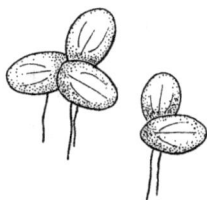

> **FIELD NOTES** Body broadly oblong to nearly spherical, obscurely 3-nerved. Tip of the rootlet rounded.

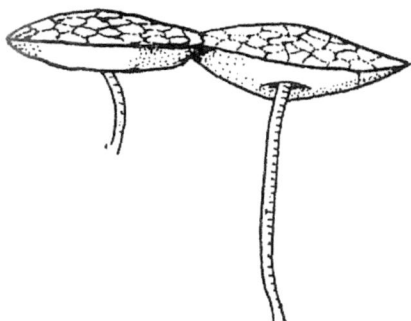

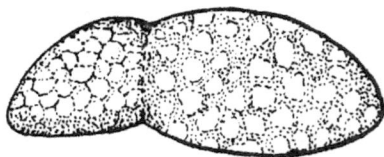

Lemna minor L.
COMMON DUCKWEED

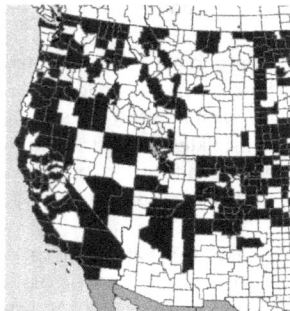

FAMILY Araceae (Arum)

FLOWERING July–August

HABITAT Quiet water, ponds.

HABIT Tiny, floating, native aquatic herb without true stems and leaves; tip of solitary rootlet rounded.

LEAVES Body (thallus or frond) obovate to elliptic, to 2 mm long, nearly as wide, symmetrical or slightly asymmetrical, flattened to weakly inflated, obscurely 3-nerved; upper surface usually slightly convex; lower surface flattened or weakly to moderately convex, pale green or occasionally reddish purple.

FLOWERS Microscopic, borne singly in pouches. **Sepals** absent. **Petals** absent. **Stamens** 1. **Pistils:** 1.

FRUIT Symmetrical, broadly ovoid; seeds longitudinally ribbed.

NOTE The plant may be solitary or more commonly clumped together in groups of 2–4.

WETLAND STATUS AW OBL | GP OBL | WMV OBL

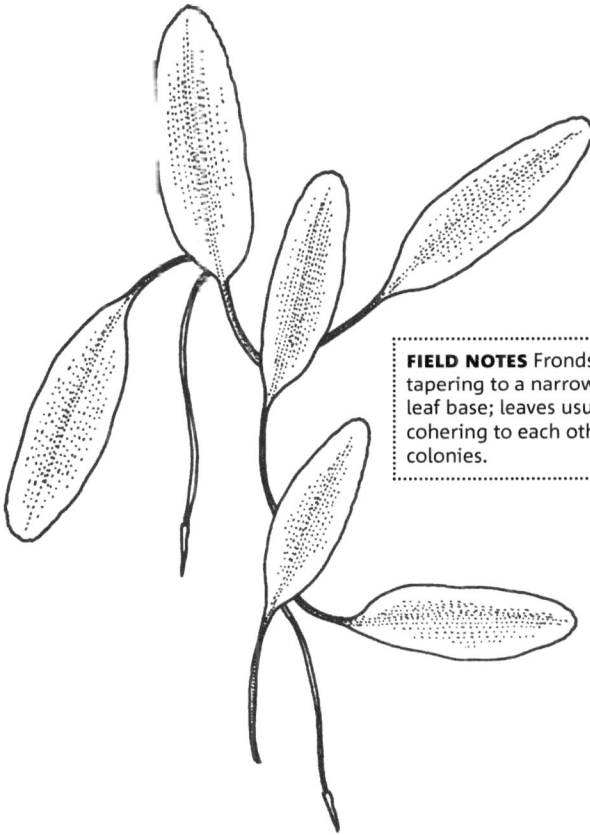

> **FIELD NOTES** Fronds elliptic, tapering to a narrow stalk at leaf base; leaves usually cohering to each other in colonies.

Lemna trisulca L.
STAR DUCKWEED

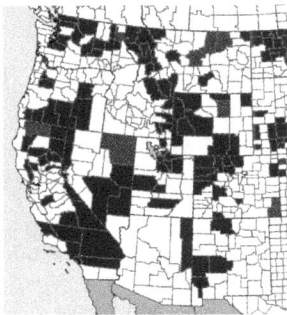

FAMILY Araceae (Arum)

FLOWERING June–October

HABITAT Quiet waters of streams, lakes, ponds, and ditches.

HABIT Native floating aquatic plant, the fronds forming colonies of many individuals; one root per frond or absent.

STEMS Absent.

LEAVES Fronds elliptic, to 13 mm long, to 6 mm wide, with a narrow stalk at the base, flat, with 3 obscure veins, very finely toothed.

FLOWERS Rarely seen; when present, 2–3 in a microscopic pouch. **Sepals** absent. **Petals** absent. **Stamens** 1 per flower. **Pistils:** 1 per flower.

FRUIT Rarely seen; when present, 1-seeded.

NOTE Although this species flowers and sets seeds on occasion, it usually reproduces asexually with new fronds developing on either side of the parent frond. The fronds are eaten by waterfowl.

WETLAND STATUS AW OBL | GP OBL | WMV OBL

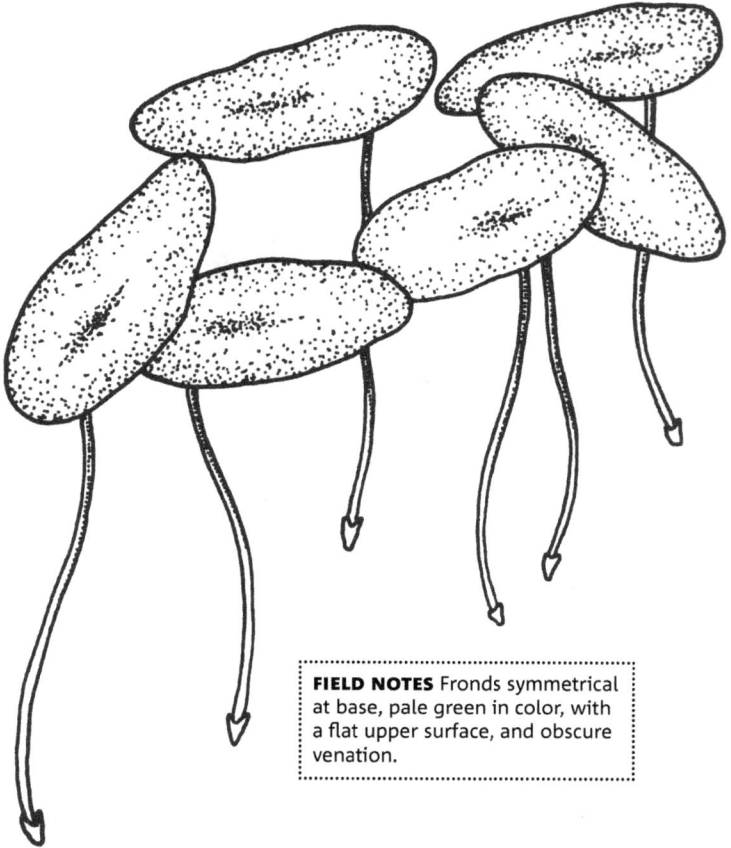

> **FIELD NOTES** Fronds symmetrical at base, pale green in color, with a flat upper surface, and obscure venation.

Lemna valdiviana Philippi
PALE DUCKWEED

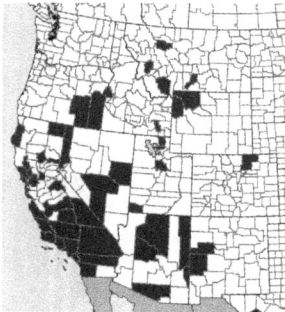

FAMILY Araceae (Arum)

FLOWERING June–September

HABITAT Quiet waters of lakes, ponds, and streams.

HABIT Native floating plant on the surface of the waters.

STEMS absent.

LEAVES Single fronds, or fronds in colonies of to 10, to 5 mm long, oblong to elliptic, usually asymmetrical at the base, flat, pale green, with obscure veins.

FLOWERS Rarely found; if present, male and female flowers borne in pouches known as spathes; male flowers 2 per pouch. **Sepals** absent. **Petals** absent. **Stamens** 1. **Pistils:** Ovary superior.

FRUIT Rarely seen, but not winged.

NOTE This duckweed is eaten by waterfowl.

WETLAND STATUS AW OBL | GP OBL | WMV OBL

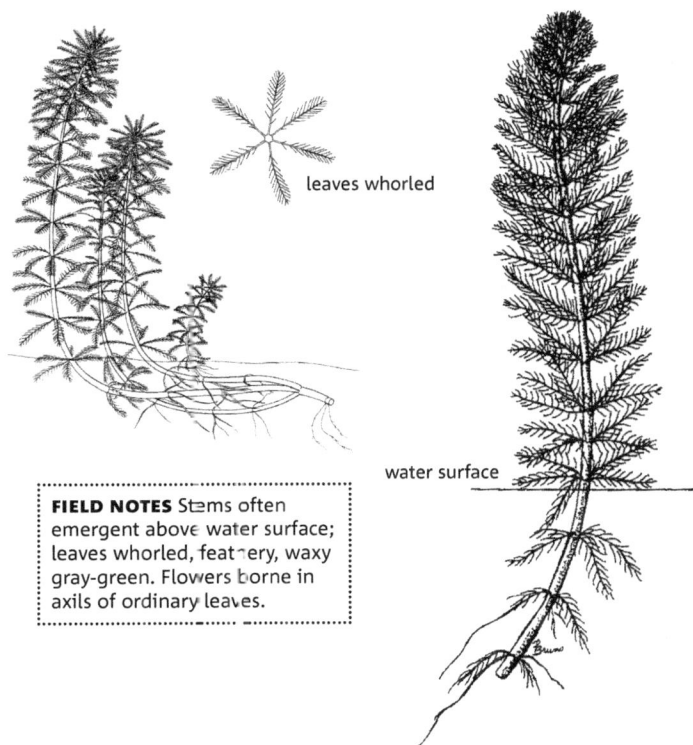

leaves whorled

water surface

> **FIELD NOTES** Stems often emergent above water surface; leaves whorled, feathery, waxy gray-green. Flowers borne in axils of ordinary leaves.

Myriophyllum aquaticum (Vell.) Verdc.
PARROT'S-FEATHER

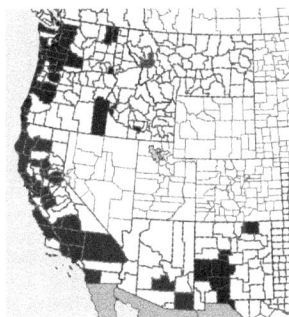

FAMILY Haloragaceae (Water-milfoil)
FLOWERING June–September
HABITAT Shallow water and creeping onto shores.
HABIT Aquatic perennial, often with a reddish hue.
STEMS Much branched, sometimes reddish, smooth, stout, bearing many leaves. The emergent stems and leaves are distinctive, growing to about 30 cm above the water surface, resembling miniature fir trees.
LEAVES Whorled, to 7 cm long, with 10–25 thread-like segments on each side, smooth, waxy blue- or gray-green in color.
FLOWERS Male and female flowers borne separately and on different plants in the axils of ordinary leaves on emergent inflorescences; each flower subtended by a 2- or 3-cleft thread-like bracteole. **Sepals** absent. **Petals** usually absent. **Stamens** 4. **Pistils:** Ovary superior, 4-lobed, smooth.
FRUIT Nearly spherical, smooth, to 3 mm in diameter, with a granular texture.
NOTE May become an aggressive weed in shallow bodies of water, and a popular aquarium plant. In North America, apparently only female plants occur and no seeds are produced; reproduction being asexual (from fragments of already rooted plants).
SYNONYMS *Myriophyllum brasiliense* Camb.
WETLAND STATUS AW OBL | GP OBL | WMV OBL

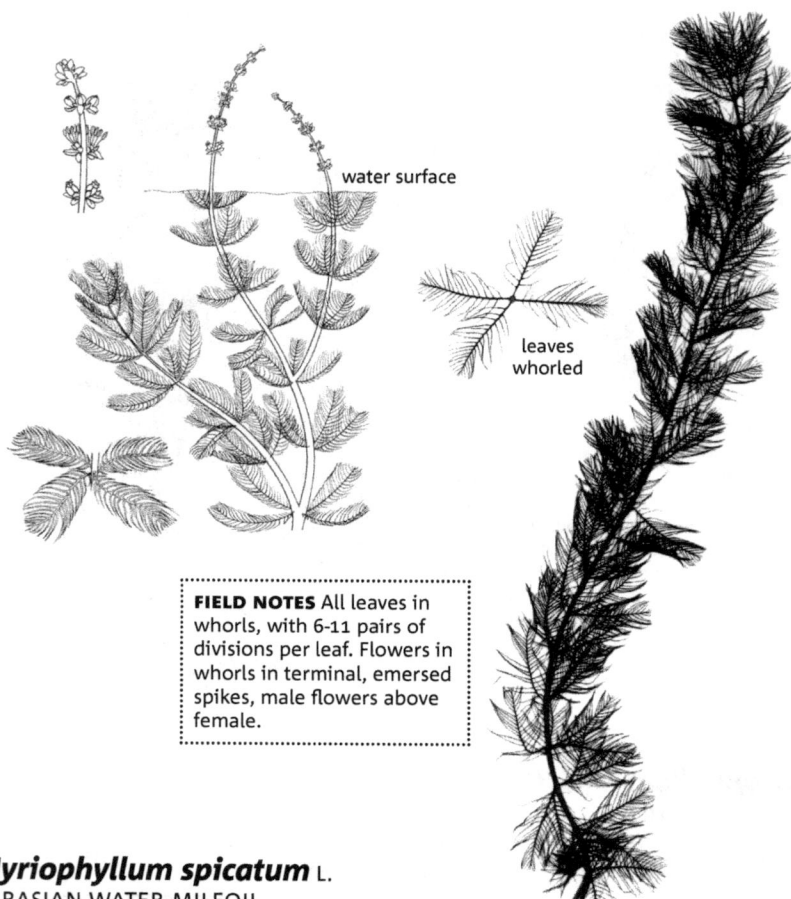

water surface

leaves
whorled

FIELD NOTES All leaves in
whorls, with 6-11 pairs of
divisions per leaf. Flowers in
whorls in terminal, emersed
spikes, male flowers above
female.

Myriophyllum spicatum L.
EURASIAN WATER-MILFOIL

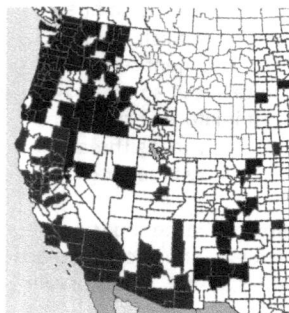

FAMILY Haloragaceae (Water-milfoil)
FLOWERING July–September
HABITAT Quiet water of lakes, ponds, and rivers.
HABIT Introduced perennial aquatic herb.
STEMS Branched or unbranched, usually red-brown but
drying white, smooth, to 1 m long.
LEAVES All in whorls, each leaf divided into 6–11 pairs
of linear segments, smooth, each leaf to 4 cm long.
FLOWERS Borne in whorls and crowded into terminal
spikes, each spike up to 10 cm long, subtended by ovate
to oblong, toothed bracts, the upper part of the spike
with male flowers only, the lower part with female flow-
ers only. **Sepals** 4, green, united below, minute. **Petals** 4, green, free from each other,
minute, or absent. **Stamens** 8. **Pistils:** Ovary superior, smooth.
FRUIT Spherical, with 4 grooves, smooth or warty, to 3 mm long.
NOTE The nutlets may be eaten by waterfowl. This Eurasian species is invasive in lakes
and slow-moving rivers, posing a threat to native aquatic species.
WETLAND STATUS AW OBL | GP OBL | WMV OBL

fruit

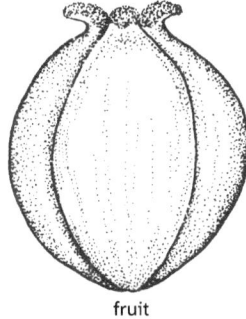

> **FIELD NOTES** Leaves, flowers, and bracts all arranged in whorls; the bracts about as long as the flowers, and deeply divided but not spiny.

Myriophyllum verticillatum L.
WHORLED WATER-MILFOIL

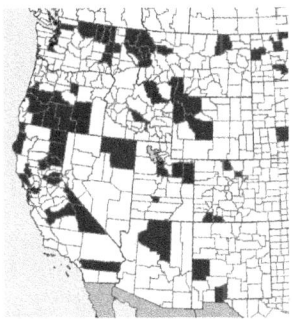

FAMILY Haloragaceae (Water-milfoil)

FLOWERING June–September

HABITAT Ponds, lakes, marshes, usually in water.

HABIT Native perennial aquatic herb with rhizomes, usually forming asexual "bulbs" (turions.)

STEMS Rather stout, mat-forming, branched or unbranched.

LEAVES All in whorls of 4 or 5, to 5 cm long, pinnately divided into 8–13 thread-like segments, smooth.

FLOWERS 3–6 in clusters in the axils of the leaves, the male and female flowers usually borne separately but on the same plant, each cluster of flowers subtended by pinnately lobed bracts about as long as the flowers. Sepals 4, green, united below, to 2 mm long. Petals 4, in the male flowers, united below, greenish, to 2 mm long, absent in the female flowers. Stamens 8. Pistils: Ovary interior; stigmas 4.

FRUIT Spherical, to 3 mm in diameter, smooth.

NOTE The fruits are eaten by waterfowl.

WETLAND STATUS AW OBL | GP OBL | WMV OBL

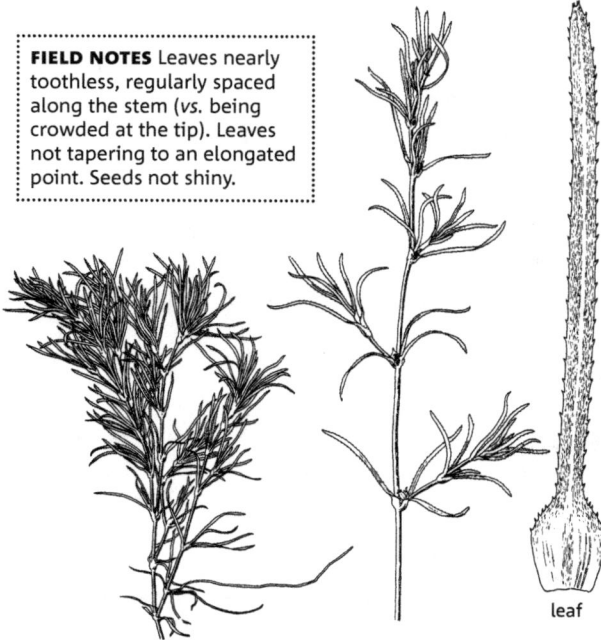

> **FIELD NOTES** Leaves nearly toothless, regularly spaced along the stem (*vs.* being crowded at the tip). Leaves not tapering to an elongated point. Seeds not shiny.

leaf

Najas guadalupensis (Spreng.) Magnus
SOUTHERN WATERNYMPH

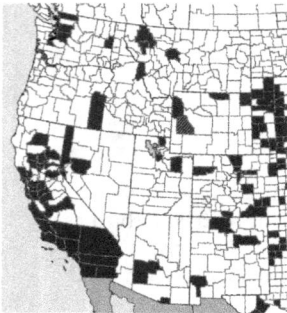

FAMILY Hydrocharitaceae (Tape-grass)

FLOWERING June–September

HABITAT Quiet or floating water in ponds, lakes, or streams.

HABIT Native submerged aquatic annual.

STEMS Very slender, much branched, essentially toothless, frequently forming mats, and sometimes rooting from the nodes.

LEAVES Very slender, alternate to subopposite, equally spaced along the stem and not crowded at the tip of the stem, essentially toothed or with a few tiny spinulose teeth, to 4 cm long, about 1 mm wide, pointed or rounded at the tip but not tapering to a long point.

FLOWERS Minute, unisexual, but with both sexes on the same plant, borne in the axils of the leaves. **Sepals** absent. **Petals** absent, although a pair of transparent membranaceous structures surround the stamen. **Stamens** 1, protruding from the two transparent structures that surround it. **Pistils:** One, with or without a transparent sheath around it; stigmas 2–4.

FRUIT Narrow and tapering to each end, to 4 mm long, not shiny, dark brown, covered by a network pattern of elongated markings.

NOTE Some plants have their stems and leaves encrusted with lime. The seeds are eaten by waterfowl.

WETLAND STATUS AW OBL | GP OBL | WMV OBL

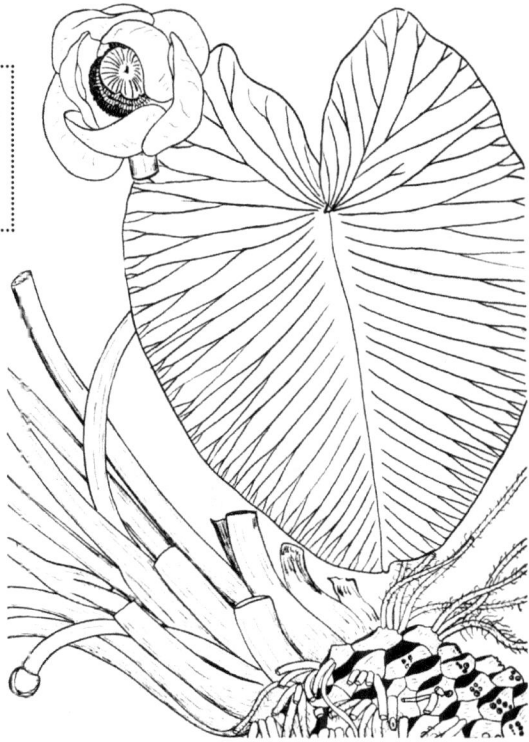

> **FIELD NOTES** Leaves leathery, shiny, deeply notched. Flowers club-shaped with usually 9 concave green to yellow petal-like sepals, and numerous small, stamen-like petals.

Nuphar polysepala Engelm.
POND-LILY

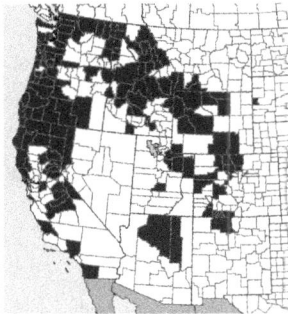

FAMILY Nymphaeaceae (Water-lily)
FLOWERING May–October
HABITAT Swamps, ponds, lakes, pools.
HABIT Native, aquatic perennial herb from thick, horizontal, cylindrical rhizomes.
STEMS Only present as rhizomes.
LEAVES Floating or emersed, simple, oblong to ovate, heart-shaped at base, the basal lobes sometimes overlapping, smooth above, smooth or finely hairy below, to 30 cm long and 20 cm wide, with terete or somewhat flattened stalks; submerged leaves generally thinner, more flaccid, and with wavier margins than the floating leaves.
FLOWERS Solitary, yellow, to 10 cm across, on long, smooth stalks that raise the flower above the water. **Sepals** (6–) 9 (–12), free from each other, concave, the outer ones green, the inner ones yellow with a green tip. **Petals** numerous, oblong, yellow, smaller than the sepals. **Stamens** numerous, in several rows. **Pistils:** Ovary disk-shaped, yellow, with up to 26 stigmatic rays.
FRUIT Leathery, ovoid, to 7 cm long, with a thick, short neck; seeds ovoid, to 8 mm long.
SYNONYMS *Nuphar lutea* ssp. *polysepala* (Engelm.) E.O. Beal
WETLAND STATUS AW OBL | GP OBL | WMV OBL

rhizome

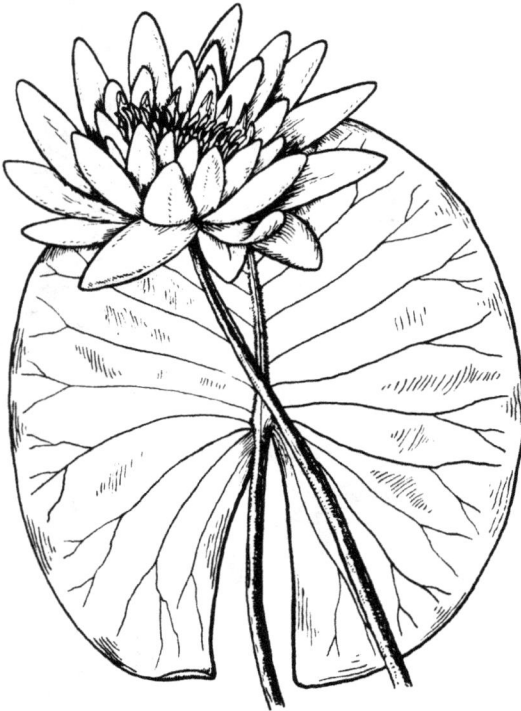

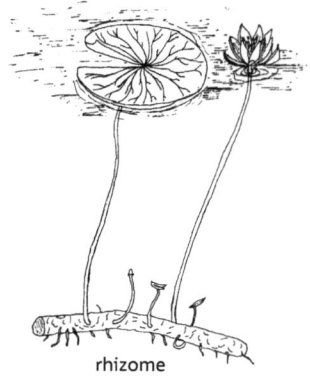

> **FIELD NOTES** Leaves circular, cleft at base, and usually purple on underside. Flowers fragrant, of 17 or more white (sometimes pink-tinged) petals.

Nymphaea odorata Ait.
AMERICAN WHITE WATER-LILY

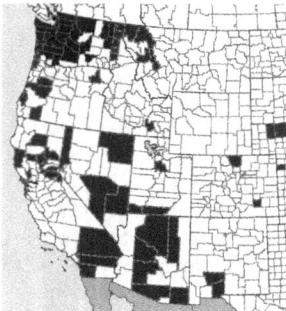

FAMILY Nymphaeaceae (Water-lily)

FLOWERING June–September

HABITAT Ponds, lakes, pools.

HABIT Aquatic perennial herb from branched rhizomes; considered adventive in western North America.

STEMS Only present as rhizomes.

LEAVES Alternate, simple, floating or less commonly emersed, spherical but with a basal cleft, up to 2 feet in diameter, green and smooth on the upper surface, usually purple and finely hairy on the lower surface, with flat, smooth stalks.

FLOWERS Solitary, showy, white, fragrant, to 20 cm across, on long stalks. **Sepals** 4, green or purplish, free from each other, narrowly ovate, usually rounded at the tip, to 10 cm long. **Petals** numerous, white, free from each other, the inner ones smaller than the outer ones. **Stamens** numerous, white, attached to the ovary. **Pistils:** Ovary superior, with a sessile, many-rayed stigma.

FRUIT Berry spherical but usually depressed at the top, containing many seeds; seeds ellipsoid to oblongoid, to 3 mm long.

NOTE The thickened rhizomes are eaten by beavers, muskrats, and other wildlife. Flower color can be variable; local populations may have varying shades of pink as well as the more common white.

WETLAND STATUS AW OBL | GP OBL | WMV OBL

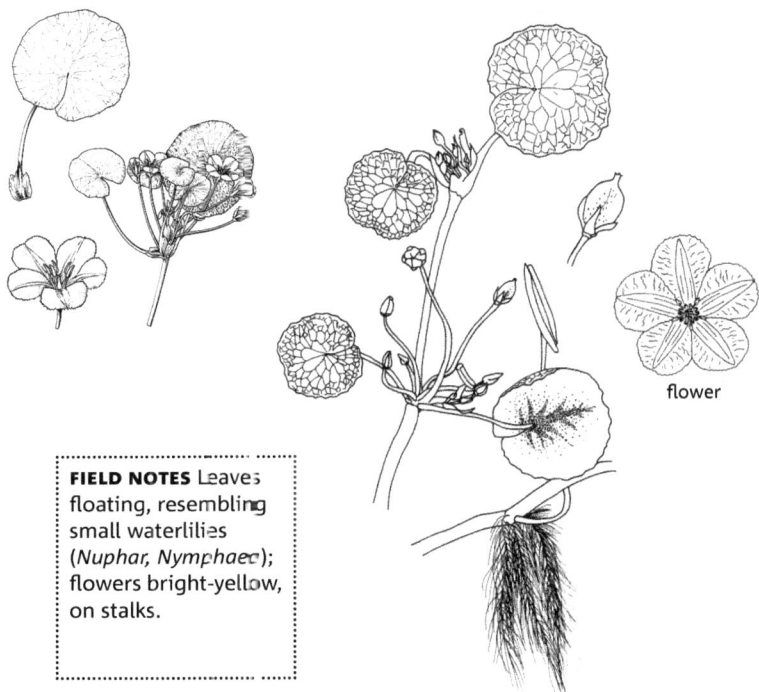

flower

> **FIELD NOTES** Leaves floating, resembling small waterlilies (*Nuphar*, *Nymphaea*); flowers bright-yellow, on stalks.

Nymphoides peltata (Gmel.) Kuntze
YELLOW FLOATINGHEART

FAMILY Menyanthaceae (Buckbean)
FLOWERING June–September.
HABITAT Quiet water of lakes and ponds.
HABIT Introduced perennial aquatic herb.
STEMS Stout, to about 1.5 m long, 2–3 mm thick, creeping and branching, rooted in bottom mud.
LEAVES From rhizomes, usually opposite and unequal; cordate to subrotund in outline, mostly 5–10 cm long and wide; margins wavy or entire; leaf underside often purplish.
FLOWERS 1–several on each stalk, these often emersed above water surface; petals 5, bright yellow, 3–5 cm in diameter when fully open; petal edges fringed.
FRUIT A beaked capsule, 1.2–2.5 cm long; seeds shiny, with stiff hairs along margins.
NOTE Native of Eurasia, invasive and capable of forming large colonies. Popular in water gardens and the source of many of its introductions when plants are disposed of into native habitats.
WETLAND STATUS AW OBL | GP OBL | WMV OBL

> **FIELD NOTES** Leaves
> wavy-edged with sharply
> fine-toothed margins.

Potamogeton crispus L.
CURLY PONDWEED

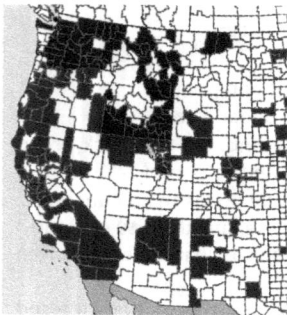

FAMILY Potamogetonaceae (Pondweed)

FLOWERING May–September

HABITAT Fresh or brackish ponds and streams.

HABIT Introduced, perennial aquatic herb from stout, creeping rhizomes.

STEMS Flattened, usually branched, to 3 mm in diameter, smooth.

LEAVES Alternate, simple, all alike, submerged, broadly linear to oblong, rounded to somewhat pointed at the tip, tapering to the nearly clasping base, to 10 cm long, to 13 mm wide, reddish green, the margins wavy and finely and irregularly toothed, 3- to 5-nerved.

FLOWERS Several, loosely arranged in spikes, the spikes cylindrical, to nearly 2.5 cm long, on stalks to 7 cm long. **Sepals** absent. **Petals** absent. **Stamens** 4. **Pistils** 4, free from each other, the ovary superior.

FRUIT Achenes ovoid, strongly and obtusely keeled with a small tooth near the base, 2-4 mm long, with a straight or incurved beak 2–3 mm long, greenish or brownish.

NOTE An invasive, introduced Eurasian species. Waterfowl may eat the achenes.

WETLAND STATUS AW OBL | GP OBL | WMV OBL

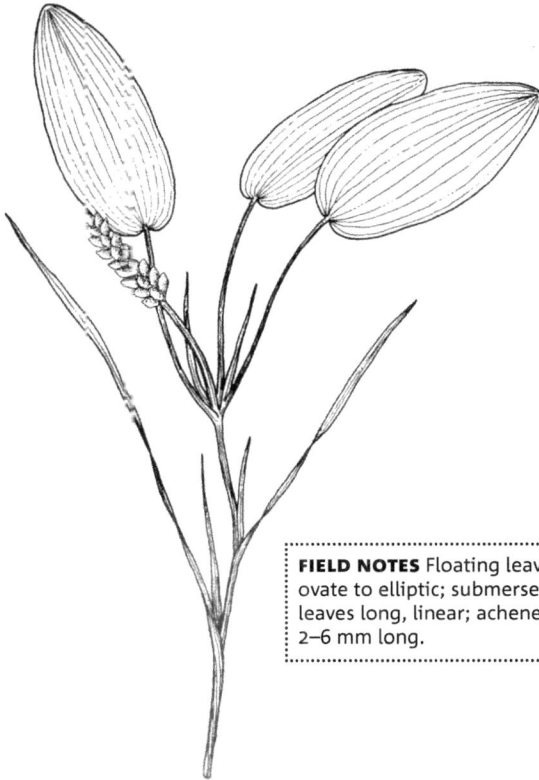

> **FIELD NOTES** Floating leaves
> ovate to elliptic; submersed
> leaves long, linear; achenes
> 2–6 mm long.

Potamogeton natans ∟
FLOATING-LEAF PONDWEED

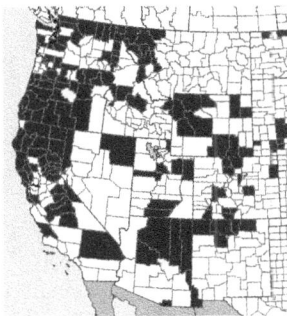

FAMILY Potamogetonaceae (Pondweed)

FLOWERING June–August

HABITAT In shallow ponds, around the edge of lakes, in slow-moving streams, often in brackish water.

HABIT Native, partially submersed perennial with a stout rhizome.

STEMS Slender, unbranched, not flattened, to 45 cm long, smooth.

LEAVES Floating leaves leathery, ovate to elliptic, to 10 cm long, to 6 cm wide, rounded at the tip, rounded or heart-shaped at the base, without teeth, smooth, with long leaf stalks; submersed leaves linear, to 20 cm long, to 2 mm wide, smooth; stipules to 10 cm long.

FLOWERS Densely crowded into spikes, the spikes to 6 cm long. **Sepals** 4, free from each other, greenish, to 3 mm long. **Petals** absent. **Stamens** 4. **Pistils** 4, free from each other, smooth.

FRUIT Achenes ellipsoid to obovoid, to 6 mm long, with a curved beak about 1 mm long.

NOTE The achenes are eaten by waterfowl.

WETLAND STATUS AW OBL | GP OBL | WMV OBL

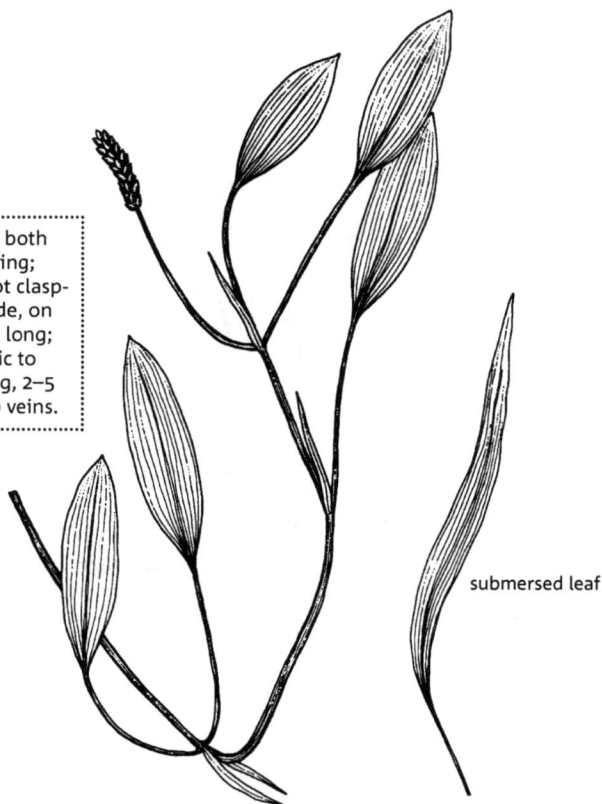

> **FIELD NOTES** Leaves both submersed and floating; submersed leaves not clasping stem, 1–4 cm wide, on stalks at least 2.5 cm long; floating leaves elliptic to oblong, 5–13 cm long, 2–5 cm wide, with 10–20 veins.

submersed leaf

Potamogeton nodosus Poir.
LONG-LEAF PONDWEED

FAMILY Potamogetonaceae (Pondweed)

FLOWERING June–August

HABITAT Quiet waters of ponds, lakes, streams, and ditches.

HABIT Native, partially submersed perennial with spreading rhizomes.

STEMS Rather stout, branched or unbranched.

LEAVES Of two kinds: submersed leaves linear-lanceolate, 4–20 cm long, 1–4 cm wide, thin and very weak, with 7–15 veins, with a leaf stalk at least 2.5 cm long; floating leaves elliptic to oblong, 2–13 cm long, 2–5 cm wide, with 10–20 veins, on stalks to 25 cm long.

FLOWERS Several crowded into spikes, the spikes to 6 cm long, on stout stalks to 15 cm long. **Sepals** 4, greenish, 3–4 mm long. **Petals** absent. **Stamens** 4. **Pistils** 4, each with a superior ovary.

FRUIT Achenes obovoid, asymmetrical, to 4 mm long, to 3 mm wide, with a short beak at the tip.

NOTE The achenes are eaten by waterfowl.

WETLAND STATUS AW OBL | GP OBL | WMV OBL

> **FIELD NOTES** Leaves all submersed, clasping stem at their base; stems usually zigzag; achenes at least 6 mm long.

Potamogeton praelongus Wulfen
WHITE-STEM PONDWEED

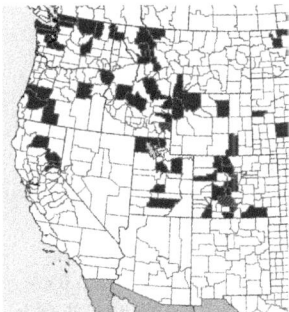

FAMILY Potamogetonaceae (Pondweed)

FLOWERING May–July

HABITAT Deep water in lakes and streams.

HABIT Native submersed perennial with a stout rhizome.

STEMS Stout, zigzag, branched, olive-green to whitish, to 30 cm long, smooth.

LEAVES Alternate, simple, all submersed, oblong-lance-olate, to 25 cm long, to 4 cm wide, more or less rounded at the tip and forming a small hood, rounded to heart-shaped at the base and clasping the stem, smooth; stipules to 10 cm long, whitish.

FLOWERS Crowded into spikes to 6 cm long, the spikes on stalks to 45 cm long. **Sepals** 4, green, free from each other, to 4 mm long. **Petals** absent. **Stamens** 4. **Pistils** 4, free from each other, smooth.

FRUIT Achenes obovoid, to 6 mm long, to 4 mm wide, smooth, with a short, persistent beak at the tip.

NOTE The achenes are eaten by waterfowl.

WETLAND STATUS AW OBL | GP OBL | WMV OBL

> **FIELD NOTES** Leaves all submersed (floating leaves absent). Submersed leaves only 2 mm wide but have 3 distinct veins. Stems slender, usually thread-like and much branched.

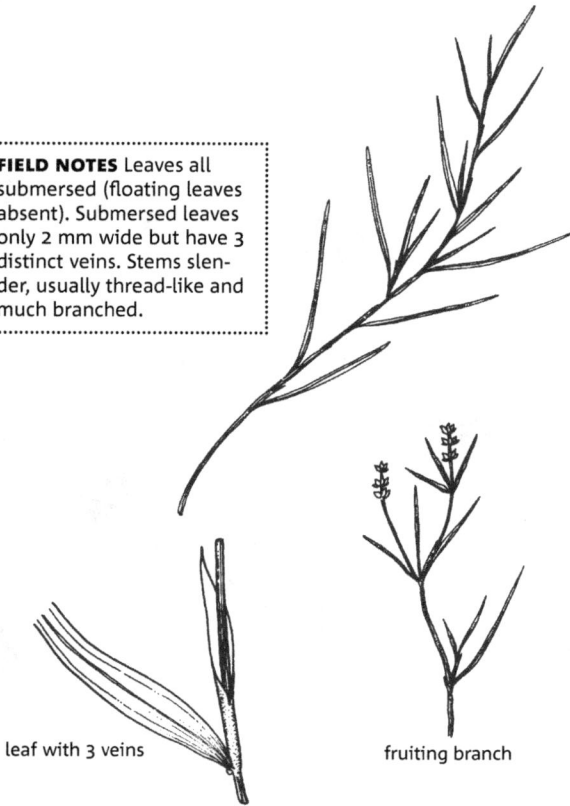

leaf with 3 veins

fruiting branch

Potamogeton pusillus L.
SMALL PONDWEED
FAMILY Potamogetonaceae (Pondweed)
FLOWERING July–September

HABITAT Quiet waters of lakes, ponds, and streams.
HABIT Native submersed perennial with slender rhizomes.
STEMS Very slender, often thread-like, much branched.
LEAVES All submersed, about 2 mm wide, with 3 conspicuous veins, smooth, with 2 minute glands where leaf is attached to the stem.
FLOWERS Several in whorls crowded into spikes, the spikes to 20 mm long, borne on smooth stalks to 6 cm long. **Sepals** 4, green, 1–2 mm long. **Petals** absent. **Stamens** 4. **Pistils** 4, each with a superior ovary.
FRUIT Achenes obovoid. asymmetrical, to 3 mm long, with a minute beak at the tip.
NOTE This species is variable in the number of flowering and fruiting heads per plant and in the length of the stalks that bear the heads. The achenes are eaten by waterfowl.
WETLAND STATUS AW OBL | GP OBL | WMV OBL

FIELD NOTES Leaves broadly rounded at base and clasping the stem. Achenes never longer than 4 mm. Stems straight, not zigzagged.

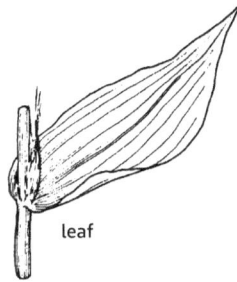

leaf

Potamogeton richardsonii (Benn.) Rydb.
RICHARDSON PONDWEED

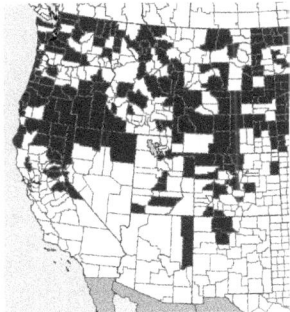

FAMILY Potamogetonaceae (Pondweed)

FLOWERING June–August

HABITAT Shallow water of lakes and streams.

HABIT Native submersed perennial herb with slender rhizomes.

STEMS Submersed, branched, rather stout, to 60 cm long, not zigzag.

LEAVES Alternate, simple, all submersed, lanceolate to broadly lanceolate, to 10 cm long, to 2.5 cm wide, pointed at the tip, broadly rounded at the base and clasping the stem, smooth; stipules 1–2.5 cm long, shredded into white fibers.

FLOWERS Crowded together into spikes, each spike to 5 cm long on stout stalks to 15 cm long. **Sepals** 4, green, free from each other, to 4 mm long. **Petals** absent. **Stamens** 4. **Pistils** 4, each with a superior ovary.

FRUIT Achenes obovoid, to 4 mm long, with a tiny erect beak at the tip.

NOTE The achenes are eaten by waterfowl.

WETLAND STATUS AW OBL | GP OBL | WMV OBL

stipule

> **FIELD NOTES** Leaves all submersed, numerous, linear to linear-lanceolate, and minutely toothed, at least near the tip.

Potamogeton robbinsii Oakes
ROBBIN'S PONDWEED

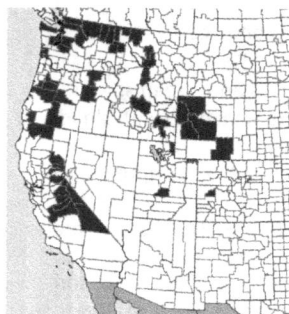

FAMILY Potamogetonaceae (Pondweed)

FLOWERING August–September

HABITAT In deep water of ponds, lakes, and streams.

HABIT Native submersed perennial with slender, creeping rhizomes.

STEMS Rather stout, much branched, to 60 cm long, smooth.

LEAVES Alternate, simple, all submersed, usually crowded, linear to linear-lanceolate, to 10 cm long, to 8 mm wide, pointed at the tip. united at the base with a stipule, finely toothed, at least near the tips.

FLOWERS Few in stiff, interrupted spikes to 2.5 cm long; stalks of spikes to 8 cm long. **Sepals** 4, reddish to greenish yellow, free from each other, 1–2 mm long. **Petals** absent. **Stamens** 4. **Pistils** 4, free from each other, smooth.

FRUIT Achenes obovoid. to 4 mm long, to 3 mm wide, smooth, with the persistent curved style at the tip.

NOTE Most specimens of this species are rarely seen in flower and even more rarely in fruit. The plants provide cover for fish.

WETLAND STATUS AW OBL | GP OBL | WMV OBL

> **FIELD NOTES** Leaves all
> submersed, to only 4 mm
> wide. Stems flattened and
> narrowly winged.

achene

Potamogeton zosteriformis Fern.
FLAT-STEM PONDWEED

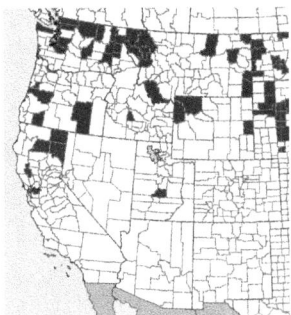

FAMILY Potamogetonaceae (Pondweed)

FLOWERING July–September

HABITAT Shallow water of ponds, lakes, and streams.

HABIT Native submersed perennial herb with slender rhizomes.

STEMS Submersed, branched, flat, narrowly winged, to 60 cm long.

LEAVES Alternate, simple, all submersed, linear, to 20 cm long, to 4 mm wide, rounded or pointed at the tip, sessile, finely veined, smooth; stipules to 4 cm long.

FLOWERS Crowded together into spikes, the spikes to 4 cm long, on flattened stalks to 10 cm long. **Sepals** 4, green, free from each other, to 2 mm long. **Petals** absent. **Stamens** 4. **Pistils** 4, each with a superior ovary.

FRUIT Achenes ellipsoid, to 6 mm long, with a straight or slightly curved beak at the tip.

NOTE The achenes are eaten by waterfowl.

WETLAND STATUS AW OBL | GP CBL | WMV OBL

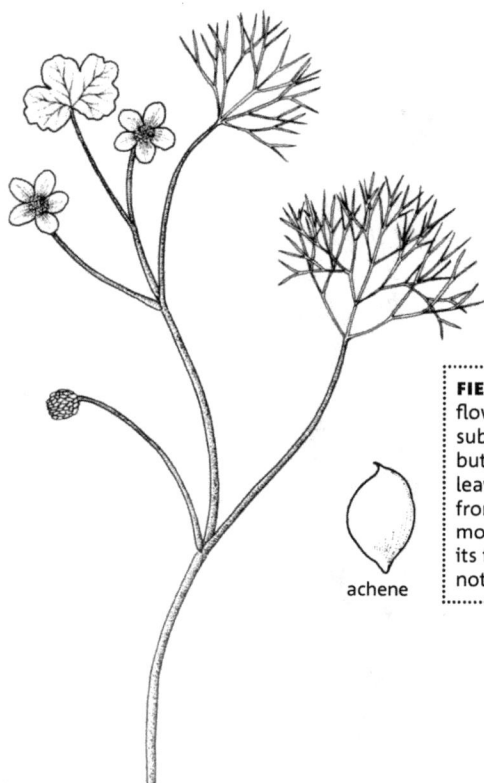

> **FIELD NOTES** Aquatic white-flowered buttercup; leaves submersed, much divided, but a few 3-lobed floating leaves may be present. Differs from the similar but less common *R. subrigidus* by having its flower stalks straight and not curved at fruiting time.

achene

Ranunculus aquatilis L.
WHITE WATER BUTTERCUP

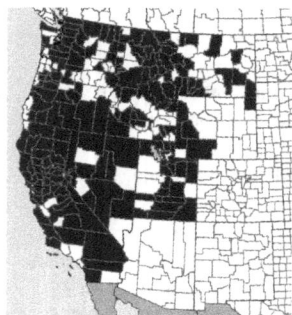

FAMILY Ranunculaceae (Buttercup)
FLOWERING April–July
HABITAT In ponds and ditches, in streams, vernal pools.
HABIT Native perennial herb with submersed stems.
STEMS Submersed, smooth or hairy, to 1 m long.
LEAVES Submersed leaves to 5 cm long, much divided into thread-like segments, smooth or hairy; floating leaves few, simple but often 3-lobed, less than 2.5 cm wide, smooth or hairy.
FLOWERS Several in clusters, to 13 mm across, each flower on a stalk to 4 cm long. **Sepals** 5, green, free from each other, to 3 mm long. **Petals** 5, white, free from each other, to 6 mm long. **Stamens** 10–25. **Pistils** many, each with a superior ovary, smooth.
FRUIT Achenes 15–25 in a cluster at the tip of a straight stalk, each achene to 2 mm long, smooth, with a minute beak.
NOTE The stems will root at the nodes if the plant is stranded on land. The submersed leaves collapse when removed from the water. The achenes are eaten by waterfowl.
WETLAND STATUS AW OBL | GP OBL | WMV OBL

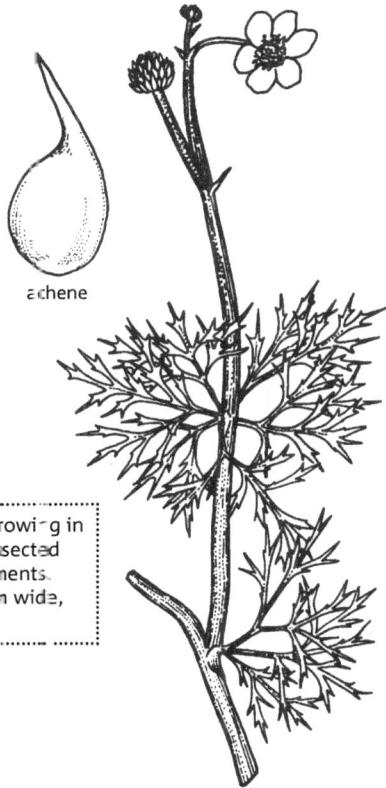

achene

FIELD NOTES When growing in water, leaves finely dissected with very narrow segments. Flowers yellow, 2–4 cm wide, with 50–80 stamens.

Ranunculus flabellaris Raf.
YELLOW WATER BUTTERCUP

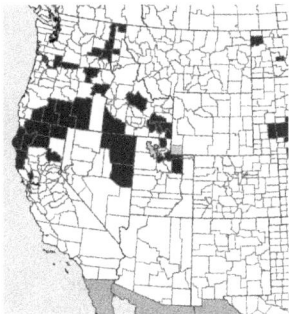

FAMILY Ranunculaceae (Buttercup)

FLOWERING April–August

HABITAT Shallow waters or in mud, in marshes, in wet ditches.

HABIT Native aquatic or mud-inhabiting perennial with thread-like roots.

STEMS Floating in water or lying on mud and rooting at the nodes, branched, to 60 cm long, smooth or hairy.

LEAVES Deeply dissected in aquatic plants, each segment extremely narrow; less divided and with broader segments when rooted in mud; usually smooth.

FLOWERS 1–few in a cluster; each flower on a stalk to 5 cm long. **Sepals** 5, free from each other, greenish yellow, 6–8 mm long. **Petals** usually 5, sometimes more, yellow, free from each other, 8–20 mm long. **Stamens** 50–80. **Pistils** many in each flower, each with a superior ovary.

FRUIT Many achenes, together in a head, the head nearly 13 mm in diameter, each achene obovoid, to 3 mm long, with a flat beak.

NOTE The achenes are eaten by birds and small mammals.

WETLAND STATUS AW FACW | GP FACW | WMV FACW

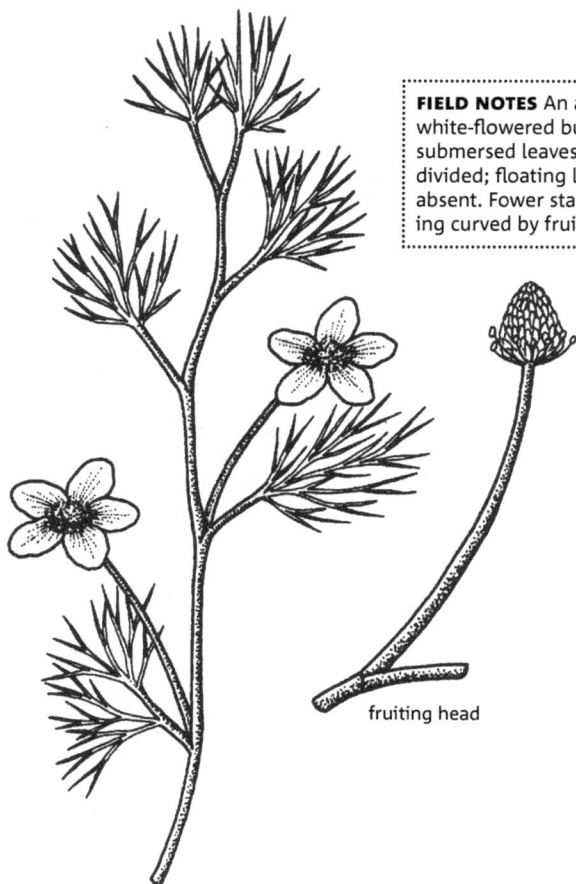

FIELD NOTES An aquatic, white-flowered buttercup; submersed leaves much divided; floating leaves absent. Fower stalks becoming curved by fruiting time.

fruiting head

Ranunculus subrigidus W. B. Drew
POND BUTTERCUP

FAMILY Ranunculaceae (Buttercup)
FLOWERING May–July
HABITAT In ponds, in slow streams.
HABIT Native perennial herb with submersed stems.
STEMS Submersed, smooth, to 60 cm long.
LEAVES All leaves submersed, to 5 cm long, much divided into thread-like segments, smooth.
FLOWERS Several in clusters, to 20 mm across, each flower on a stalk to 5 cm long. **Sepals** 5, green, free from each other, to 6 mm long. **Petals** 5, white, free from each other, to nearly 13 mm long. **Stamens** 5–10. **Pistils** many, each with a superior ovary, smooth.
FRUIT Achenes 30 or more in clusters at the tip of a curved stalk, each achene obovoid, to 2 mm long, smooth, with a minute beak.
NOTE The achenes are eaten by waterfowl.
WETLAND STATUS AW OBL | GP FACW | WMV OBL

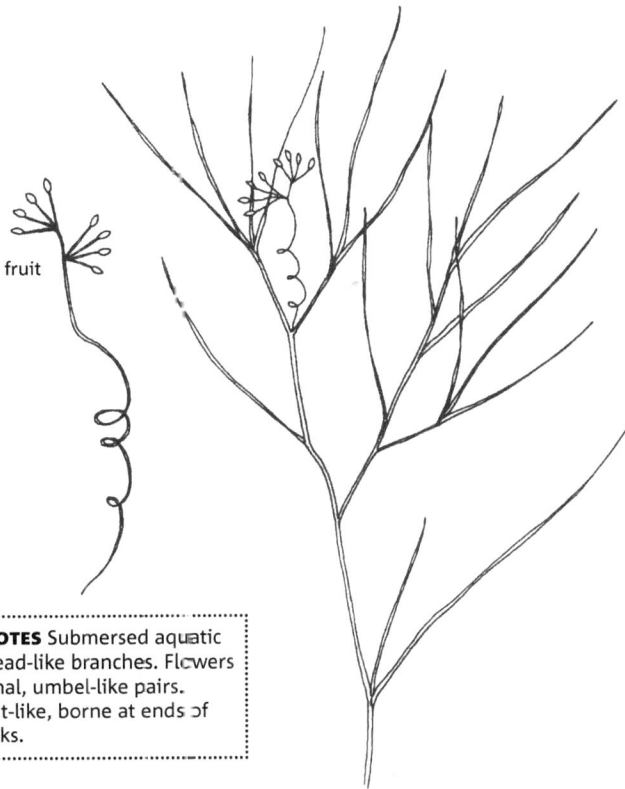

fruit

FIELD NOTES Submersed aquatic with thread-like branches. Flowers in terminal, umbel-like pairs. Fruits nut-like, borne at ends of long stalks.

Ruppia maritima L.
WIDGEON-GRASS

FAMILY Ruppiaceae (Ditch-grass Family)

FLOWERING April–July

HABITAT Alkaline or brackish water along the coasts; also in ponds, lakes, and ditches elsewhere.

HABIT Native submersed perennial with creeping rhizomes.

STEMS Thread-like, much branched, to 1 m long, smooth.

LEAVES All submersed, alternate, thread-like to linear, to 10 cm long, more or less rounded at the tip, with a sheathing stipule at the base.

FLOWERS Usually borne in an umbel-like pair at the tip of the stem, the stalks sometimes becoming spiral. **Sepals** absent. **Petals** absent. **Stamens** 2. **Pistils** usually 4 per flower, free from each other, borne on an elongated, slender stalk, giving the appearance of 4 flowers in a cluster rather than only one.

FRUIT 4, nut-like, ovoid, to 4 mm long, usually with a short, curved beak, borne on a slender stalk 2–4 cm long.

NOTE This species formerly placed in the Potamogetonaceae. All parts of this plant are eaten by waterfowl. This species also provides excellent food and protection for fish.

WETLAND STATUS AW OBL | GP OBL | WMV OBL

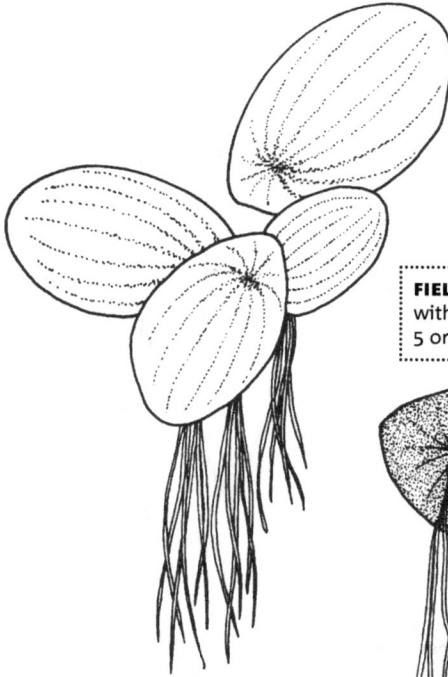

> **FIELD NOTES** Each frond with several roots and with 5 or more distinct veins.

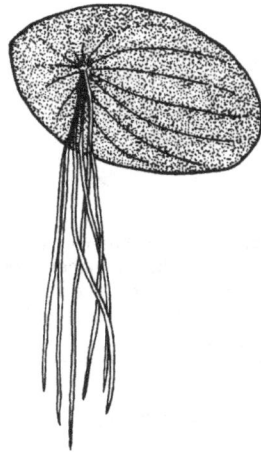

Spirodela polyrhiza (L.) Schleid.
GREATER DUCKWEED

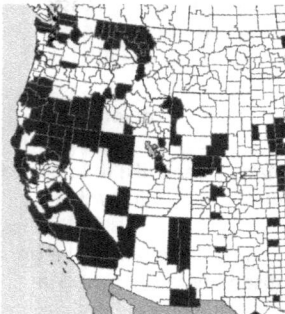

FAMILY Araceae (Arum)

FLOWERING June–September

HABITAT Quiet waters of lakes and ponds, in marshes and slow streams.

HABIT Native, plants floating on water surface.

STEMS absent.

LEAVES Single fronds or fronds cohering in clusters of 2–5, to 8 mm long, to 6 mm wide, green on the upper surface, usually red-purple on the lower surface, with at least 5 distinct veins, smooth.

FLOWERS Rarely formed; if present, the minute male and female flowers borne in pouches known as spathes; male flowers usually 2 per pouch. **Sepals** absent. **Petals** absent. **Stamens** 1. **Pistils:** Ovary superior.

FRUIT Rarely formed; if present, with a minute wing. Asexual reproductive buds, called turions, are the chief method of reproduction.

NOTE Like most duckweeds, a source of food for waterfowl.

WETLAND STATUS AW OBL | GP OBL | WMV OBL

> **FIELD NOTES** Leaves all submersed; base of each leaf attached to a stipule. Differs from sago pondweed (*S. pectinata*) by its less pointed leaves.

Stuckenia filiformis (Pers.) Böerner
FINE-LEAF PONDWEED

FAMILY Potamogetonaceae (Pondweed)

FLOWERING July–September

HABITAT Quiet waters of ponds, lakes, and streams.

HABIT Native submersed perennial with much branched rhizomes, the tips of which sometimes bear a small white tuber.

STEMS Slender, branched.

LEAVES All submersed, thread-like to linear, to 4 mm wide but usually much narrower, smooth, with 3–5 veins, with a stipule attached near its base.

FLOWERS Several in whorls of 2–8, the whorls crowded or separated on the stem, the stalk of the spike to 15 cm long. **Sepals** 4, brown, to 3 mm long. **Petals** absent. **Stamens** 4. **Pistils** 4, each with a superior ovary.

FRUIT Achenes obovoid, to 4 mm long, to 3 mm wide, with a short but prominent beak.

NOTE This is an extremely variable species with regard to leaf width and degree of congestion in the inflorescence. The achenes are eaten by waterfowl.

SYNONYMS *Potamogeton filiformis* Pers.

WETLAND STATUS AW OBL | GP OBL | WMV OBL

> **FIELD NOTES** Leaves extremely narrow, never more than 1 mm wide. Fruit yellow-brown, with a short, pointed beak.

achene

Stuckenia pectinata (L.) Böerner
SAGO PONDWEED

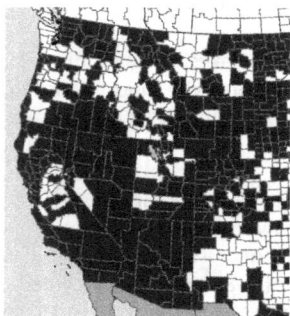

FAMILY Potamogetonaceae (Pondweed)

FLOWERING June–September

HABITAT Shallow to somewhat deep, usually fresh water.

HABIT Native submersed perennial, usually with small tubers.

STEMS Very slender, to 1 mm wide, dichotomously branched, to 1 m long.

LEAVES Alternate or subopposite, all submersed, very narrow, to 10 cm long, to 1 mm wide.

FLOWERS Very tiny, crowded into short, slender spikes, the spikes to 5 cm long, arranged in 2–7 whorls, each spike on thread-like stalks to 15 cm long. **Sepals** 4, greenish, bract-like. **Petals** absent. Stamens 4, each one attached to the base of a sepal. Pistils 4, free from each other.

FRUIT 4 in a cluster, yellow-brown, obovoid, 2–6 mm long, with a short, pointed beak.

NOTE The small tubers of this species are starchy and are used as a source of food for waterfowl and muskrats.

SYNONYMS *Potamogeton pectinatus* L.

WETLAND STATUS AW OBL | GP OBL | WMV OBL

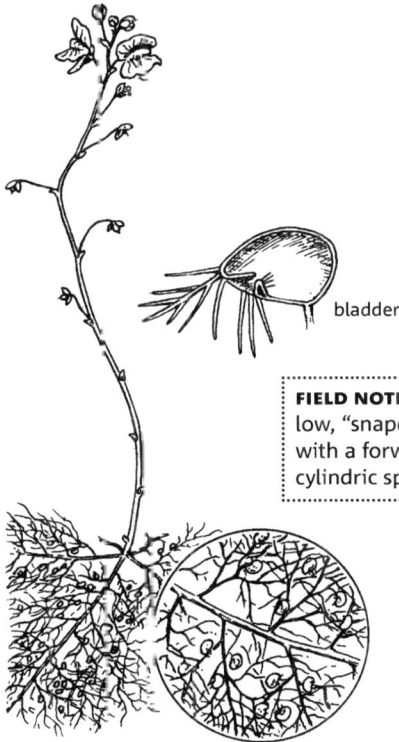

bladder

> **FIELD NOTES** Flowers yellow, "snapdragon-like," with a forward-pointing cylindric spur.

Utricularia macrorhiza Le Conte
GREATER BLADDERWORT

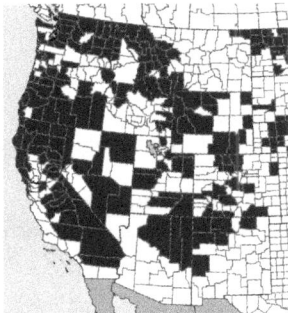

FAMILY Lentibulariaceae (Bladderwort)

FLOWERING June–September

HABITAT Quiet water of lakes, ponds, peatlands, marshes and rivers; water typically acidic.

HABIT Native perennial, floating, aquatic herb.

STEMS All submersed, sparsely branched, often forming large mats

LEAVES alternate, 1–5 cm long, 2-forked at base and repeatedly 2-forked into segments of unequal length, becoming smaller with each branching, the final segments threadlike; bladders 1–4 mm wide, borne on leaf segments.

FLOWERS yellow, 6–20 atop a stout emersed stalk 6–25 cm long; stalks bearing individual flowers curved downward in fruit. **Sepals** united to form 2 lips. **Petals** 5, pale yellow, united to form 2 lips; upper lip about equal to lower lip; spur about 2/3 as long as lower lip, often hooked upward at tip. **Stamens** 2. **Pistils:** Ovary 1, superior.

FRUIT Capsules nearly spherical, about 2 mm in diameter.

NOTE Like all bladderworts, this species is able to trap tiny aquatic organisms in its underwater bladders and utilize some of its prey for its own nutritional value.

SYNONYMS *Utricularia vulgaris* L.

WETLAND STATUS AW OBL | GP OBL | WMV OBL

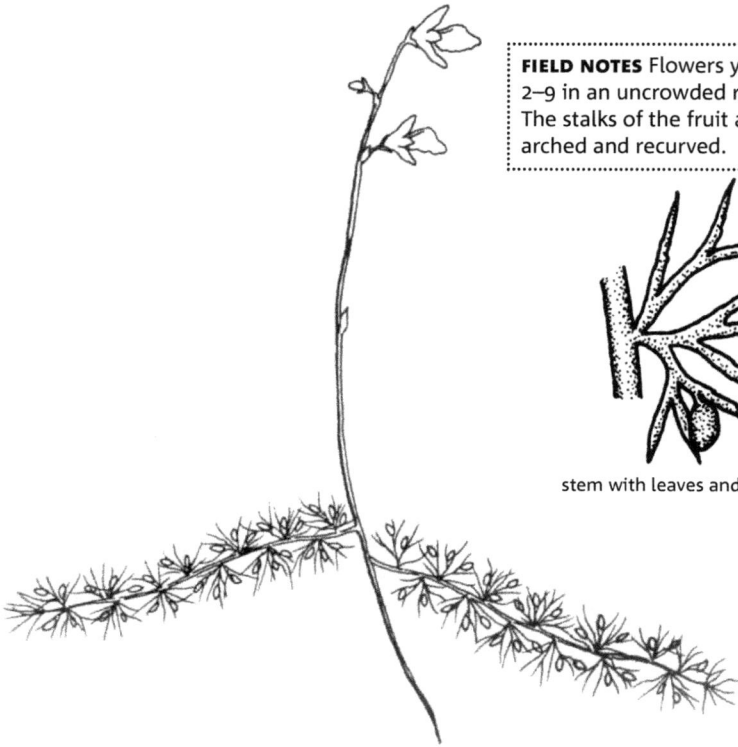

> **FIELD NOTES** Flowers yellow, 2–9 in an uncrowded raceme. The stalks of the fruit are arched and recurved.

stem with leaves and bladders

Utricularia minor L.
LESSER BLADDERWORT

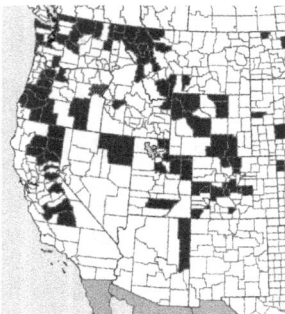

FAMILY Lentibulariaceae (Bladderwort)

FLOWERING July–September

HABITAT In streams and in shallow ponds.

HABIT Native perennial floating herb with winter buds (turions).

STEMS All submersed, thread-like, much branched.

LEAVES numerous, alternate, divided several times, each segment flat, linear, pointed at the tip; bladders to 2 mm in diameter, attached to the leaves.

FLOWERS 2–9 in an uncrowded raceme on a leafless stalk to 25 cm long: flower stalks becoming arched and recurved. **Sepals** united to form 2 lips, green, each lip to 3 mm long, the upper lip slightly longer than the lower lip. **Petals** 5, pale yellow, united to form 2 lips, to 8 mm long, with a cylindric spur to 3 mm long. **Stamens** 2, the anthers twisted. **Pistils**: Ovary 1, superior, smooth.

FRUIT Capsules nearly spherical, about 2 mm in diameter.

NOTE Bladderworts trap minute aquatic organisms in its underwater bladders and utilize some of its prey for additional nutritional.

WETLAND STATUS AW OBL | GP OBL | WMV OBL

female flower

> **FIELD NOTES** Leaves submerged, long and ribbon-like, minutely toothed along their edge.

Vallisneria americana Michx.
WILD-CELERY

FAMILY Hydrocharitaceae (Frog's-bit)
FLOWERING July–September
HABITAT Shallow waters of lakes and streams.
HABIT Native aquatic perennial herb with stolons.
STEMS Slender, smooth stolons.
LEAVES Elongated and ribbon-like, to 1 m long, to 8 mm wide, without hairs, minutely toothed along the edges, partially septate.
FLOWERS Male and female flowers borne on separate plants; male flowers numerous, crowded, subtended by a bract (spathe), breaking free while in bud and floating to the water surface on a long stalk before opening; female flowers solitary and sessile in a tubular spathe. **Sepals** 3, free from each other, greenish, ovate, pointed at the tip. **Petals** 1 in the male flower, smaller than the sepals; 3 in the female flower, nearly transparent. **Stamens** 2, the filaments united; a third sterile stamen is present, resembling a petal. **Pistils**: Ovary inferior; **stigmas** 3.
FRUIT Indehiscent, ripening under water as the stalk becomes coiled and pulls the fruit below the water surface; seeds with conspicuous cross-markings.
NOTE This species is sometimes known as eelgrass. It has the ability to clog up shallow water when it develops into dense colonies. The leaves provide cover for fish.
WETLAND STATUS AW OBL | GP OBL | WMV OBL

achenes

> **FIELD NOTES** Stems very slender; leaves opposite; achenes conspicuously beaked.

Zannichellia palustris L.
HORNED PONDWEED

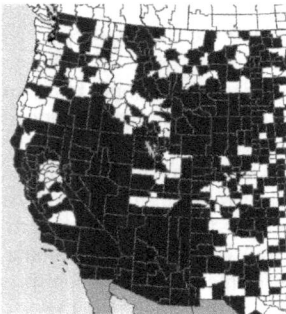

FAMILY Potamogetonaceae (Pondweed)
FLOWERING March–November
HABITAT Fresh water or brackish lakes, ponds, and streams.
HABIT Native submersed perennial with creeping rhizomes.
STEMS Thread-like, branched, to 45 cm long, smooth.
LEAVES All submersed, opposite, thread-like, to 10 cm long, pointed at the tip, with a transparent sheath at the base.
FLOWERS Male and female flowers borne separately in the same leaf axils. **Sepals** absent, although 3 transparent scales may subtend the male flower and 1 scale usually subtends the female flower. **Petals** absent. **Stamens** 1. **Pistils** 2–6, free from each other.
FRUIT 2–6 achenes; oblong, flattened, to 4 mm long, with a slender beak to 3 mm long, each achene often on a very short stalk.
NOTE Formerly placed in its own family, the Zannichelliaceae. The entire plant may be utilized as food for waterfowl and some fish.
WETLAND STATUS AW OBL | GP OBL | WMV OBL

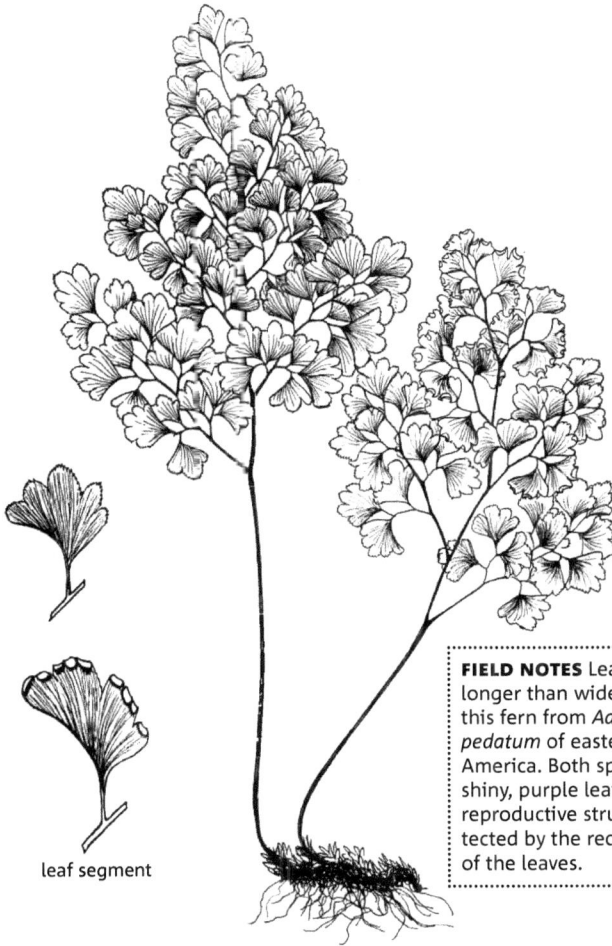

leaf segment

FIELD NOTES Leaves much longer than wide distinguish this fern from *Adiantum pedatum* of eastern North America. Both species have shiny, purple leaf stalks and reproductive structures protected by the recurved edge of the leaves.

Adiantum capillus-veneris L.
SOUTHERN MAIDENHAIR FERN

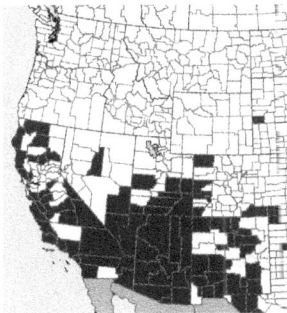

FAMILY Pteridaceae (Maidenhair Fern)
HABITAT Stream banks, limestone ledges, around springs.
HABIT Native perennial fern with creeping rhizomes.
STEMS All underground as creeping rhizomes.
LEAVES Usually drooping, 2- to 3-pinnate, the leaf segments alternate, obovate, to 30 mm long, thin, smooth, some of them round lobed and toothed, on a short stalk; main leaf stalks shiny, purple, to 20 cm long. **Sori** borne under the recurved edges of the leaves.
NOTE This fern derives its name of maidenhair from the leaf segments that have a resemblance to the leaves of the maidenhair, or ginkgo, tree.
WETLAND STATUS AW FACW | GP FACW | WMV FACW

leaf segment

> **FIELD NOTES** Leaves much divided, membranaceous. Fruiting bodies (sori) elongate.

Athyrium cyclosorum Rupr.
WESTERN LADY FERN

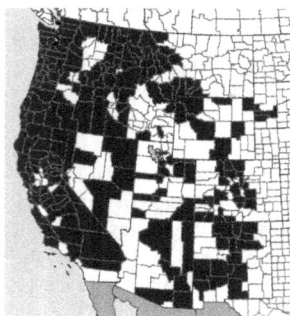

FAMILY Athyriaceae (Lady Fern)
HABITAT Rich woods, thickets, bogs, along streams.
HABIT Native, large, spreading fern with short rhizomes.
STEMS All underground as rhizomes.
LEAVES Divided as much as three times, to 75 cm long, membranaceous, usually without hairs, each segment toothed or evenly shallow lobed; leaf stalks with brown scales to 6 mm long. **Sori** elongated, about 2 mm long, scattered on the lower surface of the leaf segments.
SYNONYMS *Athyrium filix-femina* ssp. *cyclosorum* (Rupr.) C. Christens.

WETLAND STATUS AW FAC | GP FAC | WMV FAC

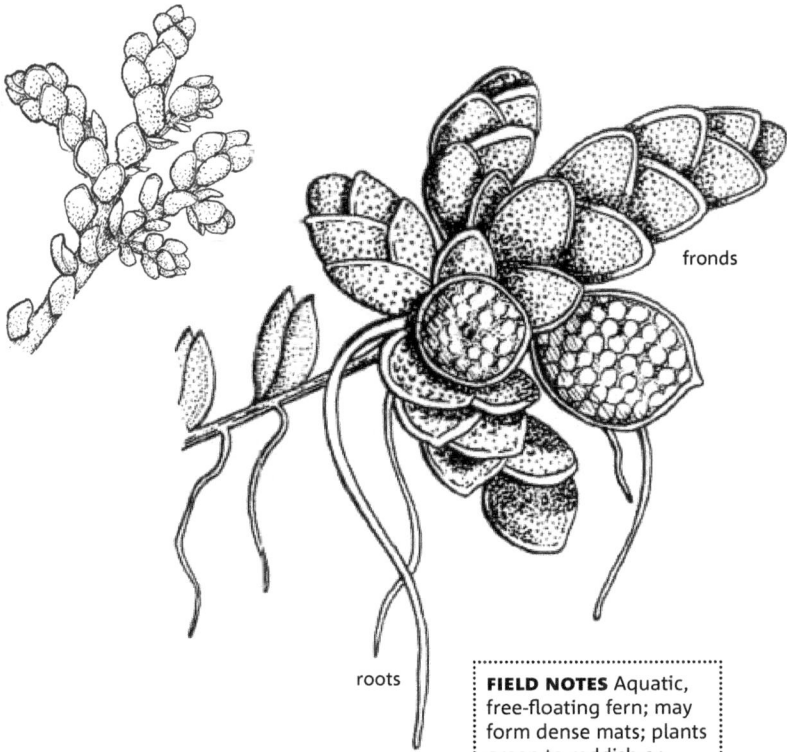

fronds

roots

> **FIELD NOTES** Aquatic, free-floating fern; may form dense mats; plants green to reddish or brown, depending on light and nutrients.

Azolla filiculoides Lam.
LARGE MOSQUITO FERN

FAMILY Salviniaceae (Water Fern)

HABITAT Covering the surface of still or slowly moving water of ponds and small streams.

HABIT Native aquatic, free-floating fern; plants green to reddish or brown, the whole plant fan-shaped, circular or irregular in outline, to about 2–3 cm in diameter.

STEMS absent.

LEAVES Fronds mostly less than 2 mm long.

NOTE *Azolla filiculoides* is able to reproduce vegetatively very quickly. It usually reaches its maximum growth in late spring, becomes fertile, then largely dies, and is replaced by other more heat-tolerant aquatics such as duckweeds (*Lemna* spp.). The only sure method of distinguishing this species from the other native North American species, *Azolla cristata*, is to examine the trichomes on the upper surfaces of the leaves (trichomes are small bumps): they are unicellular in *A. filiculoides* but septate (two-celled) in *A. cristata*.

WETLAND STATUS AW OBL | WMV OBL

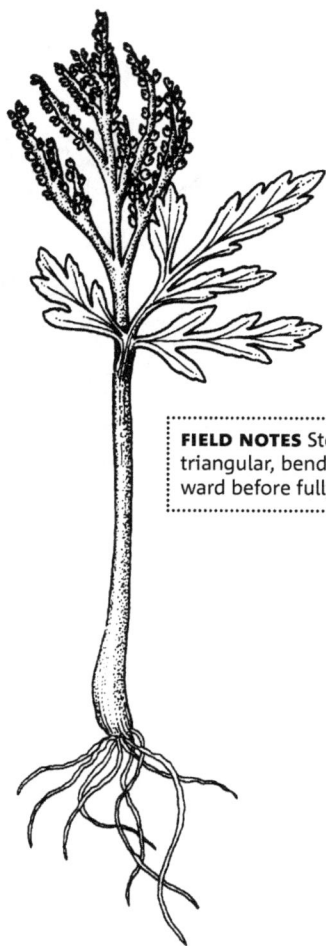

> **FIELD NOTES** Sterile leaf triangular, bending downward before fully mature.

Botrychium lanceolatum (S.G. Gmel.) Rupr.
TRIANGLE MOONWORT

FAMILY Ophioglossaceae (Adder's-tongue)

HABITAT Moist grassy or rocky areas.

HABIT Native perennial fern with an erect rhizome.

STEMS Upright, rather stout, to 30 cm tall, smooth; buds not hairy.

LEAVES Sterile blade triangular, pinnately divided, sessile, the segments linear-lanceolate to oblong, pointed at the tip, smooth, bent downward until fully mature. **Sori** spherical, borne in a terminal panicle at the top of the sterile leaf-bearing stem.

NOTE The young succulent stems of this fern may be eaten by mammals.

WETLAND STATUS AW FACW | GP FACW | WMV FACW

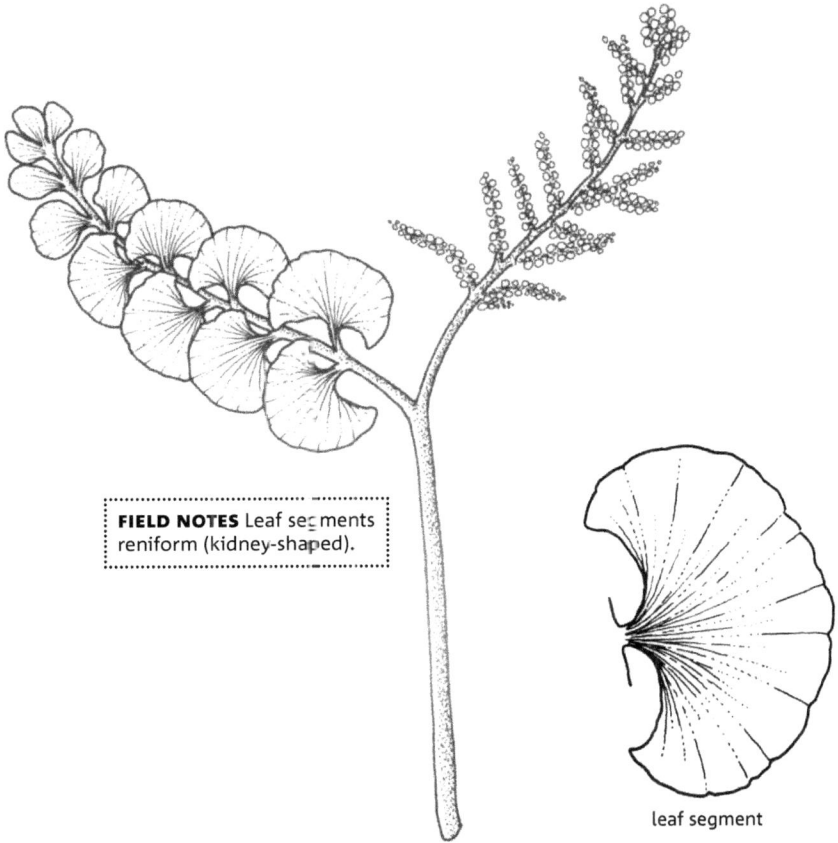

FIELD NOTES Leaf segments reniform (kidney-shaped).

leaf segment

Botrychium lunaria (L.) Swartz
MOONWORT

FAMILY Ophioglossaceae (Adder's-tongue)
HABITAT Wet meadows, moist fields.
HABIT Native perennial fern with an erect rhizome.
STEMS Upright, rather stout, to 30 cm tall, smooth; buds not hairy.
LEAVES Sterile blade borne halfway up the stem, to 13 cm long, divided into 10–16 reniform (kidney-shaped) leaflets, the leaflets shallowly toothed or lobed, smooth. **Sori** spherical, borne in a narrow spike to 10 cm long.
WETLAND STATUS AW FAC | GP FAC | WMV FAC

leaf base

> **FIELD NOTES** Plants submersed; leaves not more than 10 cm long.

megaspore

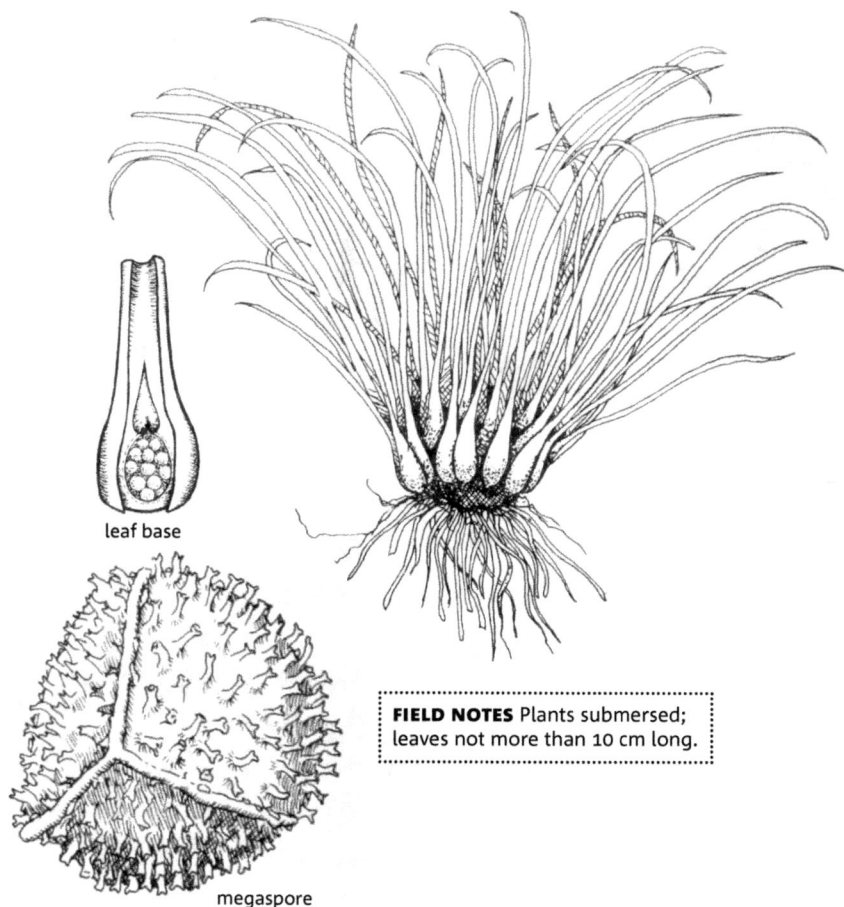

Isoetes echinospora Durieu
SPINY-SPORE QUILLWORT

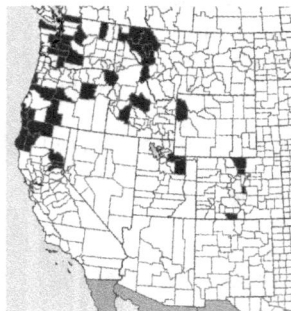

FAMILY Isoetaceae (Quillwort)

HABITAT Lakes.

HABIT Native, submersed perennial plant with a 2-lobed corm.

STEMS Underwater as a 2-lobed corm.

LEAVES Elongated, grass-like, coarse, to 30 in number, to 10 cm long, swollen at the base to contain the sporangia. **Sporangia** borne within the swollen, paler base of each leaf. **Spores** white or cream-colored with minute spines (when viewed through a microscope).

WETLAND STATUS AW OBL | GP OBL | WMV OBL

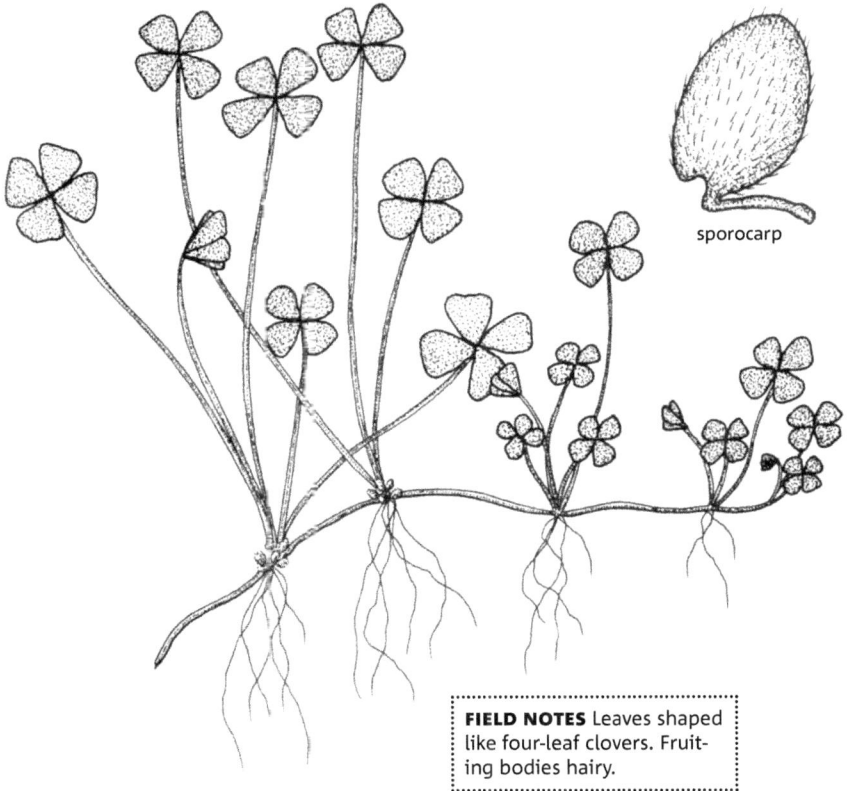

sporocarp

> **FIELD NOTES** Leaves shaped like four-leaf clovers. Fruiting bodies hairy.

Marsilea vestita Hook. & Grev.
HAIRY WATER FERN

FAMILY Marsileaceae (Water-Clover)

HABITAT Ponds, pools, creeks, roadside ditches, usually in shallow water.

HABIT Native perennial fern with slender, branching rhizomes.

STEMS All underwater as rhizomes.

LEAVES Divided into 4 leaflets resembling a four-leaf clover, each leaflet shaped like an upside-down triangle, to 2.5 cm long, to 2.5 cm wide, without teeth, smooth or hairy. **Sori** borne in hard, ellipsoid structures (sporocarps), each sporocarp solitary on stalks attached to the rhizomes, to 8 mm long, to 3 mm thick, with short, appressed hairs.

NOTE Plants stranded on land tend to be much hairier than those that live in water.

WETLAND STATUS AW OBL | GP OBL | WMV OBL

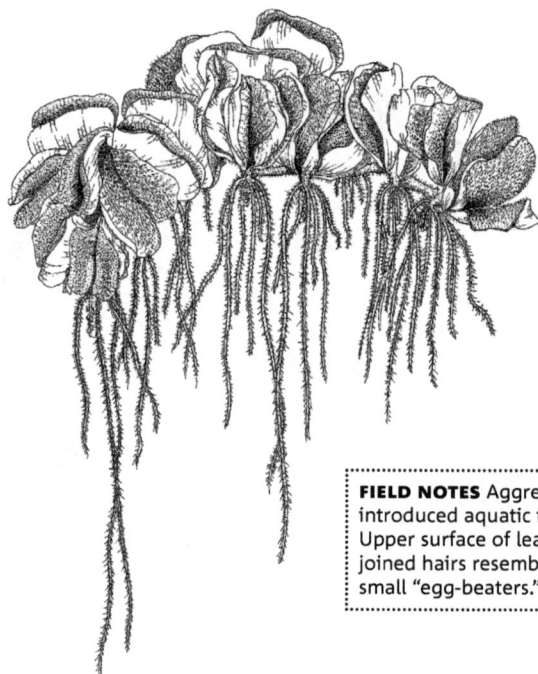

> **FIELD NOTES** Aggressive
> introduced aquatic fern.
> Upper surface of leaves with
> joined hairs resembling
> small "egg-beaters."

Salvinia molesta Mitchell
KARIBA-WEED, GIANT SALVINIA

FAMILY Salviniaceae (Water Fern)
HABITAT River channels, backwaters, floodplain pond
Habitat River channels, backwaters, floodplain ponds
HABIT Introduced, free-floating aquatic fern.
STEMS rootless (although hanging third leaf resembles roots), hairy, about 10 cm long.
LEAVES borne in threes; appear 2-ranked, but third leaf finely dissected and hanging, resembling roots; rounded to somewhat broadly elliptical, to about 3 cm long, with cordate base; upper surface with 4-pronged arching hairs joined into a dark knot at tips (resembling an egg beater), lower surface hairy. When young, the leaves are smaller and lie flat on the water surface.
SPORES Species now thought to be a sterile hybrid that spreads only by vegetative growth and fragmentation.
NOTE Native of Brazil, first introduced into North America as an ornamental, now highly invasive to the detriment of native species. When mature, *S. molesta* forms long chains of leaves to form thick mats on the surface of the water, the mats restrict oxygen and light availability for other aquatic species. A weevil, *Cyrtobagous salviniae,* has been introduced to help control the spread of this species.
WETLAND STATUS AW OBL | GP OBL | WMV OBL

leaf
underside

FIELD NOTES Leaves large, pinnate-pinnatifid. Sori elongate, borne in two rows that parallel the vein.

Woodwardia fimbriata J.E. Smith
GIANT CHAINFERN

FAMILY Blechnaceae (Chain Fern)

HABITAT Along streams, moist woods, bogs, springs.

HABIT Native perennial fern with a stout, woody, shiny, brown rhizome.

STEMS All underground as rhizomes.

LEAVES Upright, to 3 m tall, often forming a circle, pinnately divided, with each segment further deeply divided, the segments pointed at the tip, with minutely spiny teeth smooth or with resin glands. **Sori** elongated, borne in 2 rows, 1 on either side of the vein.

WETLAND STATUS AW FACW | WMV FACW

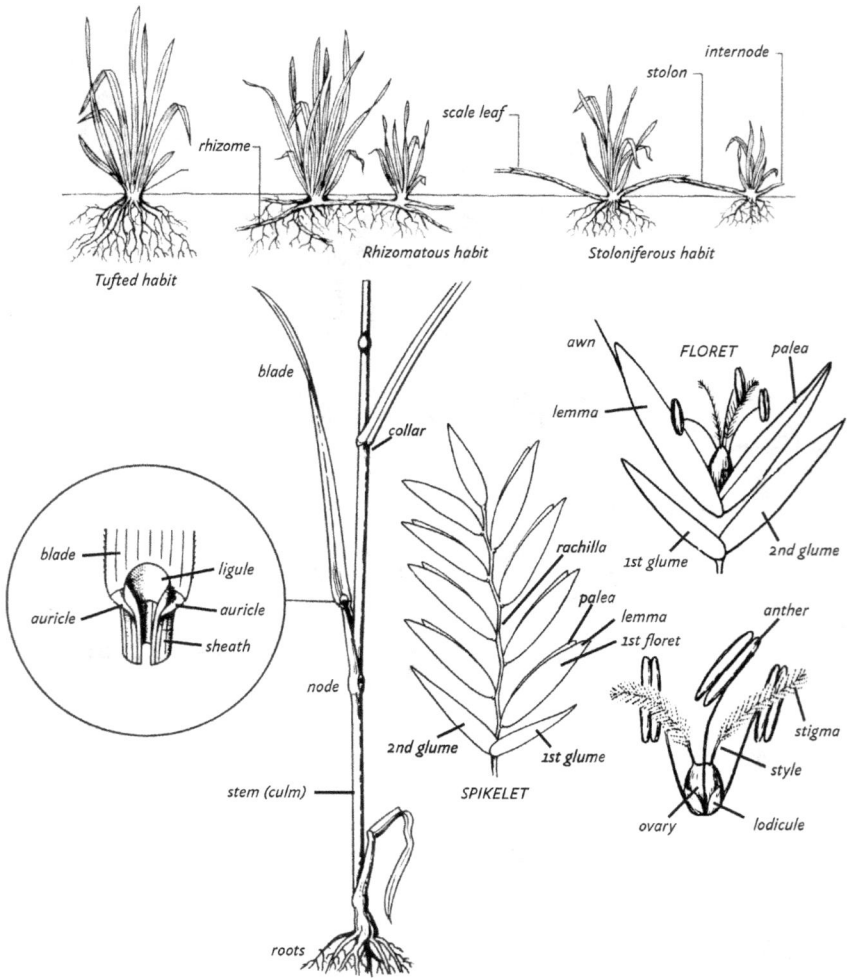

Tufted habit

Rhizomatous habit

Stoloniferous habit

rhizome

scale leaf

stolon

internode

blade

collar

awn

FLORET

palea

lemma

rachilla

palea

lemma

1st floret

anther

1st glume

2nd glume

blade

ligule

auricle

auricle

sheath

node

stigma

style

2nd glume

1st glume

stem (culm)

SPIKELET

ovary

lodicule

roots

> **FIELD NOTES** Panicles narrow, much longer than broad. Spikelets 1-flowered; glumes slightly roughened to the touch.

spikelet

Agrostis exarata Trin.
SPIKE BENTGRASS

FAMILY Poaceae (Grass)

FLOWERING July–August

HABITAT Moist places.

HABIT Native tufted perennial grass with short rhizomes.

STEMS Upright, unbranched, hollow, to 1 m tall, without hairs.

LEAVES Elongated, flat, 6–13 mm wide, rough to the touch; ligules to 8 mm long, minutely hairy.

FLOWERS Borne in spikelets, with numerous spikelets crowded into narrow, often spike-like panicles, the "spikes" sometimes interrupted; **spikelets** 1-flowered; **glumes** green or purplish, pointed or even slightly awned at the top; **lemma** awned. the awn 3–6 mm long. **Stamens** 3. **Pistils:** Ovary superior, smooth. **Grains** narrowly oblongoid, smooth.

NOTE This is an important range grass for domestic livestock.

WETLAND STATUS AW FACW | GP FACW | WMV FACW

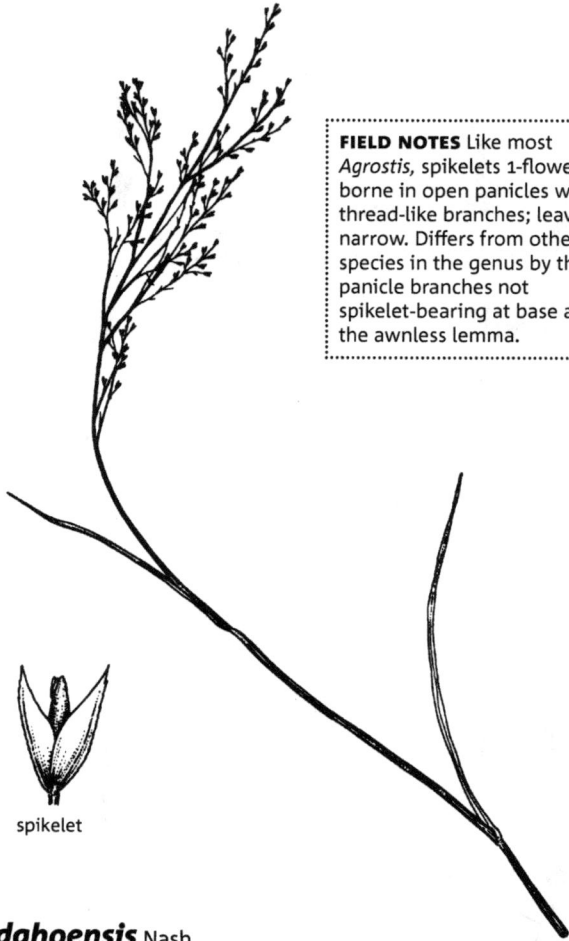

FIELD NOTES Like most *Agrostis,* spikelets 1-flowered, borne in open panicles with thread-like branches; leaves narrow. Differs from other species in the genus by the panicle branches not spikelet-bearing at base and the awnless lemma.

spikelet

Agrostis idahoensis Nash
IDAHO BENTGRASS

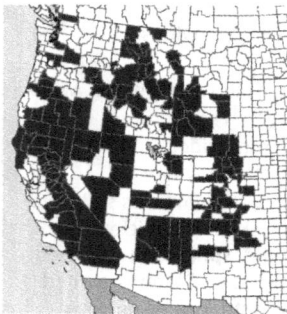

FAMILY Poaceae (Grass)
FLOWERING July–August
HABITAT Wet meadows, in bogs.
HABIT Native tufted perennial grass with fibrous roots.
STEMS Upright, slender, hollow, to 50 tall, without hairs.
LEAVES Elongated, narrow, flat or rolled into a tube, to 2 mm wide, without hairs.
FLOWERS Borne in 1-flowered spikelets, the spikelets arranged in an open panicle to 13 cm long, with the panicle branches thread-like and not bearing spikelets at their base; **spikelets** 2–3 mm long; **glumes** and **lemma** without hairs and awnless. **Stamens** 3. **Pistils:** Ovary superior. **Grains** very tiny, smooth.
WETLAND STATUS AW FACW | GP FACW | WMV FACW

> **FIELD NOTES** Spikes single, cylindrical, several times longer than broad, and not "fuzzy." Lemma awn to about 8 mm long.

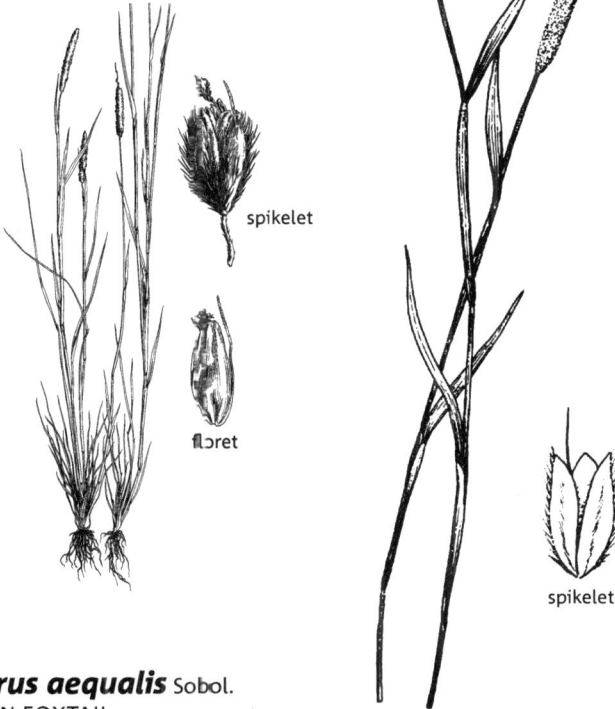

spikelet

floret

spikelet

Alopecurus aequalis Sobol.
SHORT-AWN FOXTAIL

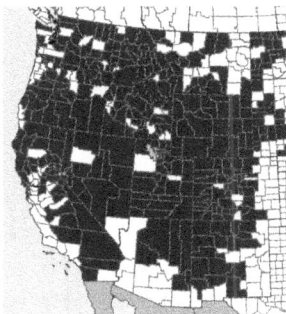

FAMILY Poaceae (Grass)
FLOWERING June–August
HABITAT Wet meadows, marshes, along streams, around ponds and lakes.
HABIT Native perennial grass with fibrous roots.
STEMS Spreading to ascending, often rooting at the nodes, to 45 cm tall, smooth.
LEAVES Elongated, to 4 mm wide, without hairs but usually rough to the touch on both surfaces; ligules to 8 mm long.
FLOWERS Borne in spikelets, with several spikelets crowded into a long cylindrical spike at the tip of the stem: spikes to 8 cm long, to 6 mm wide; **spikelets** 1-flowered: **glumes** sparsely hairy; **lemmas** smooth, with an awn arising from about the middle of the lemma, to 8 mm long. **Stamens** 3. **Pistils:** Ovary superior. **Grains** ellipsoid, smooth.
NOTE An important forage grass for domestic livestock.
WETLAND STATUS AW OBL | GP OBL | WMV OBL

> **FIELD NOTES** Head at end of stem "fuzzy," short, not more than 4 cm long.

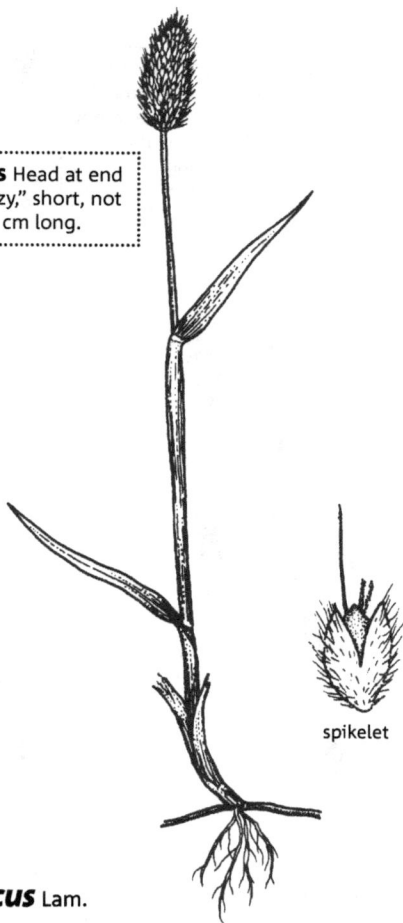

spikelet

Alopecurus magellanicus Lam.
MOUNTAIN FOXTAIL

FAMILY Poaceae (Grass)

FLOWERING June–August

HABITAT Wet meadows, along streams, mostly in the high mountains.

HABIT Native perennial grass with short rhizomes and sometimes with stolons.

STEMS Upright, unbranched. to 75 cm tall, smooth.

LEAVES Elongated, narrow, to 6 mm wide, without hairs but rough to the touch on both surfaces.

FLOWERS Borne in spikelets. with several spikelets crowded into a "fuzzy" head at the tip of the stem: heads to 4 cm long, to 13 mm wide; **spikelets** 1-flowered; **glumes** densely long-hairy: **lemmas** with an awn to 8 mm long, the awn arising from below the middle of the lemma. **Stamens** 3. **Pistils**: Ovary superior, smooth. **Grains** ellipsoid, smooth.

SYNONYMS *Alopecurus alpinus* J.E. Smith

WETLAND STATUS AW FACW | GP FACW | WMV FACW

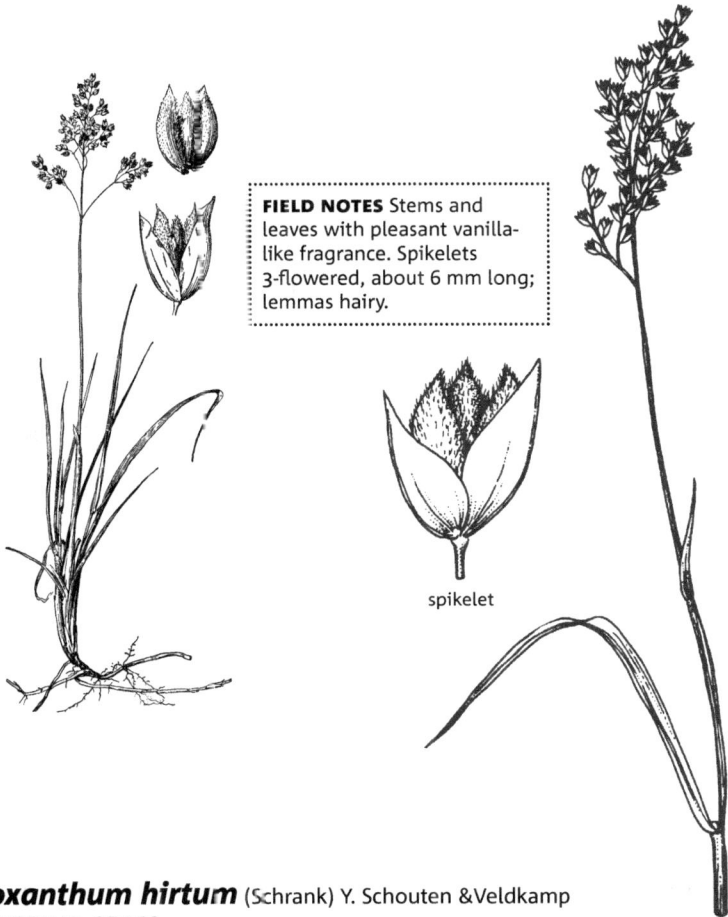

> **FIELD NOTES** Stems and
> leaves with pleasant vanilla-
> like fragrance. Spikelets
> 3-flowered, about 6 mm long;
> lemmas hairy.

spikelet

Anthoxanthum hirtum (Schrank) Y. Schouten &Veldkamp
SWEET VERNAL GRASS

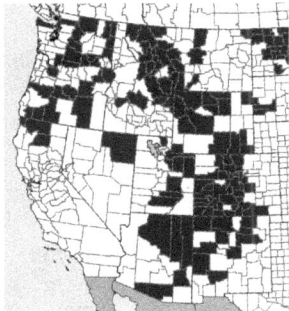

FAMILY Poaceae (Grass)

FLOWERING June–August

HABITAT Wet meadows, along streams, in bogs.

HABIT Native perennial grass with branching rhizomes.

STEMS Upright, hollow, to 60 cm tall, without hairs.

LEAVES Elongated, narrow, flat, to 6 mm wide, without hairs; ligule to 6 mm long, more or less fringed.

FLOWERS Borne in 3-flowered spikelets, the spikelets arranged in a panicle to 10 cm long; **spikelets** about 6 mm long, the lower 2 flowers male only, the upper flower with both stamens and pistils; **lemmas** hairy. **Stamens** 3. **Pistils**: Ovary superior. **Grains** small, ovoid, smooth.

NOTE This species is also known as sweetgrass or vanilla grass because of the pleasing fragrance of its stems and leaves.

SYNONYMS *Hierochloe odorata* (L.) Beauv.

WETLAND STATUS AW FACW | CP FACW | WMV FACW

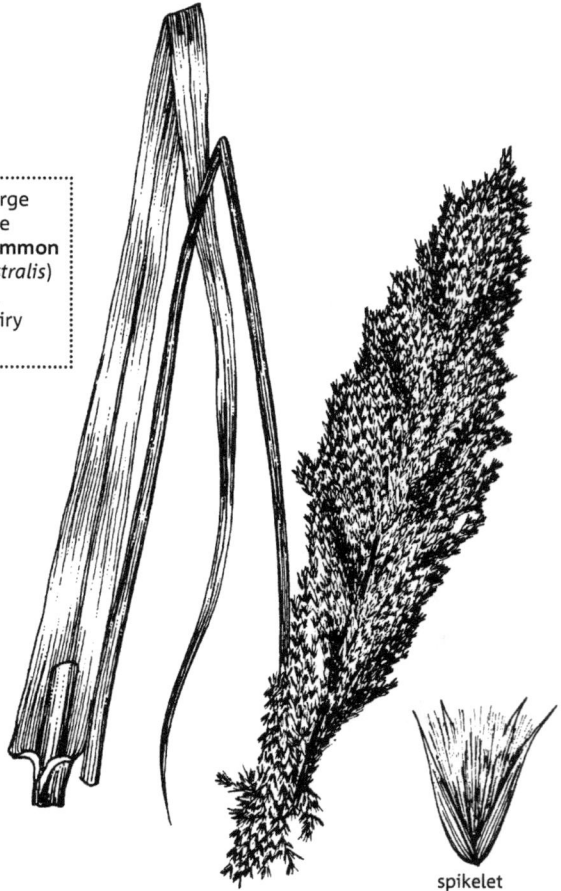

> **FIELD NOTES** Very large grass, differs from the somewhat similar **common reed** (*Phragmites australis*) by being larger in all aspects and by its hairy lemmas.

spikelet

Arundo donax L.
GIANT REED

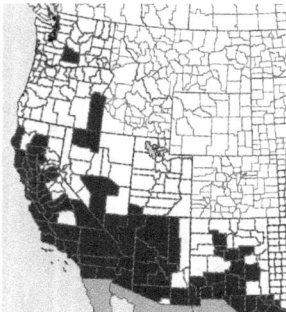

FAMILY Poaceae (Grass)

FLOWERING August–October

HABITAT Moist, disturbed soil, often along roads and in irrigation ditches.

HABIT Introduced, very large, clump-forming perennial grass with stout, knotty rhizomes.

STEMS Upright, very stout, hollow, to 6 m tall.

LEAVES Elongated, flat, to 10 cm wide, heart-shaped and with tufts of hairs at the base.

FLOWERS Borne in spikelets, the spikelets arranged in dense, plume-like spikes to 60 cm long and to 25 cm thick, each **spikelet** several-flowered, about 13 mm long; **lemmas** hairy, tipped with a short awn. **Stamens** 3. **Pistils:** Ovary superior. **Grains** ovoid, smooth.

NOTE This giant grass is native to Europe and has been planted in the United States as an ornamental which occasionally escapes into moist habitats.

WETLAND STATUS AW FACW | GP FAC | WMV FACW

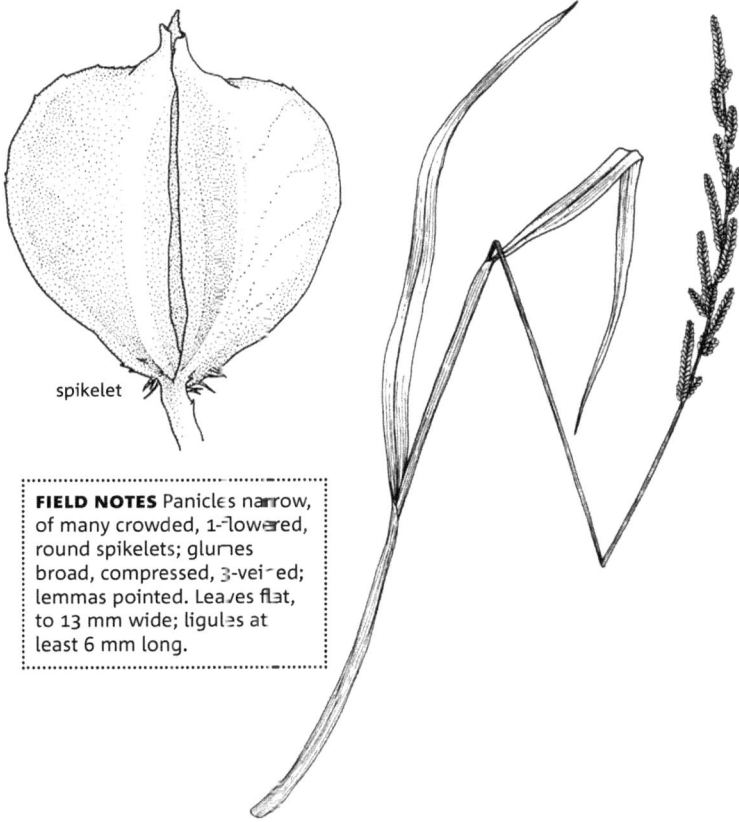

spikelet

FIELD NOTES Panicles narrow, of many crowded, 1-flowered, round spikelets; glumes broad, compressed, 3-veined; lemmas pointed. Leaves flat, to 13 mm wide; ligules at least 6 mm long.

Beckmannia syzigachne (Steud.) Fern.
AMERICAN SLOUGHGRASS

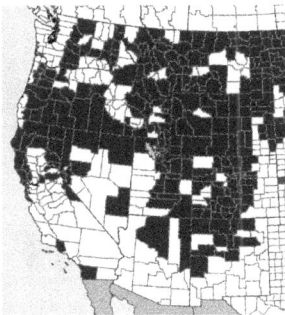

FAMILY Poaceae (Grass)

FLOWERING July–August

HABITAT Along streams, in marshes, around ponds and lakes, in wet roadside ditches.

HABIT Native annual grass with fibrous roots, but sometimes forming stolons.

STEMS Upright, hollow, usually unbranched, to 1 m tall, without hairs.

LEAVES Alternate, elongated, flat, to 13 mm wide, long-tapering to the tip, without hairs; ligules at least 6 mm long.

FLOWERS Borne in 1-flowered spikelets, the **spikelets** crowded into narrow panicles to 30 cm long, each spikelet round, to 4 mm long; **glumes** broad, compressed, 3-veined, usually smooth; **lemmas** pointed, smooth. Stamens 3. Pistils: Ovary superior, smooth. **Grains** nearly spherical, smooth.

NOTE The grains are eaten by waterfowl.

WETLAND STATUS AW OBL | GP OBL | WMV OBL

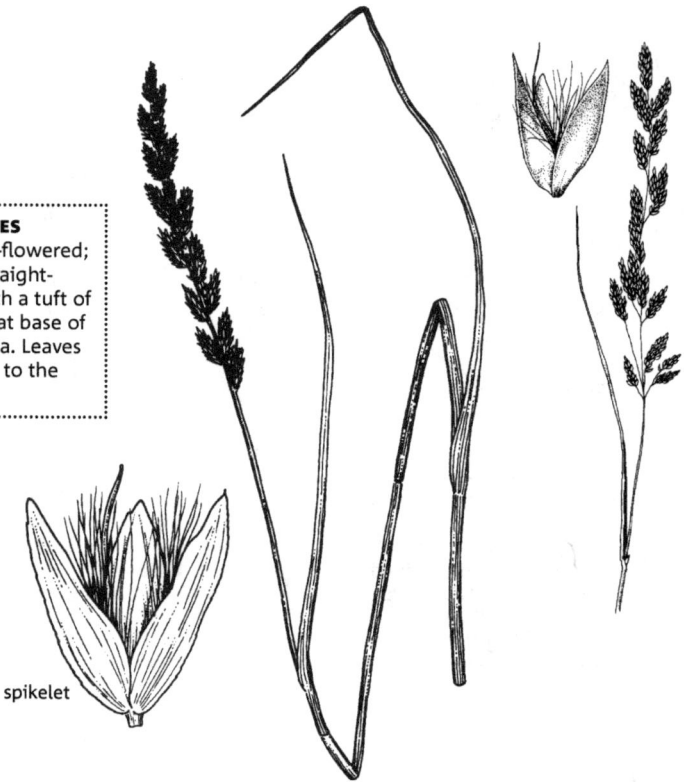

FIELD NOTES
Spikelets 1-flowered; lemmas straight-awned, with a tuft of long hairs at base of each lemma. Leaves very rough to the touch.

spikelet

Calamagrostis stricta (Timm) Koel.
NARROW-SPIKE REEDGRASS

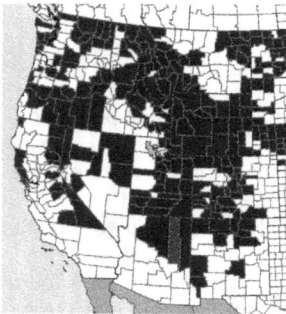

FAMILY Poaceae (Grass)
FLOWERING July–September
HABITAT Wet meadows, marshes, boggy areas.
HABIT Native tufted perennial grass with slender rhizomes.
STEMS Upright, to 1 m tall, rough to the touch.
LEAVES Elongated, narrow, flat or rolled up, 2–4 mm wide, rough to the touch.
FLOWERS 1 per spikelet, with many spikelets arranged in a narrow, dense panicle, the panicle to 15 cm long; **spikelets** to 6 mm long; **lemmas** with an awn about as long as the glumes. **Stamens** 3. **Pistils:** Ovary superior.

Grains smooth.
NOTE Bluejoint (*Calamagrostis canadensis* (Michx.) Beauv. is also a common wetland species in the western USA; distinguished from *C. stricta* by its generally larger size, more open inflorescence, and flat leaves (rather than often inrolled in *C. stricta*).
SYNONYMS *Calamagrostis inexpansa* Gray; *C. neglecta* (Ehrh.) P. Gaertn., B. Meyer & Scherb.
WETLAND STATUS AW FACW | GP FACW | WMV FACW

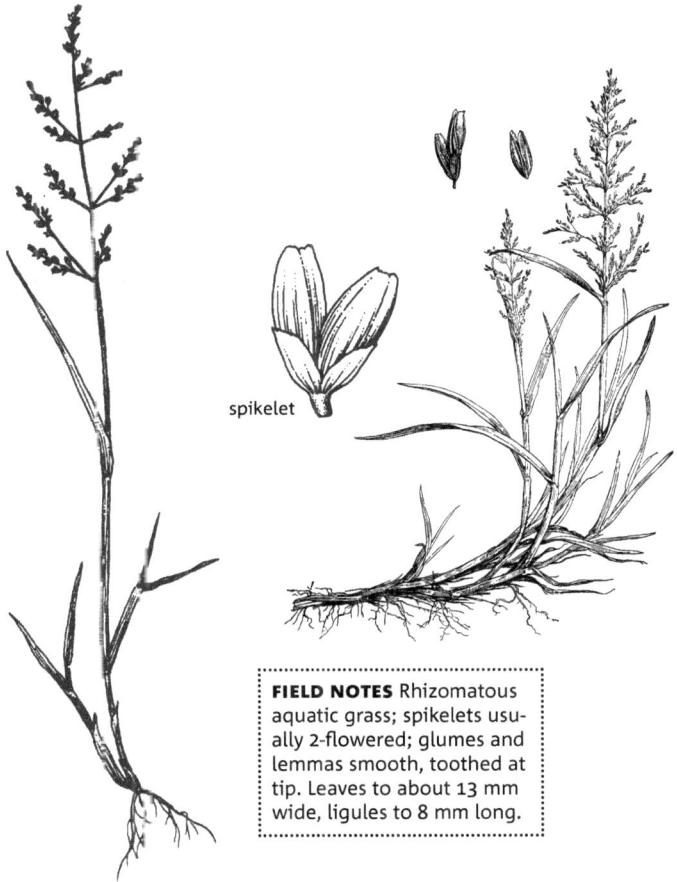

spikelet

> **FIELD NOTES** Rhizomatous aquatic grass; spikelets usually 2-flowered; glumes and lemmas smooth, toothed at tip. Leaves to about 13 mm wide, ligules to 8 mm long.

Catabrosa aquatica (L.) Beauv.
BROOKGRASS

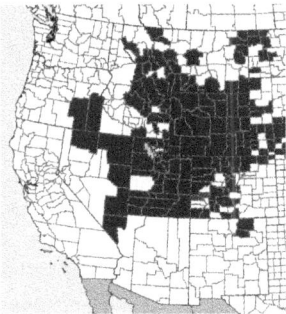

FAMILY Poaceae (Grass)
FLOWERING June–September
HABITAT Along streams, around ponds and lakes, usually in water.
HABIT Native aquatic perennial grass with rhizomes.
STEMS Decumbent and rooting at the nodes, later becoming upright, branched or unbranched, hollow, to 60 cm tall, smooth.
LEAVES Alternate, elongated, flat, to 13 mm wide, smooth; ligules to 8 mm long.
FLOWERS Borne in usually 2-flowered spikelets. the spikelets arranged in open panicles to 25 cm long; each spikelet to 4 mm long; glumes and lemmas toothed at the tip, smooth. **Stamens** 3. **Pistils**: Ovary superior, smooth. **Grains** oval, flat, brown.
NOTE The grains are eaten by waterfowl.
WETLAND STATUS AW OBL | GP OBL | WMV OBL

> **FIELD NOTES** Spikelets 1-flowered, borne in an open panicle; lemmas with a short awn. Leaves 13–20 mm wide.

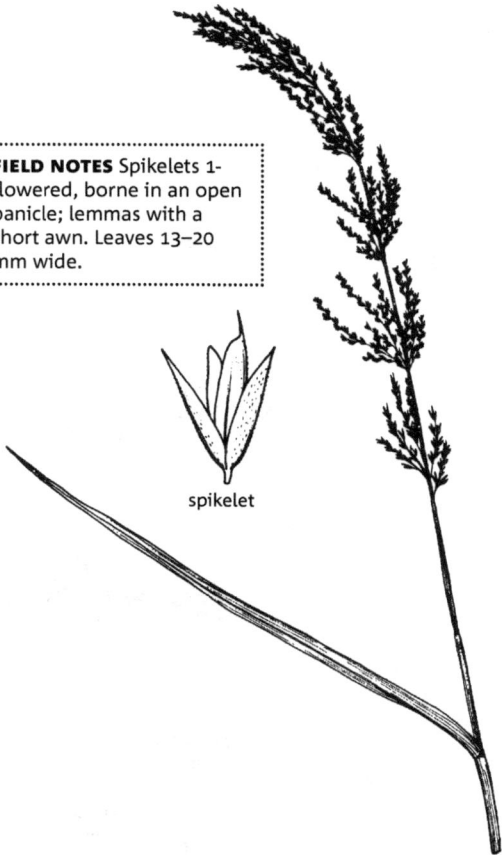

spikelet

Cinna latifolia (Trev. ex Goepp.) Griseb.
SLENDER WOOD-REEDGRASS

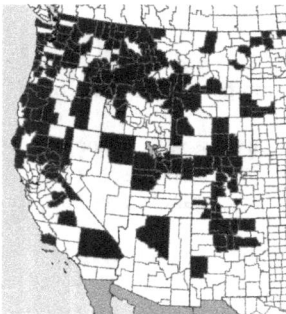

FAMILY Poaceae (Grass)

FLOWERING July–August

HABITAT Along streams, in wet meadows, damp woods, around springs.

HABIT Native perennial grass with thickened rhizomes.

STEMS Upright, unbranched, hollow, to 2 m tall, without hairs.

LEAVES Elongated, to 20 mm wide, flat, somewhat rough to the touch; ligules to 8 mm long, jagged at the tip.

FLOWERS Borne in spikelets, with many spikelets in an open panicle with thread-like branches; panicles to 30 cm long; **spikelets** 1-flowered, to 6 mm long; **glumes** narrow, lanceolate, rough to the touch; **lemmas** narrow, lanceolate, rough to the touch, with a short awn. **Stamens** 1–3. **Pistils**: Ovary superior, smooth. **Grains** ovoid, smooth.

NOTE The grains are eaten by ducks and other birds.

WETLAND STATUS AW FACW | GP OBL | WMV FACW

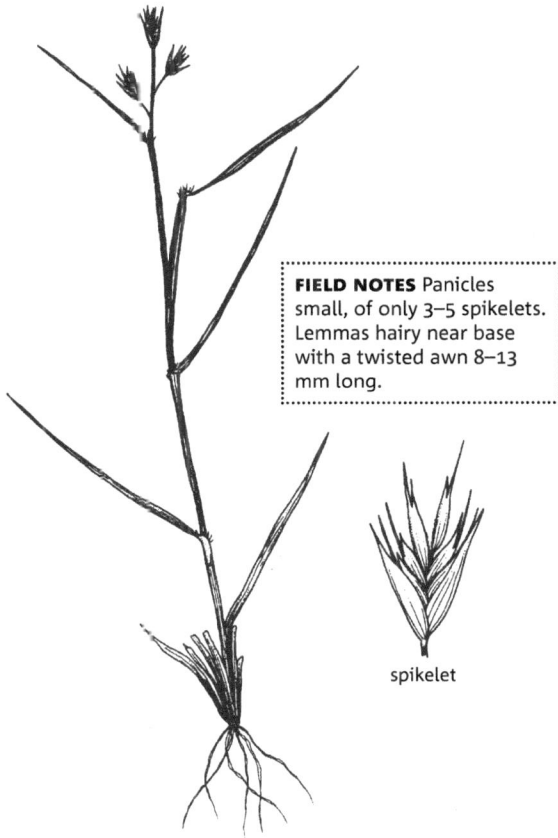

> **FIELD NOTES** Panicles small, of only 3–5 spikelets. Lemmas hairy near base with a twisted awn 8–13 mm long.

spikelet

Danthonia californica Boland.
CALIFORNIA OATGRASS

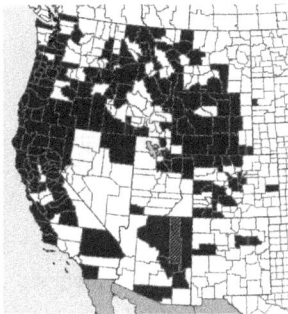

FAMILY Poaceae (Grass)
FLOWERING May–August
HABITAT Meadows and open hillsides.
HABIT Native, densely tufted perennial grass, with fibrous roots and last year's brown leaf sheaths persistent at base of plant.
STEMS Upright, hollow, to 1 m tall, smooth.
LEAVES Elongated, flat or sometimes rolled into a tube, to 25 cm long, to 4 mm wide, rough to the touch and usually with a few long hairs along the margins; sheaths with a few long hairs at the tip.
FLOWERS Borne in 5- to 8-flowered, usually purple, spikelets, with 3–5 spikelets arranged in a small panicle to 8 cm long; branches of the panicles usually hairy; **glumes** 13–20 mm long, smooth or slightly rough; **lemmas** 8–13 mm long, hairy near the base, with a twisted awn 8–13 mm long. **Stamens** 3. **Pistils:** Ovary superior, smooth. **Grains** ellipsoid, smooth.
NOTE The grains are eaten by small birds and mammals.
WETLAND STATUS AW FAC | GP FACU | WMV FAC

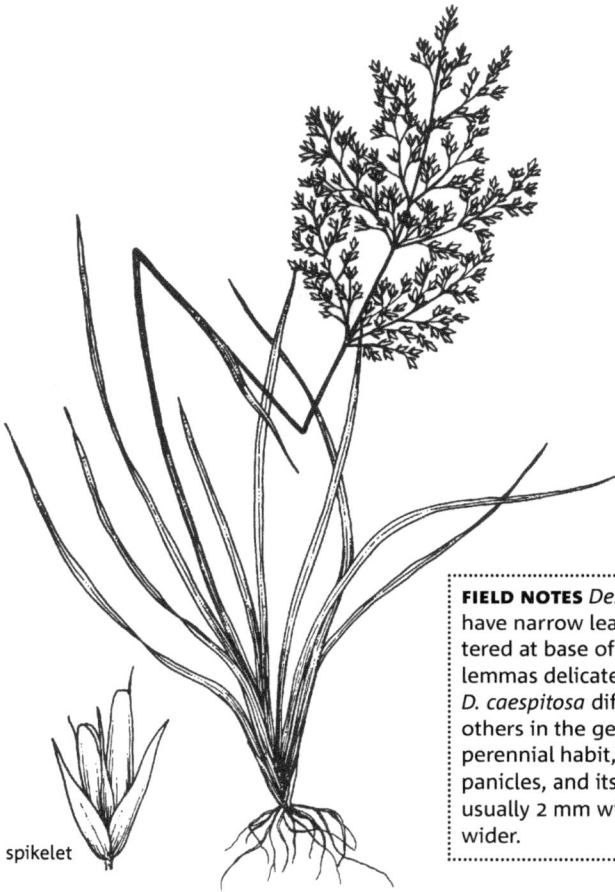

spikelet

> **FIELD NOTES** *Deschampsia* have narrow leaves clustered at base of plant and lemmas delicately awned. *D. caespitosa* differs from others in the genus by its perennial habit, its open panicles, and its leaves usually 2 mm wide or wider.

Deschampsia caespitosa (L.) Beauv.
TUFTED HAIRGRASS

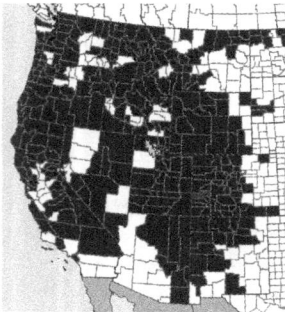

FAMILY Poaceae (Grass)

FLOWERING July–September

HABITAT Wet meadows, wet prairies, ditches, along streams, around lakes, particularly at high elevations.

HABIT Native, densely tufted perennial grass with short rhizomes.

STEMS Upright, smooth, to 1 m tall.

LEAVES Elongated, very narrow, mostly crowded near the base of the plant, flat or folded or even sometimes rolled up, 2–3 mm wide, rough to the touch.

FLOWERS Usually 2 per spikelet, with many spikelets arranged in loose, often nodding panicles with thread-like branches, the panicle to 20 cm long; **spikelets** usually purplish, shiny; **lemmas** with a slender awn to 6 mm long. Stamens 3. Pistils: Ovary superior. Grains smooth.

NOTE This species provides cover for quail and other birds.

WETLAND STATUS AW FACW | GP FACW | WMV FACW

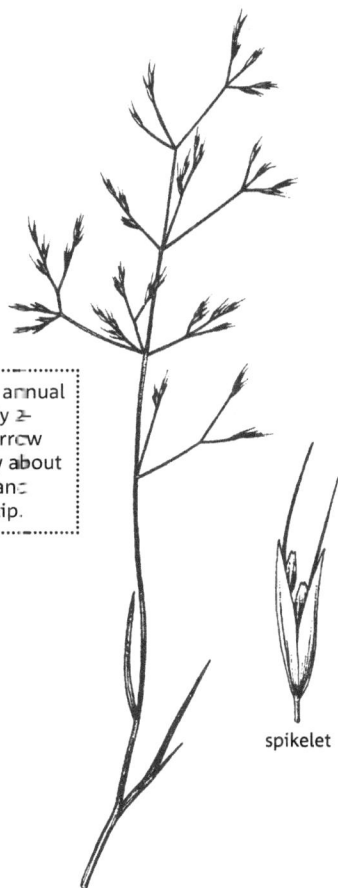

> **FIELD NOTES** Slender annual grass. Spikelets usually 2-flowered, borne in narrow panicles; lemmas only about 3 mm long, rounded and shallowly toothed at tip.

spikelet

Deschampsia danthonioides (Trin.) Munro
ANNUAL HAIRGRASS

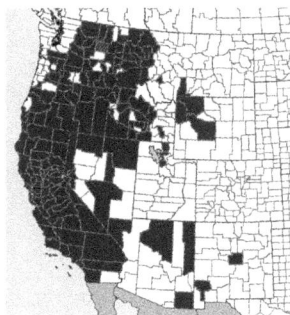

FAMILY Poaceae (Grass)

FLOWERING June–August

HABITAT Along streams, in meadows, mud flats, vernal pools.

HABIT Native tufted annual grass with fibrous roots.

STEMS Slender, upright, hollow, unbranched, to 45 cm tall, smooth.

LEAVES Elongated, thread-like, to 10 cm long, about 1 mm wide.

FLOWERS Borne in 2-flowered spikelets, the spikelets arranged in narrow panicles to 20 cm long; **glumes** to 8 mm long, pointed at the tip; **lemmas** only about 3 mm long, rounded and shallowly toothed at the tip, with hairs at the base. **Stamens** 3. **Pistils**: Ovary superior, smooth. **Grains** ellipsoid, smooth.

NOTE The grains are eaten by waterfowl.

WETLAND STATUS AW FACW | GP FACW | WMV FACW

> **FIELD NOTES** Leaves very narrow, crowded at base of plant. Panicle narrow with strongly ascending branches; lemmas with a tuft of hairs at base and a slender awn that is attached below middle of lemma.

spikelet

Deschampsia elongata (Hook.) Munro
SLENDER HAIRGRASS

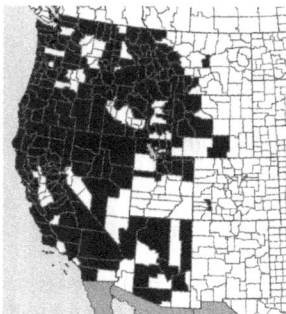

FAMILY Poaceae (Grass)

FLOWERING June–August

HABITAT Wet meadows, along streams, around ponds and lakes.

HABIT Native tufted perennial grass with fibrous roots.

STEMS Upright, slender, hollow, unbranched, to 60 cm tall, not hairy.

LEAVES Most of them crowded at the base of the plant, thread-like, to 10 cm long, about 1 mm wide; those on the stem linear, to 6 mm wide.

FLOWERS Borne in 2- or 3-flowered spikelets, the spikelets arranged in narrow panicles with ascending, thread-like branches, the panicles to 30 cm long; **glumes** to 8 mm long, hairy: **lemmas** to 3 mm long, with a tuft of hairs at the base and a slender awn to 6 mm long. **Stamens** 3. **Pistils:** Ovary superior, smooth. **Grains** ovoid, smooth.

NOTE This grass may be used as nesting cover for quail and other birds.

WETLAND STATUS AW FACW | GP FAC | WMV FACW

> **FIELD NOTES** *Diplachne* have sessile spikelets borne in two rows or two sides of the main axis (rachis). Spikelets have 5–12 flowers, with lemmas 4–8 mm long.

spikelet

Diplachne fusca (L.) Beauv. ex Roemer & J. A. Schultes
BEARDED SPRANGLETOP

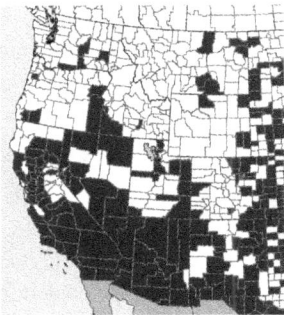

FAMILY Poaceae (Grass)

FLOWERING July–September

HABITAT Around lakes and ponds, along streams, in marshes, sometimes in shallow standing water. This species may be found in sandy or alkaline habitats.

HABIT Native tufted annual grass with fibrous roots.

STEMS Upright, branched, hollow, to 75 cm tall, smooth.

LEAVES Elongated, narrow, rolled up into a hollow tube or sometimes flat, to 8 mm wide, smooth or rough to the touch, the uppermost sometimes surpassing the inflorescence.

FLOWERS Borne in spikelets, the sessile spikelets arranged in two rows on two sides of the main axis, each spikelet with 5–12 flowers; **lemmas** 2–8 mm long, hairy, awned at the tip. **Stamens** 3. **Pistils:** Ovary superior. **Grains** small, ovoid, smooth.

NOTE The grains are eaten by waterfowl.

SYNONYMS *Leptochloa fascicularis* (Lam.) Gray

WETLAND STATUS AW FACW | GP FACW | WMV FACW

> **FIELD NOTES** Spikelets uni-
> sexual, 5- to 15-flowered,
> with the two sexes borne
> on separate plants. Plants
> typically sod-forming, often
> in salty and alkaline areas.

spikelet

Distichlis spicata (L.) Greene
INLAND SALTGRASS

FAMILY Poaceae (Grass)

FLOWERING May–August

HABITAT Salt marshes, alkaline flats, along roads.

HABIT Native perennial grass from extensive, much
branched rhizomes.

STEMS Mat-forming, but with ascending stems to 40 cm
tall, smooth, stiff.

LEAVES Narrow, elongated, flat to rolled into a tube,
long-tapering to the tip, to 3 mm wide, rough to the
touch.

FLOWERS 5–15 flowers in a spikelet, with male flowers
in separate spikelets on separate plants from the
female; spikelets many in a short panicle to 8 cm long; each **spikelet** to 20 mm long;
glumes and **lemmas** straw-colored, without awns. **Stamens** 3. **Pistils:** Ovary superior.
Grains smooth.

NOTE The female plants usually have shorter stems than the male plants.

SYNONYMS *Distichlis stricta* (Torr.) Scribn.

WETLAND STATUS AW FAC | GP FACW | WMV FACW

FIELD NOTES Spikelets mostly 3- to 5-flowered. Rhizomes absent. Glumes very short-awned, and glumes at least 2/3 as long as the spikelets.

spikelet

Elymus trachycaulus (Link) Gould ex Shinners
SLENDER WHEATGRASS

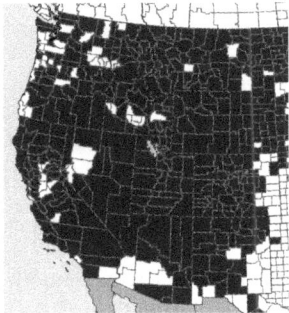

FAMILY Poaceae (Grass)
FLOWERING June–August
HABITAT Along streams, moist woods, roadside ditches, meadows.
HABIT Native perennial grass with fibrous roots.
STEMS Upright, unbranched, hollow, to 1 m tall, smooth, although the sheaths may be hairy.
LEAVES Elongated, flat, to 8 mm wide, hairy or at least rough to the touch on the upper surface; ligules extremely short and ciliate.
FLOWERS Borne in spikelets on either side of the axis, sometimes overlapping, mostly 3- to 5-flowered, to 20 mm long; **glumes** pointed at the tip or with a very short awn, the glumes as least 2/3 as long as the spikelets: **lemmas** to 13 mm long, awnless or with an awn to 4 cm long. **Stamens** 3. **Pistils:** Ovary superior, smooth. **Grains** oblongoid, usually hairy at the tip.
SYNONYMS *Agropyron trachycaulum* (Link) Malte ex H.F. Lewis
NOTE This species is a valuable range grass for domesticated livestock. This species is known to hybridize with other grasses so that intermediate plants may be found.
WETLAND STATUS AW FACU | GP FACU | WMV FAC

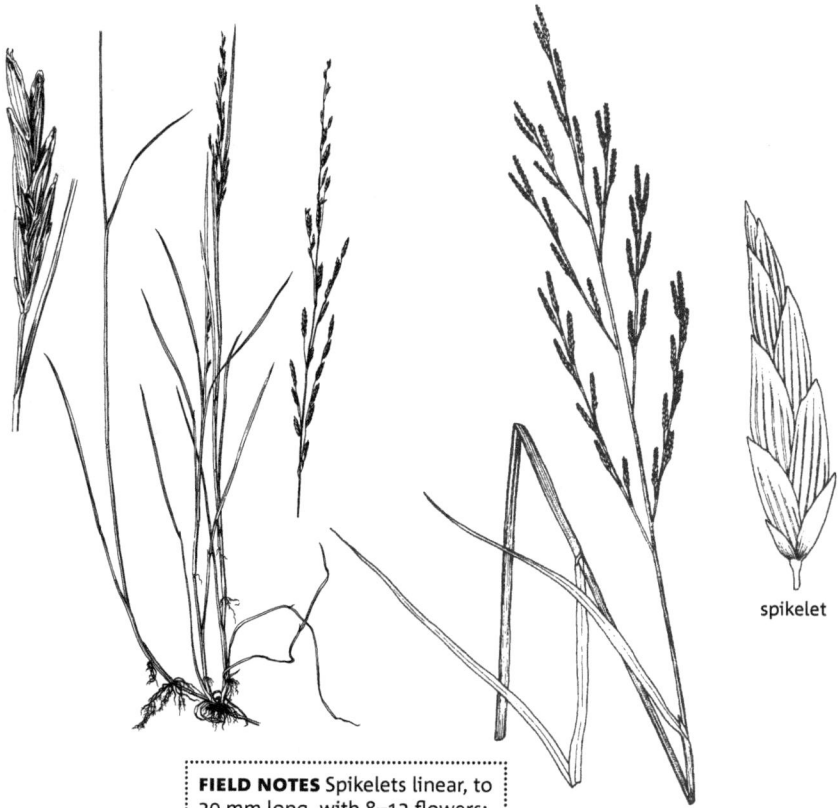

spikelet

> **FIELD NOTES** Spikelets linear, to
> 20 mm long, with 8–12 flowers;
> lemmas do not have hairs.

Glyceria borealis (Nash) Batchelder
SMALL FLOATING MANNA GRASS

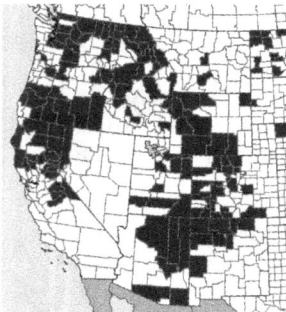

FAMILY Poaceae (Grass)

FLOWERING June–August

HABITAT In shallow water of ponds and lakes and in wet meadows in the mountains.

HABIT Native perennial grass with rhizomes.

STEMS Eventually upright, unbranched, hollow, to 1.5 m tall, without hairs.

LEAVES Elongated, flat or folded, 6–8 mm wide, smooth or rough to the touch; ligules to 13 mm long.

FLOWERS Borne in spikelets, with several spikelets in narrow panicles to 45 cm long; **spikelets** 8- to 12-flowered, 13–20 mm long; **glumes** lanceolate, smooth; **lemmas** without hairs, 7-nerved. **Stamens** 3. **Pistils:** Ovary superior, smooth. **Grains** ellipsoid, smooth.

NOTE The grains are eaten by waterfowl.

WETLAND STATUS AW OBL | GP OBL | WMV OBL

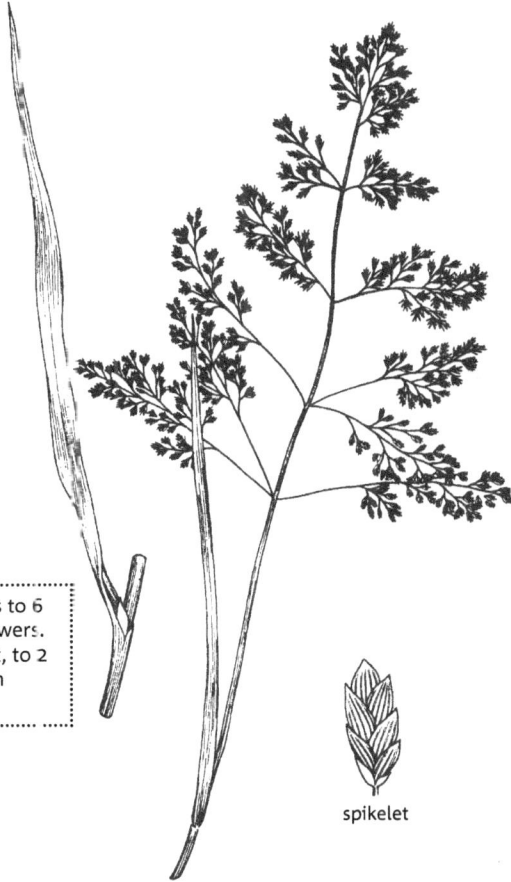

FIELD NOTES Spikelets to 6 mm long, with 4–8 flowers. Stems rather succulent, to 2 m tall; leaves 4–12 mm wide.

spikelet

Glyceria elata (Nash ex Rydb.) M.E. Jones
TALL MANNA GRASS

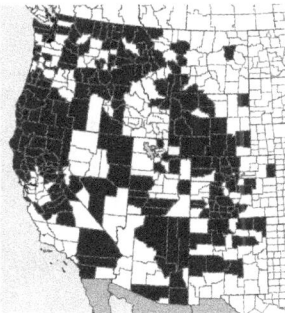

FAMILY Poaceae (Grass)
FLOWERING June–August
HABITAT Wet meadows, moist woods, in shallow streams, ponds and lakes, ditches, around springs, in the mountains.
HABIT Perennial herb with creeping rhizomes.
STEMS Upright, usually unbranched, rather succulent, to 2 m tall, smooth.
LEAVES Elongated, 4–12 mm wide, rough to the touch; ligules to 6 mm long, finely hairy.
FLOWERS 4–8 in a spikelet, with many spikelets arranged in an open panicle to 25 cm long; **spikelets** to 6 mm long, somewhat flattened; **lemmas** with 7 nerves. **Stamens** 2. **Pistils:** Ovary superior, smooth. **Grains** narrow-ovoid, smooth.
NOTE The seeds are eaten by waterfowl.
WETLAND STATUS AW OBL | GF OBL | WMV OBL

FIELD NOTES Spikelets
many-flowered, awnless,
less than 6 mm long.

spikelet

Glyceria striata (Lam.) A. S. Hitchc.
FOWL MANNA GRASS

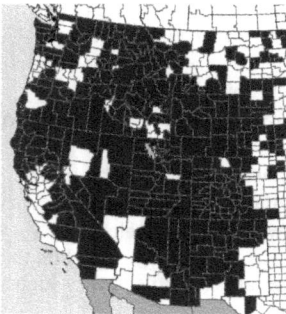

FAMILY Poaceae (Grass)

FLOWERING May–August.

HABITAT Wet meadows, low woods, bogs, roadside ditches, swamps.

HABIT Native, tufted perennial grass with fibrous roots or short rhizomes.

STEMS Erect, smooth, to 1.2 m tall.

LEAVES Flat or sometimes folded lengthwise, somewhat rough to the touch, to 8 mm wide.

FLOWERS 3–7 per spikelet, with many spikelets in slender panicles to 30 cm long; **spikelets** less than 6 mm long. **Stamens** 3. **Pistils**: Ovary superior. **Grains** shiny, reddish, 1–2 mm long.

WETLAND STATUS AW OBL | GP OBL | WMV OBL

> **FIELD NOTES** Leaves and stems velvety-hairy. Spikelets 2-flowered; glumes hairy, lemmas ciliate; the awn of one of the lemmas is hooked.

spikelet

Holcus lanatus L.
COMMON VELVET GRASS

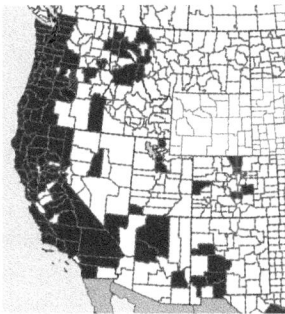

FAMILY Poaceae (Grass)

FLOWERING May–September

HABITAT Moist meadows, ditches, other disturbed areas.

HABIT Introduced perennial grass with fibrous roots.

STEMS Upright, unbranched, hollow, to 60 cm tall, velvety-hairy.

LEAVES Elongated, flat, 8–13 mm wide, velvety-hairy; ligules 3–4 mm long, finely hairy.

FLOWERS Borne in spikelets, with many spikelets crowded in rather narrow panicles, sometimes appearing spike-like, to 15 cm long; **spikelets** 2-flowered, to 4 mm long; **glumes** hairy; **lemmas** ciliate, the upper one with a hooked awn. **Stamens** 3. **Pistils**: Ovary superior, smooth. **Grains** ovoid, smooth.

NOTE European native. The leaves and stems are grazed by deer and domestic animals.

WETLAND STATUS AW FAC | GP FAC | WMV FAC

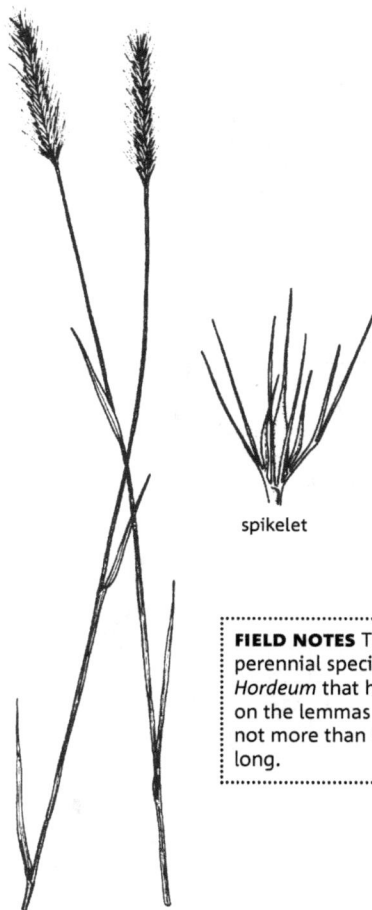

spikelet

> **FIELD NOTES** The only perennial species of *Hordeum* that has awns on the lemmas usually not more than 13 mm long.

Hordeum brachyantherum Nevskii
MEADOW BARLEY

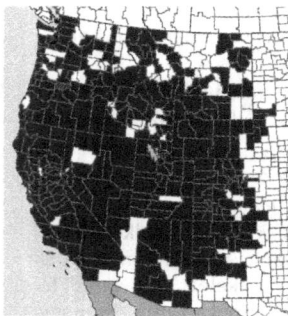

FAMILY Poaceae (Grass)

FLOWERING June–August

HABITAT Wet meadows, along streams, in disturbed areas.

HABIT Native tufted perennial grass with fibrous roots.

STEMS Upright or spreading, hollow, to 60 cm tall, smooth or sparsely hairy.

LEAVES Elongated, flat, to 8 mm wide, usually hairy and rough to the touch.

FLOWERS Borne in clusters of 3 spikelets, with many clusters forming a spike to 8 cm long; **glumes** awn-like to 2.5 cm long; **lemmas** with an awn not more than 13 mm long. **Stamens** 3. **Pistils:** Ovary superior. **Grains** oblongoid. hairy at the tip.

NOTE A valuable forage grass for domestic livestock, particularly in the higher ranges.

WETLAND STATUS AW FACW | GP FAC | WMV FACW

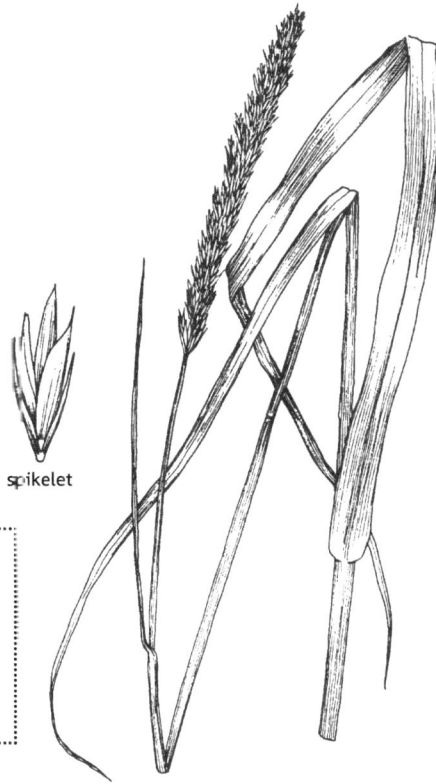

spikelet

> **FIELD NOTES** Large, clump-forming grass; rhizomes usually absent. Glumes awn-like and not broadened above the base. Spikelets 3–6 at each node in the inflorescence. Leaves 6–20 mm wide.

Leymus cinereus (Scribn. & Merr.) A. Löve
BASIN WILD-RYE

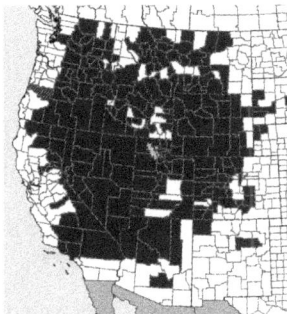

FAMILY Poaceae (Grass)

FLOWERING June–August

HABITAT Along streams, in meadows, along roads, edge of woods, sagebrush areas.

HABIT Native, clump-forming perennial grass, usually without rhizomes.

STEMS Upright, unbranched, stout, hollow, to 2 m tall, smooth or rough-hairy.

LEAVES Elongated, 6–20 mm wide, hairy; ligules to 8 mm long, membranaceous.

FLOWERS Borne in spikelets, with 3–6 spikelets at a node, forming a spike to 20 cm long, to 13 mm broad; **glumes** awn-like, not broadened above the base; **lemma** usually hairy, with or without an awn to 6 mm long. **Stamens** 3. **Pistils:** Ovary superior, smooth. **Grains** narrowly oblongoid, hairy at the tip.

NOTE This grass may be browsed by livestock. It is a good soil stabilizer and is used in new roadcuts. Black ergot, a fungus, is often present in the spikelets and can prove harmful to livestock.

SYNONYMS *Elymus cinereus* Scribn. & Merr.

WETLAND STATUS AW FAC | GP JPL | WMV FAC

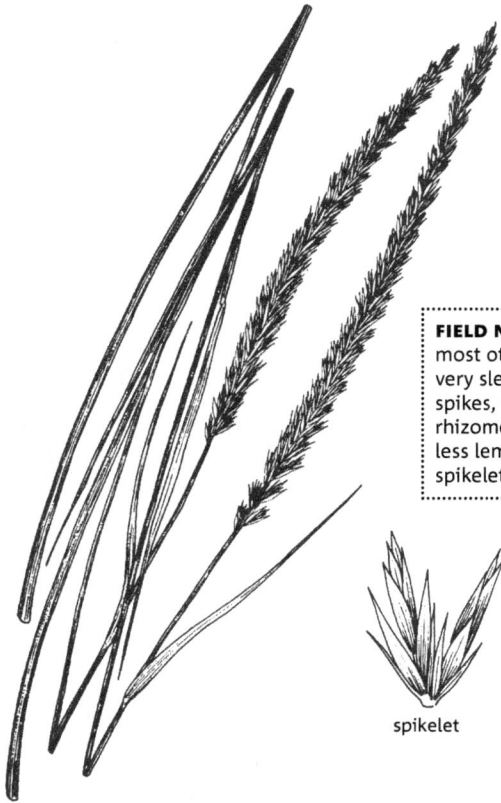

> **FIELD NOTES** Differs from most other *Leymus* by its very slender and short spikes, its long-creeping rhizomes, its usually awnless lemmas, and its paired spikelets 13–20 mm long.

spikelet

Leymus triticoides (Buckl.) Pilger
CREEPING WILD-RYE

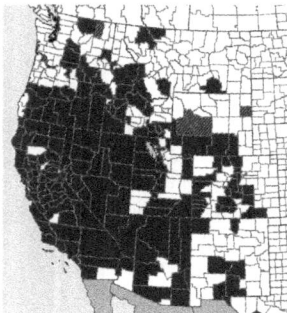

FAMILY Poaceae (Grass)

FLOWERING May–August

HABITAT Meadows, damp ravines, often in salty areas.

HABIT Native perennial grass from much branched rhizomes.

STEMS Upright, hollow, usually unbranched, to 1 m tall, smooth or slightly rough to the touch.

LEAVES Elongated, flat to rolled into a narrow tube, to 8 mm wide, rough to the touch; ligules extremely short, with tiny cilia.

FLOWERS Borne in spikelets, the spikelets usually paired and forming a slender spike to 25 cm long; **spikelets** 3- to 8-flowered, 13–20 mm long; **glumes** slender-pointed, 2–8 mm long; **lemmas** less than 13 mm long, usually awnless. **Stamens** 3. **Pistils:** Ovary superior. **Grains** narrowly oblongoid, smooth except for a few hairs at the tip.

NOTE This plant produces viable seeds very irregularly. As a result, it has minimal value as a food for wildlife.

SYNONYMS *Elymus triticoides* Buckley

WETLAND STATUS AW FAC | GP FAC | WMV FAC

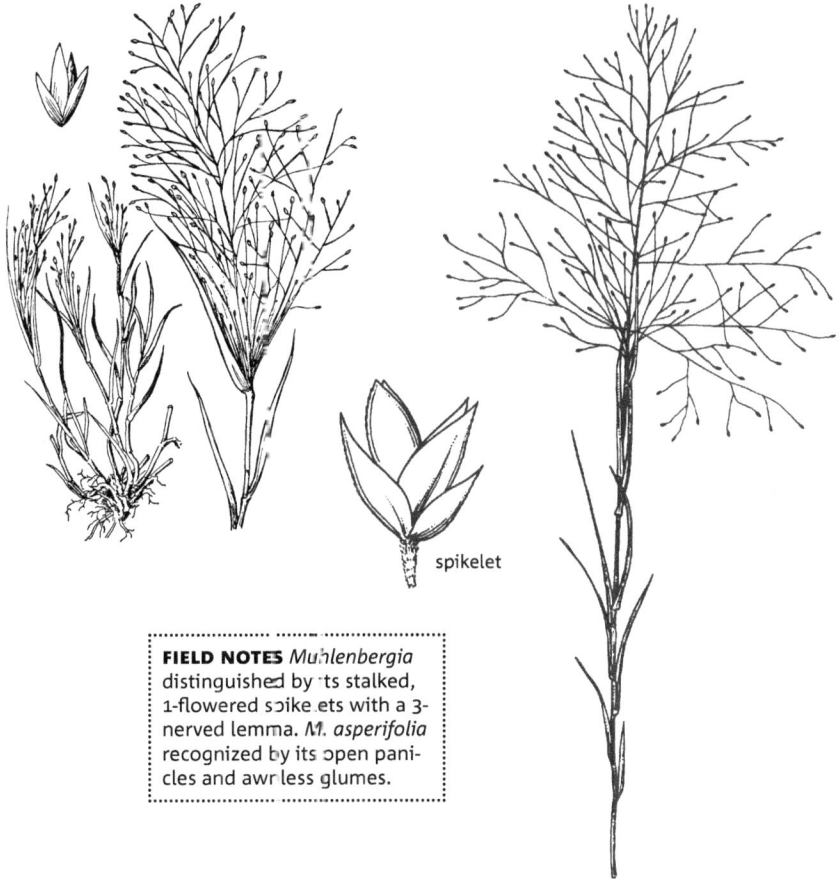

spikelet

> **FIELD NOTES** *Muhlenbergia* distinguished by its stalked, 1-flowered spikelets with a 3-nerved lemma. *M. asperifolia* recognized by its open panicles and awnless glumes.

Muhlenbergia asperifolia (Nees & Meyen ex Trin.) Parodi
ALKALI MUHLY

FAMILY Poaceae (Grass)
FLOWERING July–September
HABITAT Damp meadows, around ponds, along streams, alkaline flats, disturbed areas.
HABIT Native perennial grass with elongated, scaly rhizomes.
STEMS Upright, branched, to 120 cm tall, smooth, somewhat flattened.
LEAVES Narrow, elongated, flat, to 3 mm wide, usually rough to the touch; sheaths smooth.
FLOWERS Borne in 1-flowered spikelets; spikelets several in open panicles at the tips of thread-like branches; panicle to 25 cm long; **spikelets** to 2 mm long; **glumes** awnless. **Stamens** 3. **Pistils:** Ovary superior. **Grains** smooth, about 1 mm long.
NOTE This species, sometimes known as scratchgrass, may invade lawns.
WETLAND STATUS AW FACW | GP FACW | WMV FACW

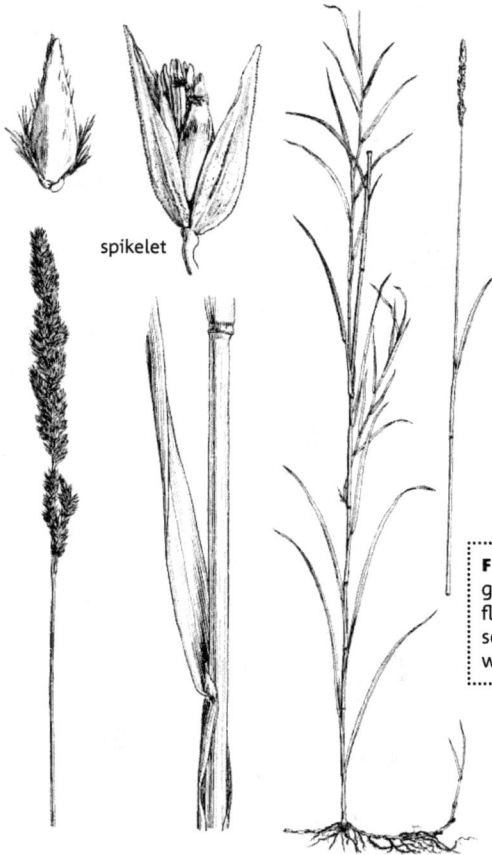

spikelet

> **FIELD NOTES** Aggressive grass; spikelets with 1 flower and 3 or 4 empty scales. Leaves to 20 mm wide.

Phalaris arundinacea L.
REED CANARY GRASS
FAMILY Poaceae (Grass)

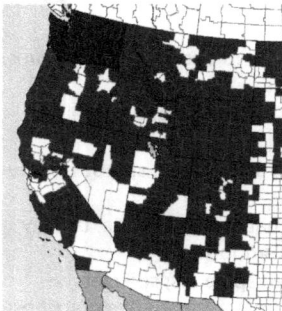

FLOWERING August–September

HABITAT Wet meadows, along streams.

HABIT Adventive, tufted perennial grass with fibrous roots; may form extensive stands to the exclusion of most other species.

STEMS Upright, usually branched from below, hollow, to 1.5 m tall, without hairs.

LEAVES Elongated, narrow, flat, to 20 mm wide, smooth or slightly rough to the touch.

FLOWERS Borne in 1-flowered spikelets, with many spikelets in an open panicle; panicle to 25 cm long; **spikelets** 2–5 mm long, smooth; first (lowest) **glume** about 1 mm long, less than half as long as second glume; **lemma** 3–4 mm long, ellipsoid, shiny. **Stamens** 3. **Pistils:** Ovary superior. **Grains** ovoid, smooth.

NOTE The grains are eaten by birds.

WETLAND STATUS AW FACW | GP FACW | WMV FACW

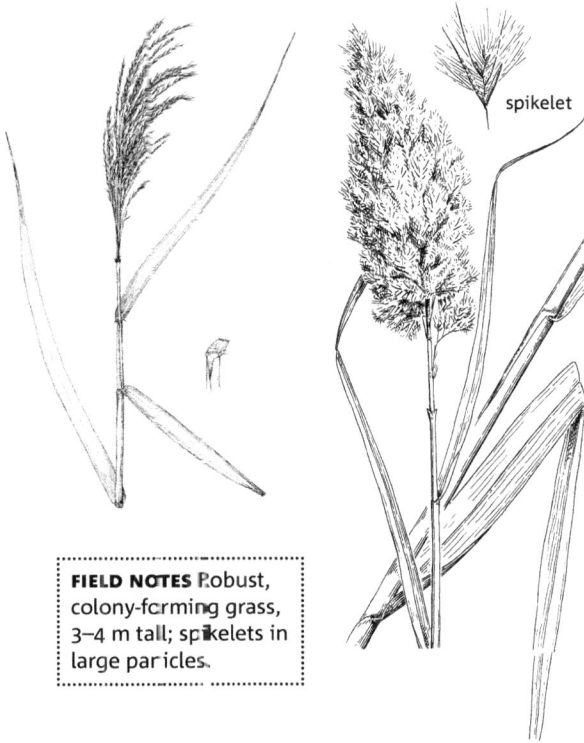

spikelet

> **FIELD NOTES** Robust, colony-forming grass, 3–4 m tall; spikelets in large panicles.

Phragmites australis (Cav.) Trin. ex Steud.
COMMON REED

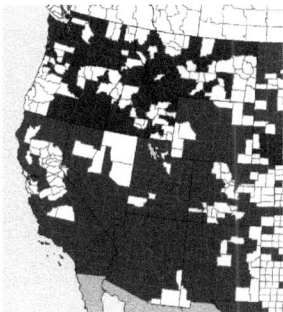

FAMILY Poaceae (Grass)

FLOWERING July–September.

HABITAT Along streams, around ponds, sloughs, reclaimed stripmine areas.

HABIT Native, robust perennial grass with stout, creeping rhizomes, forming dense colonies.

STEMS Erect, smooth, to 3–4 m tall.

LEAVES Flat, elongated, smooth, to 6 cm wide.

FLOWERS 3–7 flowers per spikelet, with many spikelets arranged in a large, dense, much-branched panicle to 40 cm long; **spikelets** 20 mm long, bearing numerous silky hairs. **Stamens** 3. **Pistils**: Ovary superior.

SYNONYMS *Phragmites communis* Trin.

WETLAND STATUS AW FACW | GP FACW | WMV FACW

FIELD NOTES Rhizomes absent; lemmas with a web at base; panicles relatively small; spikelets to 8 mm long.

spikelet

Poa leptocoma Trin.
MARSH BLUEGRASS

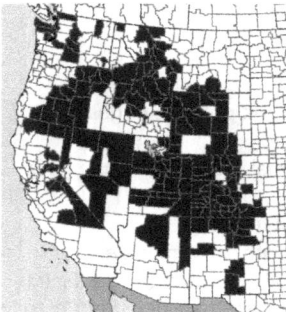

FAMILY Poaceae (Grass)
FLOWERING July–August
HABITAT Along streams, wet meadows, bogs, high in the mountains.
HABIT Native perennial grass with fibrous roots.
STEMS Spreading, often rooting at the nodes, to 60 cm tall, smooth.
LEAVES Elongated, narrow, flat, to 4 mm wide, slightly folded at the tip. rough to the touch.
FLOWERS Borne in spikelets. with the spikelets arranged in a small panicle to 15 cm long; **spikelets** 2- to 6-flowered, flattened, purplish, to 8 mm long; **lemmas** hairy, webbed at the base. **Stamens** 3. **Pistils:** Ovary superior, smooth. **Grains** ellipsoid, smooth.
NOTE This grass is palatable to domestic livestock.
WETLAND STATUS AW FACW | GP OBL | WMV FACW

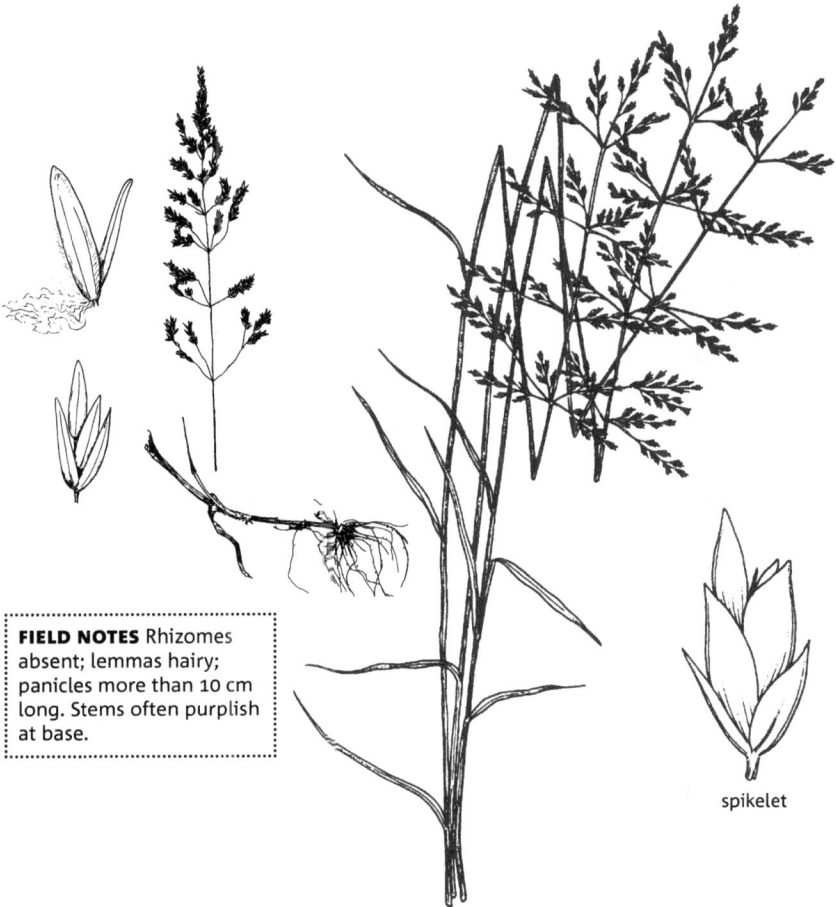

FIELD NOTES Rhizomes absent; lemmas hairy; panicles more than 10 cm long. Stems often purplish at base.

spikelet

Poa palustris L.
FOWL BLUEGRASS

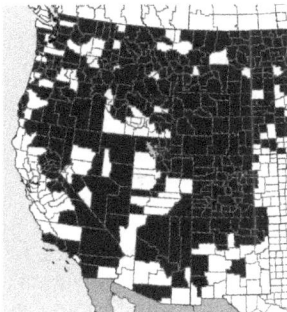

FAMILY Poaceae (Grass)
FLOWERING May–August
HABITAT Wet meadows, roadside ditches, other moist areas.
HABIT Native perennial grass with fibrous roots.
STEMS Spreading, often rooting at the nodes, to 1 m tall, smooth, often purplish at the base.
LEAVES Alternate, elongated, narrow, flat or folded, to 20 cm long, boat-shaped at the tip, rough to the touch on the upper surface.
FLOWERS Borne in spikelets. with the spikelets arranged in an open panicle to 30 cm long; **spikelets** 2- to 4-flowered, to 6 mm long; **lemmas** hairy, bronze at the tip. **Stamens** 3. **Pistils**: Ovary superior, smooth. **Grains** obovoid. smooth.
WETLAND STATUS AW FAC | GP FACW | WMV FAC

spikelet

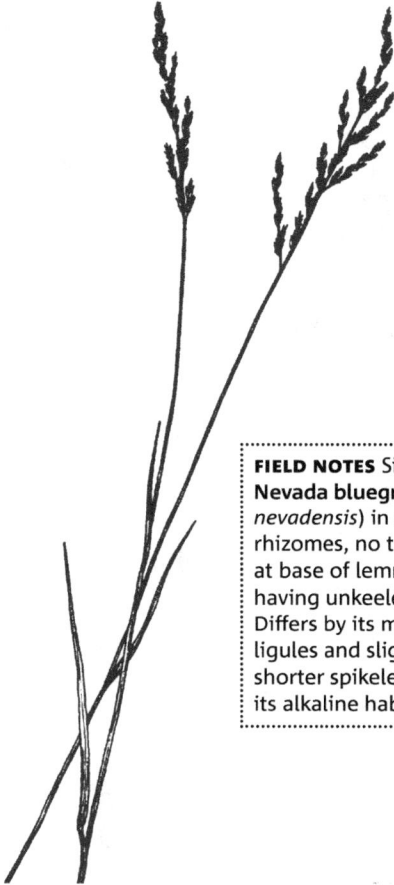

> **FIELD NOTES** Similar to **Nevada bluegrass** (*P. nevadensis*) in absence of rhizomes, no tufted hairs at base of lemmas, and having unkeeled lemmas. Differs by its much shorter ligules and slightly shorter spikelets, and by its alkaline habitat.

Poa secunda J. Presl
ALKALI BLUEGRASS

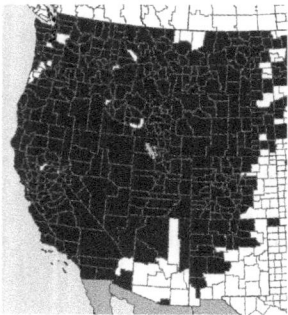

FAMILY Poaceae (Grass)
FLOWERING May–July
HABITAT Alkaline meadows, from valleys to mountains.
HABIT Native tufted perennial grass without rhizomes.
STEMS Upright, unbranched. to 60 cm tall, without hairs.
LEAVES Elongated, narrow, rolled up into a tube, to 20 cm long, to 2 mm wide, without hairs; ligules less than 3 mm long.
FLOWERS Borne in 2- to 5-flowered spikelets, the **spikelets** 6–8 mm long, arranged in a narrow panicle to 15 cm long; **lemmas** sometimes minutely hairy but not with a tuft of hairs at the base, unkeeled, to 6 mm long. **Stamens** 3. **Pistils**: Ovary superior. **Grains** ellipsoid, smooth.
SYNONYMS *Poa juncifolia* Scribn., *Poa nevadensis* Vasey ex Scribn.
WETLAND STATUS AW FACU | GP FACU | WMV FACU

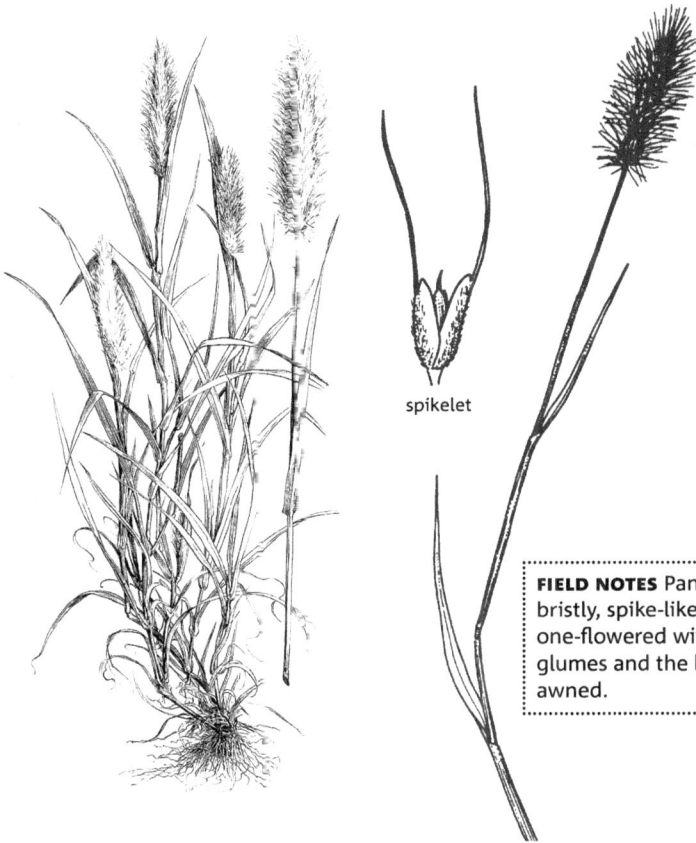

spikelet

> **FIELD NOTES** Panicles soft-bristly, spike-like; spikelets one-flowered with both the glumes and the lemmas awned.

Polypogon monspeliensis (L.) DesfAnnual
RABBIT-FOOT GRASS

FAMILY Poaceae (Grass)

FLOWERING May–September

HABITAT Wet soil in ditches, marshes, along streams and rivers, around lakes and ponds.

HABIT Introduced annual grass with fibrous roots.

STEMS Solitary or several in clumps, upright, to 60 cm tall, hollow.

LEAVES Elongated, narrow, to 20 cm long, to 8 mm wide, rough along the edges, otherwise smooth, conspicuously ridged on the upper surface; ligule to 6 mm long.

FLOWERS Borne in spikelets, the spikelets arranged in a dense panicle that resembles a spike. **Spikelets** 1-flowered; the **glumes** about 2 mm long, awned from the notched tip, the awn to 8 mm long, the **lemma** with an awn about 2 mm long. **Stamens** 3. **Fistils:** Ovary superior. **Grains** ovoid, smooth.

NOTE This native of Europe is regularly found in moist, disturbed areas. The grains are eaten by waterfowl.

WETLAND STATUS AW FACW | GP FACW | WMV FACW

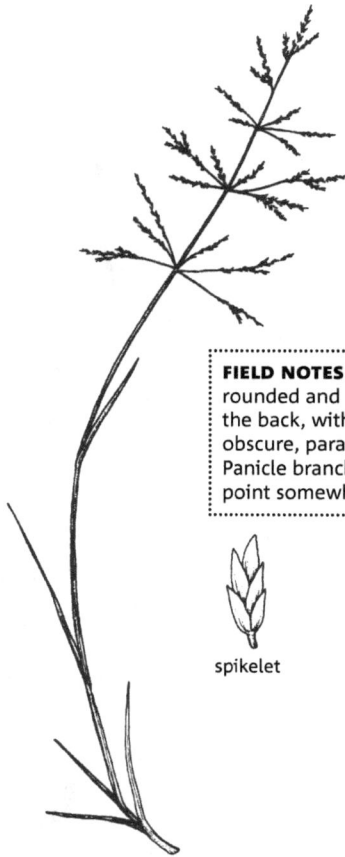

> **FIELD NOTES** Lemmas
> rounded and not keeled on
> the back, with usually 5
> obscure, parallel nerves.
> Panicle branches typically
> point somewhat downward.

spikelet

Puccinellia distans (L.) Parlat.
WEEPING ALKALI GRASS

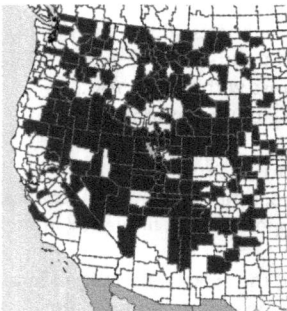

FAMILY Poaceae (Grass)

FLOWERING June–August

HABITAT Moist habitats, nearly all in alkaline soils.

HABIT Introduced perennial grass with fibrous roots.

STEMS Spreading at first, becoming upright, unbranched. hollow, to 1.2 m tall, without hairs but sometimes somewhat rough to the touch.

LEAVES Elongated, flat but becoming inrolled, to 4 mm wide, smooth or slightly rough to the touch; ligules very short, rounded at the tip.

FLOWERS Borne in spikelets, the spikelets arranged in open panicles, with some of the branches pointing downward, the panicles to 15 cm long; **spikelets** 2- to 6-flowered, to 6 mm long. **Stamens** 3. **Pistils:** Ovary superior, smooth. **Grains** oblongoid, smooth.

NOTE European native, widespread over much of the United States; widely used as a forage plant for domesticated animals.

WETLAND STATUS AW FACW | GP FACW | WMV FACW

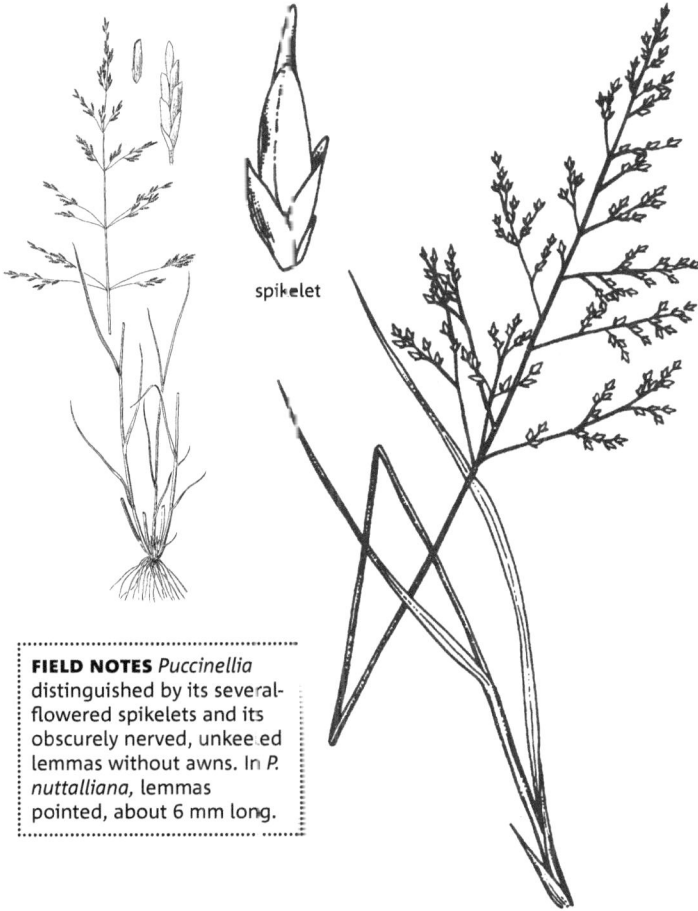

spikelet

FIELD NOTES *Puccinellia* distinguished by its several-flowered spikelets and its obscurely nerved, unkeeled lemmas without awns. In *P. nuttalliana*, lemmas pointed, about 6 mm long.

Puccinellia nuttalliana (I. A. Schultes) A. S. Hitchc.
NUTTALL'S ALKALI GRASS

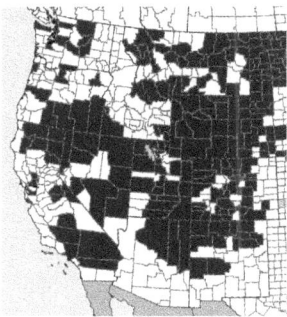

FAMILY Poaceae (Grass)
FLOWERING June–August
HABITAT Alkaline flats.
HABIT Native tufted perennial grass with fibrous roots.
STEMS Upright, to 60 cm tall, smooth or rough to the touch.
LEAVES Narrow, elongated, rolled up into a tube, to 2 mm wide.
FLOWERS 2–7 in a spikelet, with several spikelets forming an open panicle with erect to spreading branches; **spikelets** to 6 mm long, pointed at the tip; **lemmas** rounded on the back, awnless, with obscure nerves.
Stamens 3. Pistils: Ovary superior. Grains smooth, about 2 mm long.
SYNONYMS *Puccinellia airoides* (Nutt.) S. Wats. & Coult.
WETLAND STATUS AW FACW | GP OBL | WMV FACW

> **FIELD NOTES** Spikelets compressed, 1-flowered densely arranged in 2 rows on 2 sides of the axis; second glume awnless or with a minute awn. Leaves to 6 mm wide.

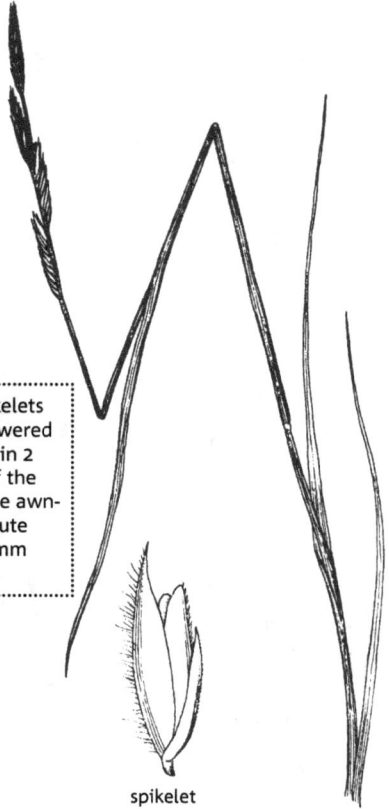

spikelet

Spartina gracilis Trin.
ALKALI CORDGRASS

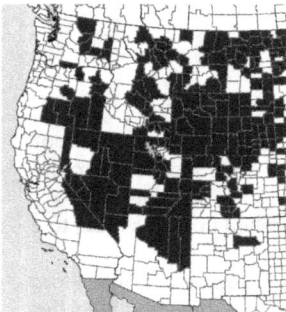

FAMILY Poaceae (Grass)

FLOWERING June–September

HABITAT Along streams, around lakes and ponds, in wet meadows, frequently in alkaline habitats.

HABIT Native perennial grass with well-developed rhizomes.

STEMS Upright, unbranched, hollow, to 1 m tall, smooth.

LEAVES Elongated, narrow, flat, to 20 cm long, to 6 mm wide, without hairs but rough to the touch on the upper surface.

FLOWERS Borne in 1-flowered spikelets arranged in compressed spikes, with many spikes forming a panicle to 20 cm long, each spike to 8 cm long; **glumes** to 8 mm long, pointed at the tip; **lemmas** lanceolate, ciliate, not awned. **Stamens** 3. **Pistils:** Ovary superior. **Grains** obovoid, smooth.

SYNONYMS *Sporobolus hookerianus* P. M. Peterson & Saarela

WETLAND STATUS AW FACW | GP FACW | WMV FACW

> **FIELD NOTES** Plants forming large clumps, topped by broad, open panicles. Spikelets 1-flowered; glumes and lemmas without awns.

spikelet

Sporobolus airoides (Torr.) Torr.
ALKALI SACATON

FAMILY Poaceae (Grass)

FLOWERING June–August

HABITAT Alkaline meadows.

HABIT Native, stout, perennial grass forming dense tufts.

STEMS Upright, unbranched. to 1.2 m tall, usually hollow, smooth.

LEAVES Alternate, elongated, narrow, crowded at the base of the plant and recurved, flat to inrolled. to 4 mm wide. rough to the touch, hairy near the base; ligule a ring of hairs.

FLOWERS Borne in 1-flowered spikelets. with many spikelets forming a broad, open panicle to 40 cm long; **glumes** pointed at the tip. smooth, awnless, to 3 mm long; **lemmas** pointed at the tip, smooth, awnless, to 3 mm long. **Stamens** 3. **Pistils:** Ovary superior, smooth. **Grains** obovoid, smooth.

NOTE This is an important forage grass for domestic livestock.

WETLAND STATUS AW FAC | GP FAC | WMV FAC

FIELD NOTES Plants with rhizomes. Lemmas conspicuously nerved. Leaves sometimes as much as 2 cm wide.

spikelet

Torreyochloa pallida (Torr.) Church
WEAK MANNA GRASS

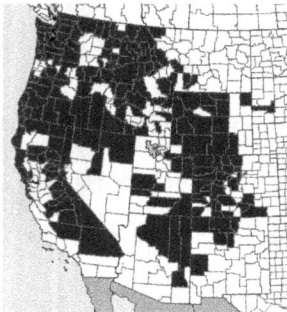

FAMILY Poaceae (Grass)
FLOWERING June–August
HABITAT Along streams, marshes, often in standing water.
HABIT Native perennial grass with stout rhizomes.
STEMS Spreading at first but becoming upright, unbranched, hollow, to 1.2 m tall, smooth; sheaths on stems with cross-markings.
LEAVES Elongated, flat, to 20 mm wide, without hairs but usually slightly rough to the touch.
FLOWERS Borne in spikelets, the spikelets arranged in open panicles to 50 cm long, the branches of the panicle ascending, spreading, or pointing downward. **Stamens** 3. **Pistils**: Ovary superior, smooth. **Grains** oblongoid, smooth.
SYNONYMS *Puccinellia pauciflora* (J. Presl) Munz
WETLAND STATUS AW OBL | GP OBL | WMV OBL

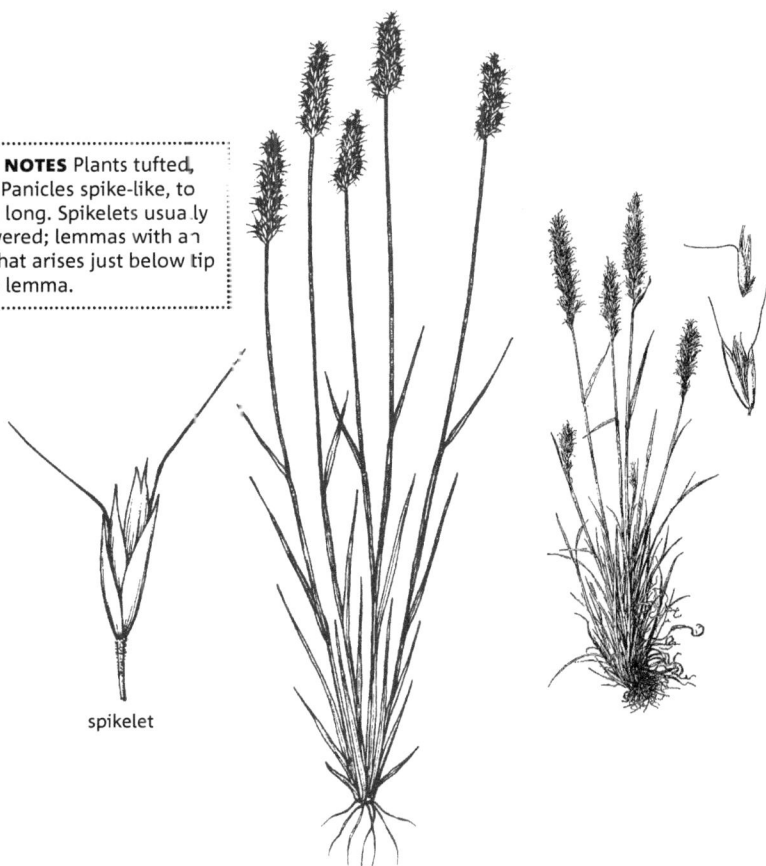

FIELD NOTES Plants tufted, hairy. Panicles spike-like, to 15 cm long. Spikelets usually 2-flowered; lemmas with an awn that arises just below tip of the lemma.

spikelet

Trisetum spicatum (L.) Richter
SPIKED FALSE-OATS

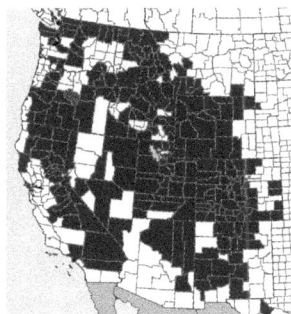

FAMILY Poaceae (Grass)
FLOWERING July–September
HABITAT Wet meadows, particularly in the high mountains.
HABIT Native tufted perennial grass with fibrous roots.
STEMS Upright, unbranched, hollow, to 60 cm tall, smooth or hairy.
LEAVES Elongated, flat or folded, to 6 mm wide, usually hairy; ligules to 4 mm long, ciliate, jagged at the tip.
FLOWERS Borne in spikelets, the spikelets crowded into spike-like panicles to 15 cm long; **spikelets** usually 2-flowered (sometimes 3), 2–6 mm long, purplish to silvery; **lemmas** notched at the tip, with an awn arising just below the notch, the awn 6–8 mm long. **Stamens** 3. **Pistils**: Ovary superior, smooth. **Grains** ellipsoid, smooth.
NOTE This grass is an important forage species for domestic cattle in the high mountains.
WETLAND STATUS AW FACU | CP FACU | WMV UPL

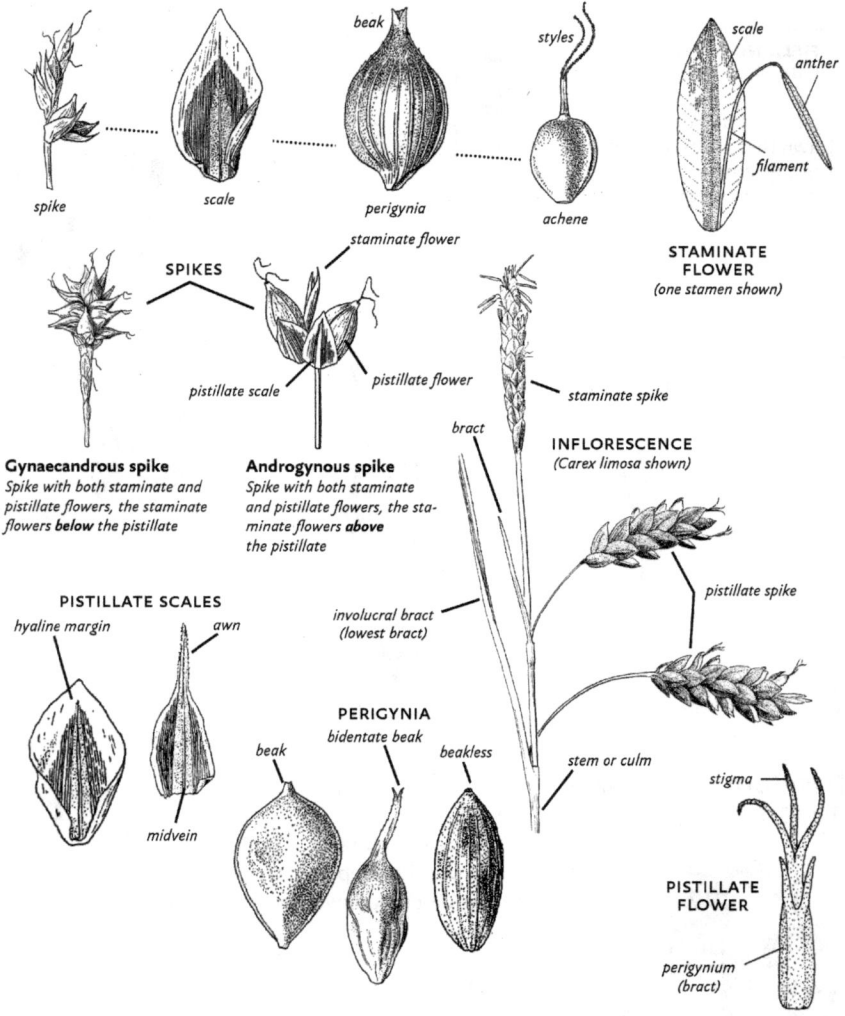

beak

styles

scale

anther

filament

spike

scale

perigynia

achene

STAMINATE FLOWER
(one stamen shown)

staminate flower

SPIKES

staminate spike

bract

INFLORESCENCE
(Carex limosa shown)

pistillate scale

pistillate flower

Gynaecandrous spike
*Spike with both staminate and pistillate flowers, the staminate flowers **below** the pistillate*

Androgynous spike
*Spike with both staminate and pistillate flowers, the staminate flowers **above** the pistillate*

pistillate spike

PISTILLATE SCALES

hyaline margin

awn

involucral bract
(lowest bract)

PERIGYNIA

bidentate beak

beak

beakless

stem or culm

stigma

midvein

PISTILLATE FLOWER

perigynium
(bract)

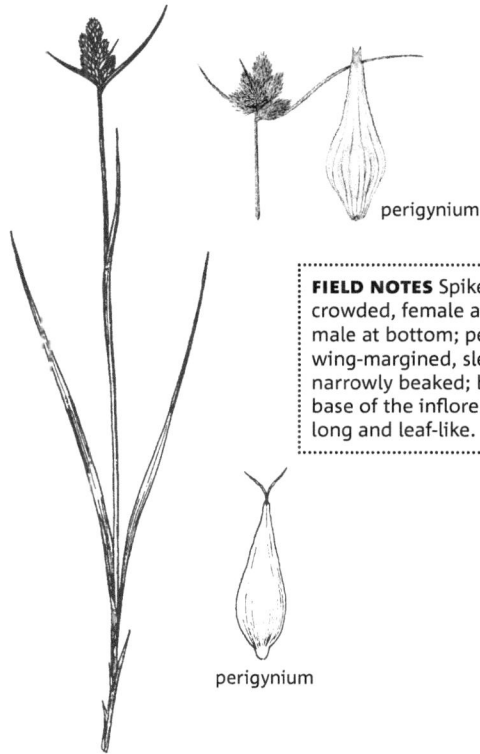

perigynium

> **FIELD NOTES** Spikes crowded, female at top, male at bottom; perigynia wing-margined, slender, narrowly beaked; bract at base of the inflorescence long and leaf-like.

perigynium

Carex athrostachya Olney
SLENDER-BEAK SEDGE

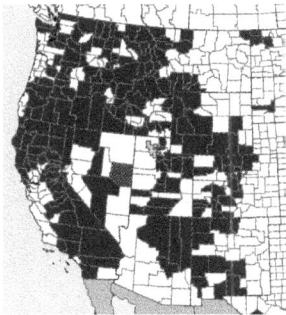

FAMILY Cyperaceae (Sedge)

FLOWERING May–July

HABITAT Marshes, meadows, woodlands.

HABIT Native, densely tufted perennial sedge with fibrous roots.

STEMS Upright, unbranched, to 60 cm tall, smooth.

LEAVES Elongated, often all near the base of the plant, flat, to 4 mm wide, without hairs.

FLOWERS Male and female borne in the same spikes, with the female flowers above the male ones, with several spikes crowded in the inflorescence, each spike 6–13 mm long, the inflorescence subtended by a leaf-like bract much longer than the inflorescence; **scales** narrower and usually slightly shorter than the perigynia, brownish, tapering to a point. **Stamens** 3. Pistils enclosed in a perigynium, the **perigynium** flattened, pale green to tan, lanceolate, with a narrow, minutely toothed wing on either side, without or with only a few nerves, to 6 mm long, to 2 mm wide; **stigmas** 2.

FRUIT Achenes lenticular, to 2 mm long.

NOTE The achenes are eaten by birds. This is a good forage species for cattle and horses.

WETLAND STATUS AW FACW | GP FACW | WMV FACW

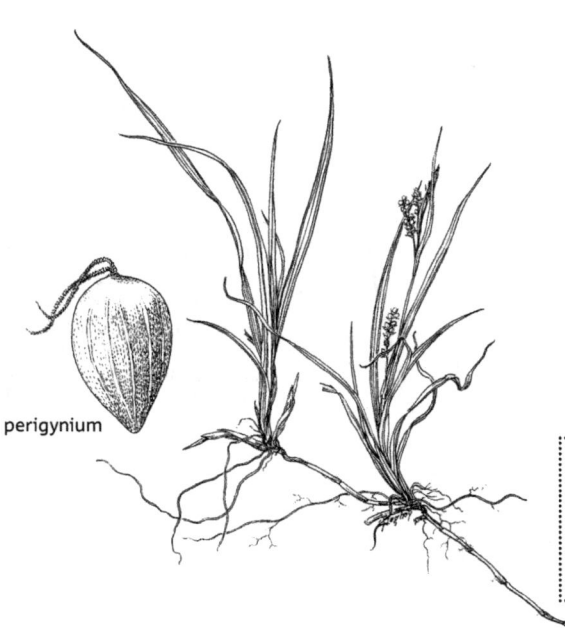

perigynium

> **FIELD NOTES** Male spike usually separate from the female; perigynia golden when mature; scales tinged reddish brown; plants short, usually less than 40 cm tall.

Carex aurea Nutt.
GOLDEN-FRUIT SEDGE

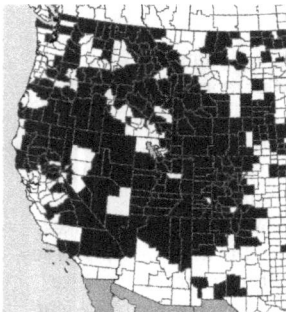

FAMILY Cyperaceae (Sedge)

FLOWERING May–July

HABITAT Wet meadows, marshes, moist woods.

HABIT Native perennial sedge with whitish, creeping rhizomes.

STEMS Upright, unbranched, slender, to 40 cm tall, smooth.

LEAVES Elongated, flat, to 4 mm wide, smooth, sometimes longer than the stem.

FLOWERS Male and female borne in separate spikes; male spike solitary, terminal, 6–20 mm long; female spikes 2–5, 6–20 mm long, the lowest one usually subtended by a leaf-like bract longer than the inflorescence; **scales** half as long as to as long as the perigynia, rounded or with a short point at the tip, brown to straw-colored, usually with a green center and a transparent border. **Stamens** 3. Pistils enclosed in a perigynium, the **perigynium** ellipsoid to nearly spherical, rounded and without a beak at the tip, to 3 mm long, with or without nerves, golden or yellow-brown when mature; **stigmas** 2.

FRUIT Achenes lenticular, to 2 mm long.

NOTE The achenes are eaten by waterfowl.

WETLAND STATUS AW OBL | GP OBL | WMV FACW

FIELD NOTES Spikes with female flowers at top, male spikes at bottom, and none of the spikes overlapping. Achenes triangular, stigmas 3, perigynium scarcely beaked.

perigynium

Carex bella Bailey
SHOWY SEDGE

FAMILY Cyperaceae (Sedge)
FLOWERING May–August
HABITAT Wet meadows, along streams, moist woods, usually in the higher mountains.
HABIT Native perennial sedge with fibrous roots; rhizomes absent.
STEMS Upright, to 60 cm tall, smooth.
LEAVES Elongated, flat, to 8 mm wide, smooth.
FLOWERS Crowded into spikelets, the spikelets to 4 cm long, subtended by a bract without a sheath at its base; male flowers borne at the base of each spikelet; lower spikelets tending to droop. **Scales** ovate, pointed at the tip, reddish brown to blackish brown with a transparent edge and a green mid-vein, to 3 mm long. **Stamens** 3. Pistils enclosed in a perigynium; each perigynium obovoid. to 6 mm long, with a very short, terminal beak; **stigmas** 3.
FRUIT Achenes triangular, to nearly 6 mm long, smooth.
NOTE The achenes are eaten by birds and small mammals.
WETLAND STATUS AW FACU | GP FAC | WMV FACU

perigynium   achene

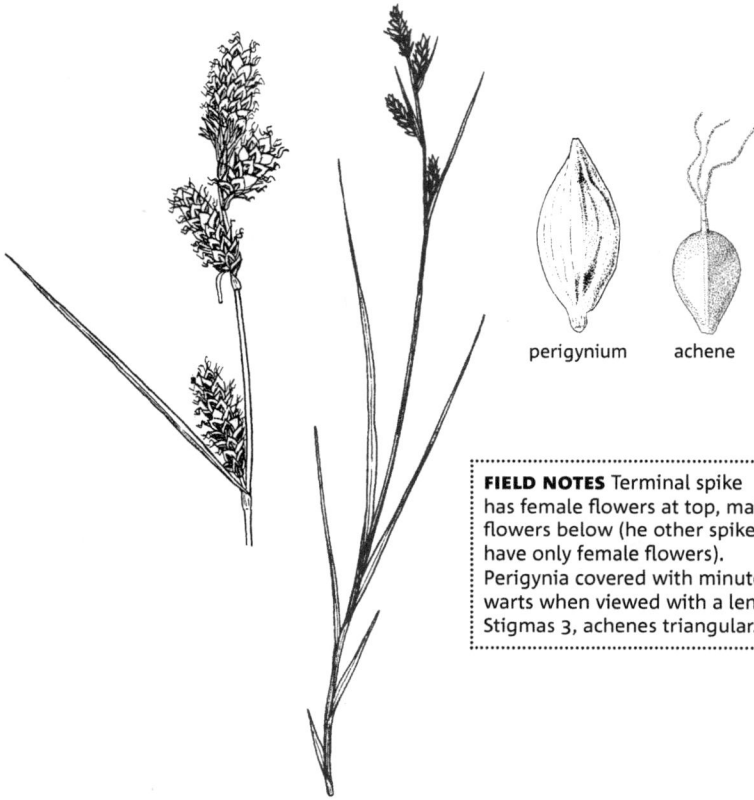

> **FIELD NOTES** Terminal spike
> has female flowers at top, male
> flowers below (he other spikes
> have only female flowers).
> Perigynia covered with minute
> warts when viewed with a lens.
> Stigmas 3, achenes triangular.

Carex buxbaumii Wahlenb.
BROWN BOG SEDGE

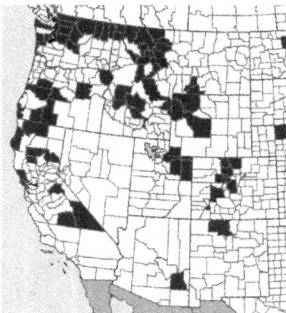

FAMILY Cyperaceae (Sedge)

FLOWERING May–August

HABITAT Swamps, bogs, around lakes, wet meadows.

HABIT Native perennial sedge with creeping rhizomes;
last year's leaves not persisting at base of the plant.

STEMS Upright, unbranched. to 1 m tall, without hairs.

LEAVES Elongated, to 4 mm wide, without hairs, usually
all shorter than the stems.

FLOWERS Borne in 2–5 spikes, the terminal spike to 4
cm long, with female flowers at the top and male flowers below, the other spikes to 4 cm long, with only
female flowers; **scales** longer than the perigynia, brown
to purple-black, with a paler midvein, tapering to a slender awn at the tip. **Stamens** 3.
Pistils enclosed in a perigynium, the **perigynium** ellipsoid to obovoid, to 5 mm long,
covered with minute warts, beakless or with a very short beak, pale gray-green; **stigmas** 3.

FRUIT Achenes triangular, to 2 mm long, smooth.

NOTE The achenes are eaten by waterfowl.

WETLAND STATUS AW OBL | GP OBL | WMV OBL

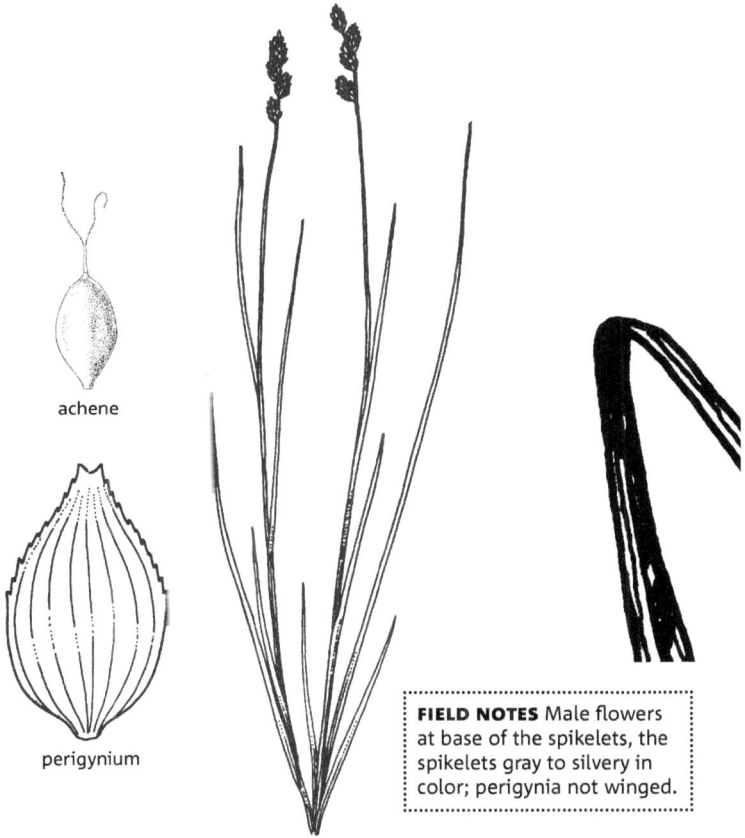

achene

perigynium

> **FIELD NOTES** Male flowers at base of the spikelets, the spikelets gray to silvery in color; perigynia not winged.

Carex canescens L.
HOARY SEDGE

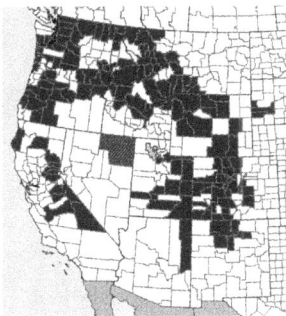

FAMILY Cyperaceae (Sedge)

FLOWERING May–August

HABITAT Wet meadows, swamps, particularly in the mountains.

HABIT Native tufted perennial sedge with short rhizomes.

STEMS Upright, unbranched, to 60 cm tall, smooth.

LEAVES Elongated, mostly near the base of the plant, rarely longer that the flowering stems, flat, to 4 mm wide, sometimes bluish, smooth.

FLOWERS Male and female flowers borne separately but in the same spikelet, the male at the base of each spikelet; spikelets pale to silvery, 4–8 in a crowded cluster, each spikelet less than 13 mm long. **Scales** pale and often transparent except for the green midvein, not awned, shorter than the perigynia. **Stamens** 3. Pistils enclosed in a perigynium; **perigynium** straw-colored to silvery, to 3 mm long, not beaked, wingless, ellipsoid to elliptic-ovoid.

FRUIT Achenes lenticular, to 2 mm long.

WETLAND STATUS AW OBL | GP OBL | WMV OBL

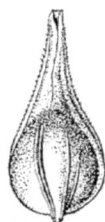

perigynium

female scale

> **FIELD NOTES** Spikelets with male flowers at tip in a crowded, compact inflorescence. Styles 2; achenes lenticular, rhizomes absent.

Carex diandra Schrank Lesser
PANICLED SEDGE

FAMILY Cyperaceae (Sedge)

FLOWERING May–August

HABITAT Wet meadows, bogs, swamps, around lakes and ponds.

HABIT Native tufted perennial sedge from fibrous roots.

STEMS Upright, triangular, to 1 m tall, without hairs.

LEAVES Alternate, elongated, flat, to 4 mm wide, usually as long as or slightly longer than the stem, without hairs; sheaths red-dotted.

FLOWERS Borne in spikelets with the male flowers at the tip of each spikelet; spikelets several, crowded into a compact inflorescence to 5 cm long. **Scales** lanceolate, pale brown, pointed at the tip but rarely awned. **Stamens** 3. Pistils enclosed in a perigynium; each **perigynium** ovoid to lance-ovoid, to 3 mm long, dark brown, tapering to or contracted to a short beak; beak minutely toothed; **stigmas** 2.

FRUIT Achenes lenticular, about 1 mm long.

NOTE The achenes are eaten by waterfowl.

WETLAND STATUS AW OBL | GP OBL | WMV OBL

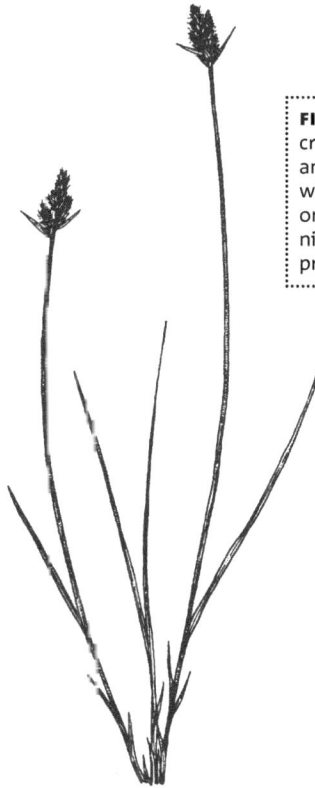

> **FIELD NOTES** Spikes crowded, to 20 mm long, and either male or female, with the two sexes usually on separate plants. Perigynia to 5 mm long, with a prominent beak.

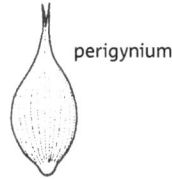

perigynium

Carex douglasii Boott
DOUGLAS' SEDGE

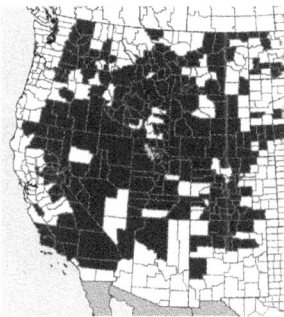

FAMILY Cyperaceae (Sedge)

FLOWERING May–August

HABITAT Wet or dry prairies, ditches, tolerating alkaline conditions.

HABIT Native perennial sedge with long, slender rhizomes.

LEAVES Upright, unbranched, to 120 cm tall, without hairs.

FLOWERS Male and female flowers borne in spikes, usually on different plants or, if on the same plant, the male flowers at the top of the spike and the female flowers at the bottom; spikes severely crowded into a head to 5 cm long, each spike to 20 mm long; **scales** lanceolate to ovate-lanceolate, pointed or even awned at the tip, pale brown with a green midvein and transparent margins, longer than the perigynia. **Stamens** 3. Pistils enclosed in a perigynium, each **perigynium** ellipsoid to ellipsoid-ovoid, to 5 mm long, tapering to a prominent and minutely toothed beak at the tip, straw-colored to pale brown; **stigmas** 2.

FRUIT Achenes lenticular, to 2 mm long, smooth.

NOTE The achenes are eaten by birds. This species is a forage plant for livestock.

WETLAND STATUS AW FAC | GP FAC | WMV FAC

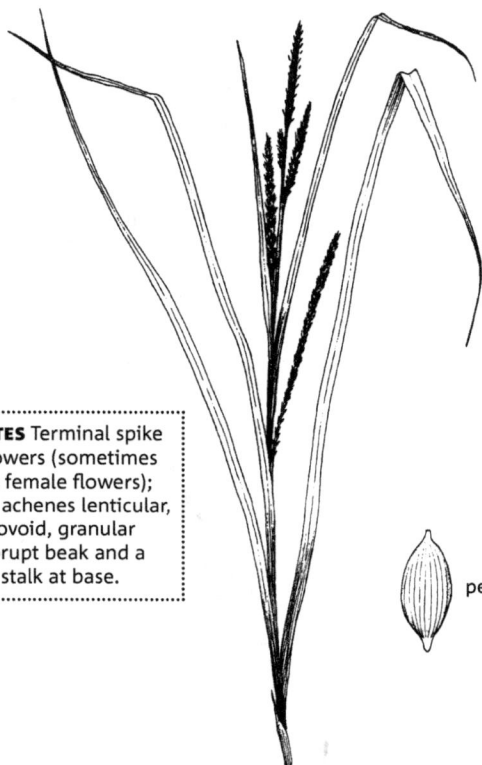

FIELD NOTES Terminal spike of male flowers (sometimes with a few female flowers); stigmas 2, achenes lenticular, perigynia ovoid, granular with an abrupt beak and a very short stalk at base.

perigynium

Carex kelloggii W. Boott
KELLOGG'S SEDGE

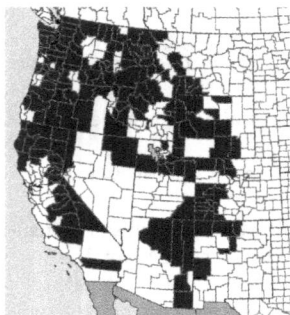

FAMILY Cyperaceae (Sedge)

FLOWERING May–June

HABITAT In and along streams and the edges of lakes; wet meadows, particularly in the mountains.

HABIT Native clump-forming perennial sedge with thickened rootstocks.

STEMS Upright, to 45 cm tall, shorter than the leaves, without hairs but somewhat rough to the touch near the inflorescence.

LEAVES Elongated, flat and sometimes channeled, to 3 mm wide, smooth.

FLOWERS Borne in 4–6 spikes, the terminal spike with mostly all male flowers, the others with all female flowers and sessile; spikes to 5 cm long, to 6 mm wide; **scales** oblong to ovate, purple-brown to black, narrower and slightly shorter than the perigynia. **Stamens** 3. Pistils enclosed in a perigynium, the **perigynium** ovoid, pale green, granular, finely nerved, to 3 mm long, abruptly beaked with a short, dark-colored beak and a short stalk at the base; **stigmas** 2.

FRUIT Achenes lenticular, 1–2 mm long, smooth.

NOTE The achenes are eaten by waterfowl.

WETLAND STATUS AW OBL | GP OBL | WMV OBL

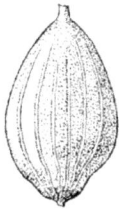

perigynium

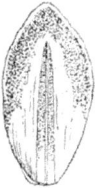

female scale

perigynium

FIELD NOTES Male and female spikes separate; stigmas 2; achenes lenticular; scales brown to blackish; perigynia with nerves on both faces. Differs from the similar *Carex nebrascensis* by its narrower leaves (less than 4 mm wide), and the beak of the perigynium which is not 2-toothed.

Carex lenticularis Michx.
SHORE SEDGE

FAMILY Cyperaceae (Sedge)
FLOWERING June–September
HABITAT Around lakes and ponds, along streams, wet meadows, sometimes in the mountains.
HABIT Native, densely tufted perennial sedge; without rhizomes or with very short rhizomes.
STEMS Upright, to 75 cm tall, without hairs.
LEAVES Elongated, narrow, to 4 mm wide, without hairs, usually not surpassing the inflorescence.
FLOWERS Crowded into dense spikes, the terminal spike usually with all male flowers or occasionally with a few female flowers at the bottom, the other 2–5 spikes with only female flowers, the spikes to 5 cm long, to 6 mm thick, the lowest subtended by a leafy bract often surpassing the inflorescence; **scales** brown to blackish, shorter than the perigynia. **Stamens** 3. Pistils enclosed in a perigynium; **perigynia** ellipsoid to ovoid, somewhat flattened, to 3 mm long with a small, entire beak at the tip, with 3–7 nerves on each face; **stigmas** 2.
FRUIT Achenes lenticular, about 2 mm long, smooth.
NOTE The achenes are eaten by waterfowl.
WETLAND STATUS AW OBL | GP OBL | WMV OBL

perigynium

female scale

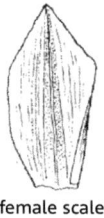

FIELD NOTES Plants slender and delicate, with a single slender spike at tip of each stem; male flowers at top of the spike. Perigynia beak absent.

Carex leptalea Wahlenb.
BRISTLY-STALK SEDGE

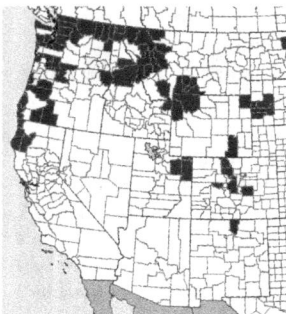

FAMILY Cyperaceae (Sedge)
FLOWERING May–August
HABITAT Swamps, bogs, fens, around lakes and ponds.
HABIT Native perennial sedge with branched rhizomes.
STEMS Upright, very slender, to 120 cm tall, smooth.
LEAVES Elongated, very narrow, flat, to 1 mm wide, smooth, nearly always shorter than the stem.
FLOWERS Male and female flowers borne in a solitary terminal slender spike with the male flowers at the top, the spike to 20 mm long, not subtended by a bract.
Scales ovate to lanceolate, rounded or pointed or even awned at the tip, greenish or brownish, shorter than the perigynia. Stamens 3. Pistils enclosed in a perigynium; each **perigynium** ellipsoid, rounded and not beaked at the tip, narrowed to a spongy base. 2–6 mm long, pale green to straw-colored; **stigmas** 3.
FRUIT Achenes triangular, about 2 mm long, smooth.
NOTE The achenes are eaten by waterfowl.
WETLAND STATUS AW OBL | GP OBL | WMV OBL

> **FIELD NOTES** Male spike solitary on a long stalk; female spikes 1–3, the lowest ones drooping; stigmas 3; achene triangular; perigynium covered by numerous minute dots. Rhizomes long, creeping, with roots covered by yellow wool. Leaves 1–2 mm wide.

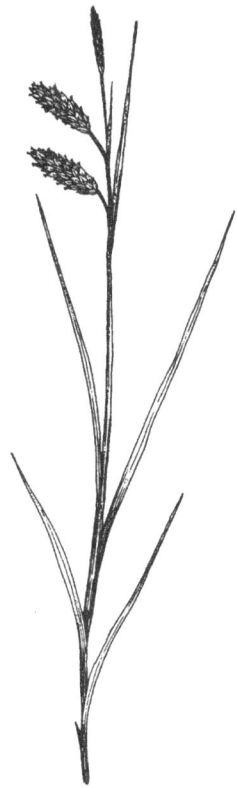

perigynium

Carex limosa L.
MUD SEDGE

FAMILY Cyperaceae (Sedge)
FLOWERING May–June
HABITAT Sphagnum bogs.
HABIT Native perennial sedge with long-creeping rhizomes, and roots covered by a yellow wool.
STEMS Upright, slender, triangular, to 50 cm tall, without hairs.
LEAVES Very narrow and elongated, shorter than the stem, to 2 mm wide, flat but channelled, without hairs.
FLOWERS Borne in spikes, the male spike solitary and terminal, the other 1–3 spikes female; male spike to 6 mm thick, on an upright stalk; female spikes oblongoid, to 30 mm long, to 8 mm thick, at least the uppermost on slender, drooping stalks; scales ovate, shorter than the perigynia, brown. **Stamens** 3. Pistils enclosed in a perigynium, the **perigynium** ovoid, abruptly beaked at the tip, with a short stalk at the base, flattened, greenish or straw-colored, to 4 mm long, to 2 mm wide, smooth, covered by numerous dots; **stigmas** 3.
FRUIT Achenes triangular, about 2 mm long.
NOTE The achenes are sometimes eaten by small mammals.
WETLAND STATUS AW OBL | CP OEL | WMV OBL

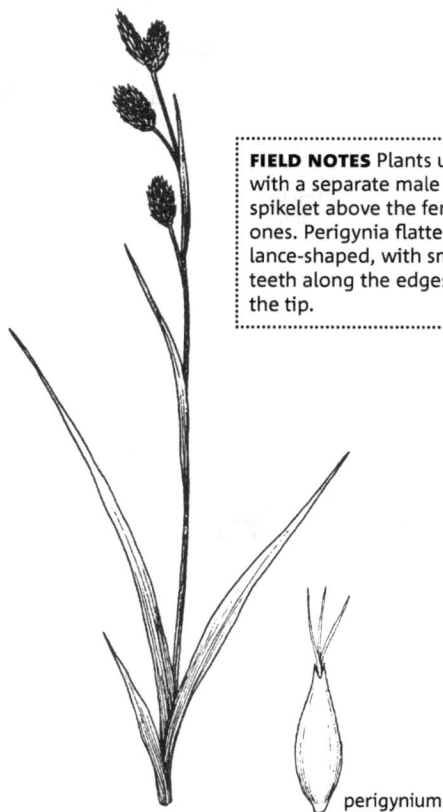

> **FIELD NOTES** Plants usually with a separate male spikelet above the female ones. Perigynia flattened, lance-shaped, with small teeth along the edges near the tip.

perigynium

Carex luzulina Olney
WOOD-RUSH SEDGE

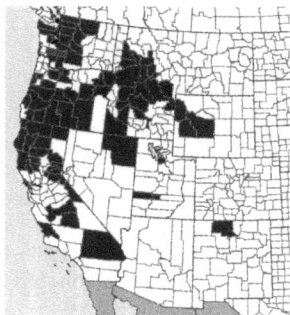

FAMILY Cyperaceae (Sedge)
FLOWERING May–July
HABITAT Wet meadows, bogs, other moist areas.
HABIT Native perennial sedge with fibrous roots.
STEMS Upright, to 60 cm tall, smooth.
LEAVES Elongated, nearly all crowded near the base of the plant and not longer than the flowering stems, to 15 cm long, a little less than 13 mm wide, smooth.
FLOWERS Borne in spikelets. with 2–6 spikelets not overlapping at the tip of the stem, the inflorescence subtended by a short bract: terminal spike usually male, to 4 cm long; female spikes similar but a little thicker, to 4 cm long. **Scales** lanceolate, brown, never longer than the perigynia. smooth or slightly hairy along the mid-vein. **Stamens** 3. Pistils enclosed in perigynia; each **perigynium** lanceoloid, to 6 mm long, smooth or rarely with a few hairs when young, flattened, ciliate and minutely toothed near the tip, tapering to a tiny beak; **stigmas** 3.
FRUIT Achenes triangular, smooth, to 2 mm long.
NOTE The achenes are eaten by birds and small mammals.
WETLAND STATUS AW OBL | GP OBL | WMV OBL

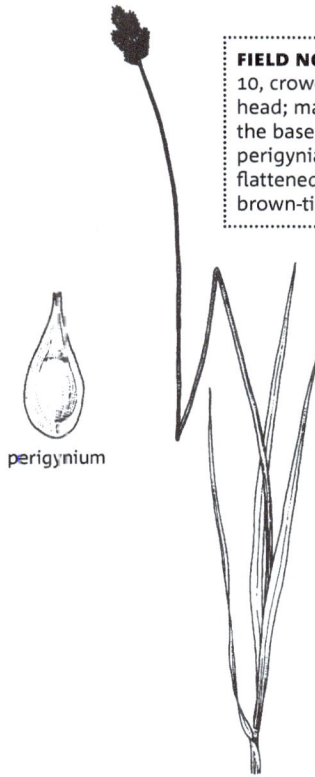

> **FIELD NOTES** Spikelets 3–10, crowded into a dense head; male flowers borne at the base of each spikelet; perigynia straw-colored, flattened, usually with a brown-tipped beak.

perigynium

Carex microptera Mackenzie
SMALL-WING SEDGE

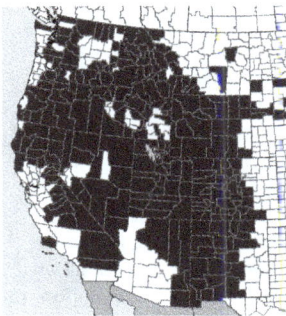

FAMILY Cyperaceae (Sedge)
FLOWERING May–August
HABITAT Wet meadows, fens.
HABIT Native tufted perennial sedge without creeping rhizomes.
STEMS Upright, triangular, to 60 cm tall, smooth or rough to the touch.
LEAVES Narrow, elongated, to 6 mm wide, without hairs, usually shorter than the stems.
FLOWERS Male and female flowers borne separately in the same spikelet, the male flowers at the base of the spikelet; spikelets 3–10, crowded into a dense head to 2.5 cm long. **Scales** lanceolate, brownish, shorter than the perigynia. Stamens 3. Pistils enclosed in a perigynium; each **perigynium** straw-colored, but usually with a brown-tipped beak, lanceolate to ovate, 2–6 mm long, smooth, flattened; **stigmas** 2.
FRUIT Achenes lenticular, about 2 mm long.
NOTE This species is distinguished with difficulty from the closely related *Carex festivella*, *C. ebenea*, and *C. haydeniana*. The achenes are eaten by birds and small mammals.
WETLAND STATUS AW FAC | GP FAC | WMV FACU

> **FIELD NOTES** Leaves bluish, with 1–2 terminal male spikes, 2 stigmas, and a short-beaked perigynia with 5–10 veins on each face.

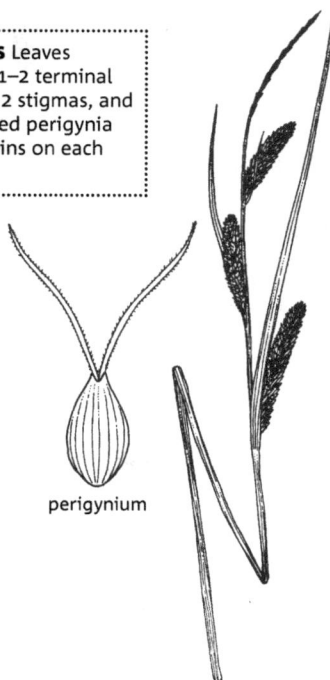

perigynium

Carex nebrascensis Dewey
NEBRASKA SEDGE

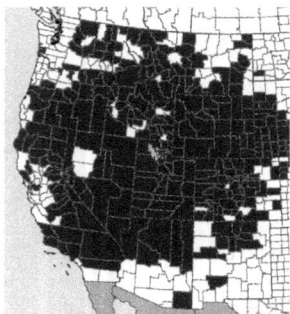

FAMILY Cyperaceae (Sedge)

FLOWERING May–September

HABITAT Swamps, wet meadows, around lakes and ponds.

HABIT Native perennial sedge from creeping rhizomes.

STEMS Stems upright, rather stout, triangular, to 1 m tall, smooth or somewhat roughened.

LEAVES Alternate, elongated, shorter or longer than the stem, to 13 mm wide, smooth or sometimes roughened along the edges, often bluish; sheaths yellow-brown.

FLOWERS Borne in spikes, the male flowers usually in separate spikes from the female flowers, but on the same plant; male spikes 1 or 2, above the female spikes, narrowly cylindrical, to 5 cm long; female spikes 2–5, to 8 cm long, the lowest spike subtended by a leafy bract as long as or longer than the inflorescence. **Scales** lanceolate, pointed at the tip, longer or shorter than the perigynia, with a green or pale midvein. **Stamens** 3. Pistils enclosed in a perigynium; each **perigynium** ellipsoid to obovoid, abruptly tapering to a very short, 2-cleft beak, straw-colored to brown, to 5 mm long, with 5–10 veins on each face; **stigmas** 2.

FRUIT Achenes lenticular, smooth, to 3 mm long.

NOTE The uppermost female spike sometimes has a few male flowers near its tip. This sedge is sometimes an important forage species for livestock. The achenes are eaten by waterfowl.

WETLAND STATUS AW OBL | GP OBL | WMV OBL

WESTERN WETLAND FLORA

perigynium

female scale

perigynium

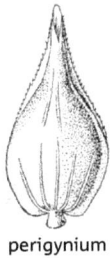

> **FIELD NOTES** Spikelets with male flowers at tip, at least the lowest spikelet not crowded with the rest; styles 2; achenes lenticular; perigynia at least 2 mm long. Rhizomes coarse, black, and scaly.

Carex praegracilis W. Boott
CLUSTERED FIELD SEDGE

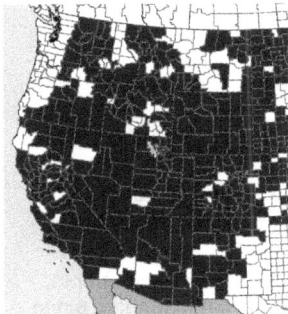

FAMILY Cyperaceae (Sedge)

FLOWERING May–September

HABITAT Most moist habitats from sea level to the mountains, often in alkaline soils.

HABIT Native perennial sedge with coarse, black, scaly rhizomes.

STEMS Upright, 1 or few together, to 60 cm tall, smooth, triangular.

LEAVES Alternate, elongated, mostly all near the base of the plant, flat, to 4 mm wide, not hairy.

FLOWERS Borne in spikelets, with the male flowers at the tip of each spikelet; spikelets 6–25. crowded into a cylindrical head, except the lowest spikelet not crowded into the head; each spikelet less than 13 mm long. **Scales** pale brown, usually with a green mid-vein, as long as or longer than the perigynium. **Stamers** 3. Pistils enclosed in a perigynium: each **perigynium** pale brown to brown to brown-black, at least 2 mm long, ovoid to ellipsoid, with a short, prominent beak, usually with minute teeth on the beak; **stigmas** 2.

FRUIT Achenes lenticular, to 2 mm long.

NOTE This species is valuable as a forage plant for livestock. It occurs from the prairies and plains to moderate elevations in the mountains. It is characteristic of alkaline flats.

WETLAND STATUS AW FACW | GP FACW | WMV FACW

spikes

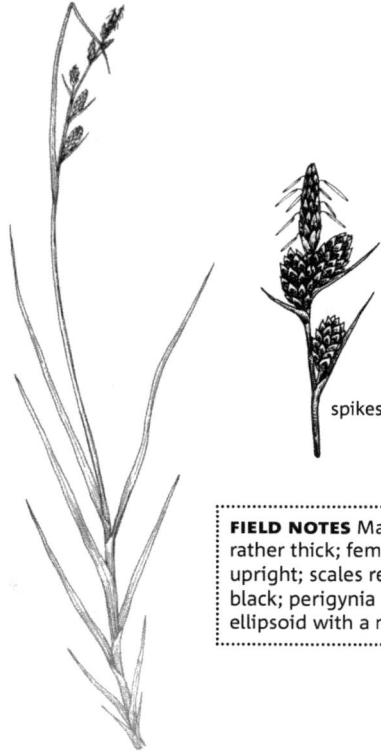

> **FIELD NOTES** Male spike solitary, rather thick; female spikes 2–5, upright; scales red-brown to black; perigynia plump, green, ellipsoid with a minute beak.

Carex raynoldsii Dewey
RAYNOLDS' SEDGE

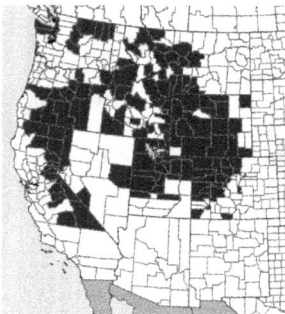

FAMILY Cyperaceae (Sedge)

FLOWERING June–August

HABITAT Meadows, open slopes, particularly in the mountains.

HABIT Native tufted perennial sedge with short, stout rhizomes.

STEMS Upright, triangular, to 60 cm tall, not hairy.

LEAVES Alternate, elongated, narrow, to 8 mm wide, flat, not hairy.

FLOWERS Male and female borne in separate spikes; male spike solitary, terminal, to 20 mm long, to 6 mm thick; female spikes 2–5, all upright, sessile except for the lowermost, to 2.5 cm long, to 10 mm thick. **Scales** ovate, shorter and narrower than the perigynia. red-brown to black. **Stamens** 3. Pistils enclosed in a perigynium; **perigynia** ellipsoid, green, plump, to 5 mm long, smooth, minutely beaked, with 2 conspicuous and several obscure nerves; **stigmas** 3.

FRUIT Achenes triangular, to 3 mm long, smooth.

NOTE The leaf bases from the preceding year are often persistent at the base of the plant. At maturity, the green perigynia contrast markedly with the subtending red-brown to black scales. The achenes are eaten by waterfowl.

WETLAND STATUS AW FACU | GP FACU | WMV FACU

> **FIELD NOTES** Male and
> female flowers borne in
> separate spikes; scales dark
> colored; perigynium ellip-
> soid, abruptly contracted
> into a very short beak.

spikes

Carex saxatilis L.
RUSSET SEDGE

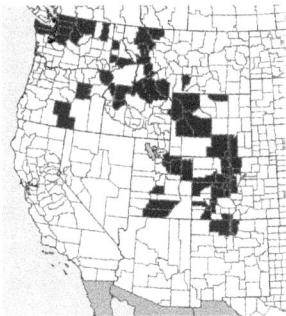

FAMILY Cyperaceae (Sedge)
FLOWERING May–August
HABITAT Edge of ponds, along streams, wet meadows, fens.
HABIT Native perennial sedge with extensively creeping rhizomes.
STEMS Upright, slender, triangular, to 60 cm tall, not hairy.
LEAVES Alternate, elongated, narrow, usually not surpassing the stem (except for the lowermost bract), to 4 mm wide, not hairy, usually septate, particularly on the sheaths.

FLOWERS Male and female borne in separate spikes; male spikes 1–2, terminal, to 4 cm long, to 6 mm thick, dark in color; female spikes 1–2, to 4 cm long, to 13 mm thick, the lowest on a stalk and sometimes nodding. **Scales** lanceolate to elliptic, shorter and narrower than the perigynia, usually dark-colored, rarely straw-colored. **Stamens** 3. Pistils enclosed in a perigynium; **perigynia** crowded, ascending, usually dark-colored, rarely straw-colored, ellipsoid, to 6 mm long, abruptly contracted into a very short beak, smooth; **stigmas** 2.
FRUIT Achenes lenticular, about 2 mm long, smooth.
NOTE The achenes are sometimes eaten by waterfowl.
WETLAND STATUS AW OBL | CP OBL | WMV OBL

> **FIELD NOTES** Scales black, subtending an often blackish perigynia with a minute beak.

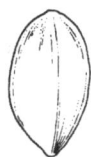

perigynium

Carex scopulorum Holm
HOLM'S ROCKY MOUNTAIN SEDGE

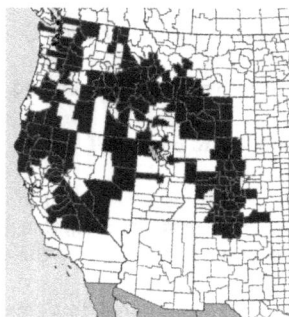

FAMILY Cyperaceae (Sedge)

FLOWERING May–August

HABITAT Along streams, wet meadows, around lakes and ponds, open slopes, particularly in the mountains and sometimes above timberline.

HABIT Native perennial sedge from elongated rhizomes, often sod-forming.

STEMS Upright, triangular, to 40 cm tall, without hairs.

LEAVES Alternate, elongated, narrow, to 6 mm wide, flat, not hairy, not surpassing the stems.

FLOWERS Male and female flowers borne separately in 3–6 spikes, the terminal spike nearly always male only, the lowest spike nearly always female only, the other spikes usually with male flowers above and female flowers below: spikes to 30 mm long, upright, all but the lowest one sessile. **Scales** black-purple, shorter than the perigynia. **Stamens** 3. Pistils enclosed in a perigynium; **perigynia** ellipsoid, usually purple-black, to 4 mm long, minutely beaked, smooth, without conspicuous veins; **stigmas** 2.

FRUIT Achenes lenticular, to 2 mm long.

NOTE In some areas, this species seems to grade into the closely related *Carex aquatilis*. The achenes are eaten by waterfowl.

WETLAND STATUS AW FACW | GP FACW | WMV OBL

> **FIELD NOTES** Male flowers at tip of spikelets; stigmas 2; achenes lenticular; perigynia brown, shiny, less than 3 mm long, with a minute beak.

perigynium

Carex simulata Mackenzie
SHORT-BEAK SEDGE

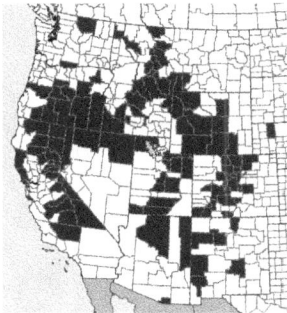

FAMILY Cyperaceae (Sedge)

FLOWERING May–August

HABITAT Wet meadows, along streams, in bogs, swamps.

HABIT Native perennial sedge with long, brown rhizomes.

STEMS Upright, triangular, to 45 cm tall, longer than the leaves.

LEAVES Alternate, elongated, narrow, to 4 mm wide, flat, not hairy.

FLOWERS Male and female borne in the same spikelet, with the male flowers at the tip; spikelets 8–25, less than 13 mm long, more or less crowded into a head to 4 cm long; bracts subtending the spikelets not leaf-like. **Scales** lanceolate, brown except for the paler edges, longer than the perigynia. **Stamens** 3. Pistils enclosed in a perigynium; **perigynia** very small, ovoid to ellipsoid, to 3 mm long, minutely beaked, brown, shiny, smooth, minutely toothed, especially near the tip; stigmas 2.

FRUIT Achenes lenticular, ellipsoid to obovoid, to 2 mm long, smooth.

NOTE Some plants may have all male spikelets or all female spikelets. The achenes are sometimes eaten by birds.

WETLAND STATUS AW OBL | GP OBL | WMV OBL

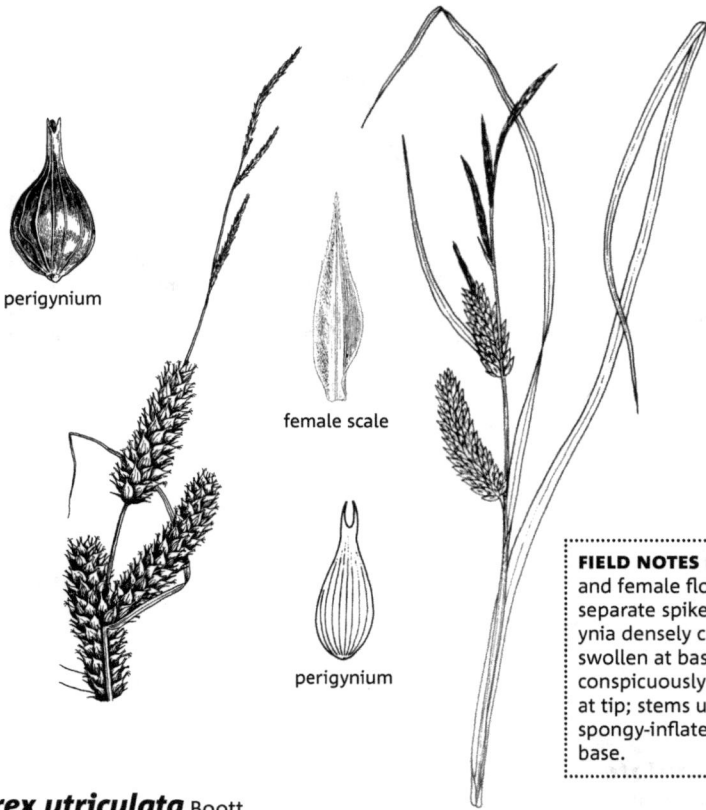

perigynium

female scale

perigynium

> **FIELD NOTES** Male and female flowers in separate spikes; perigynia densely crowded, swollen at base and conspicuously beaked at tip; stems usually spongy-inflated at base.

Carex utriculata Boott
BEAKED SEDGE

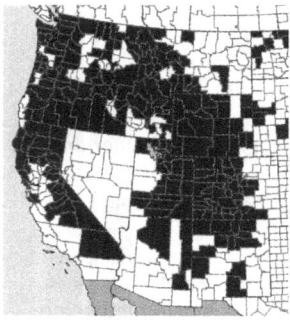

FAMILY Cyperaceae (Sedge)

FLOWERING May–August

HABITAT Along streams, around lakes and ponds, wet meadows, swamps, from sea level to the mountains.

HABIT Native perennial sedge from stout, elongated rhizomes.

STEMS Upright, triangular, to 1 m tall, smooth, usually spongy-inflated at the base.

LEAVES Several, elongated, flat, to 13 mm wide, not hairy, with evident cross-walls.

FLOWERS Borne in spikes, the male in 2–4 upper spikes, to 8 cm long, the female in several lower, thicker spikes to 10 cm long; **scales** pointed or awned at the tip, usually straw-colored, shorter than the perigynia. **Stamens** 3. Pistils enclosed in a perigynium; each **perigynium** pale brown, smooth, shiny, to 4 mm long with a conspicuous beak about 1/4 the length, ovoid, broadly rounded below, with 8–16 conspicuous nerves; **stigmas** 3.

FRUIT Achenes triangular, to 2 mm long, with a persistent, twisted style.

NOTE The achenes are eaten by waterfowl. Formerly often termed *Carex rostrata* J. Stokes but that species more northern uncommon in Washington, Idaho and Montana.

WETLAND STATUS AW OBL | GP OBL | WMV OBL

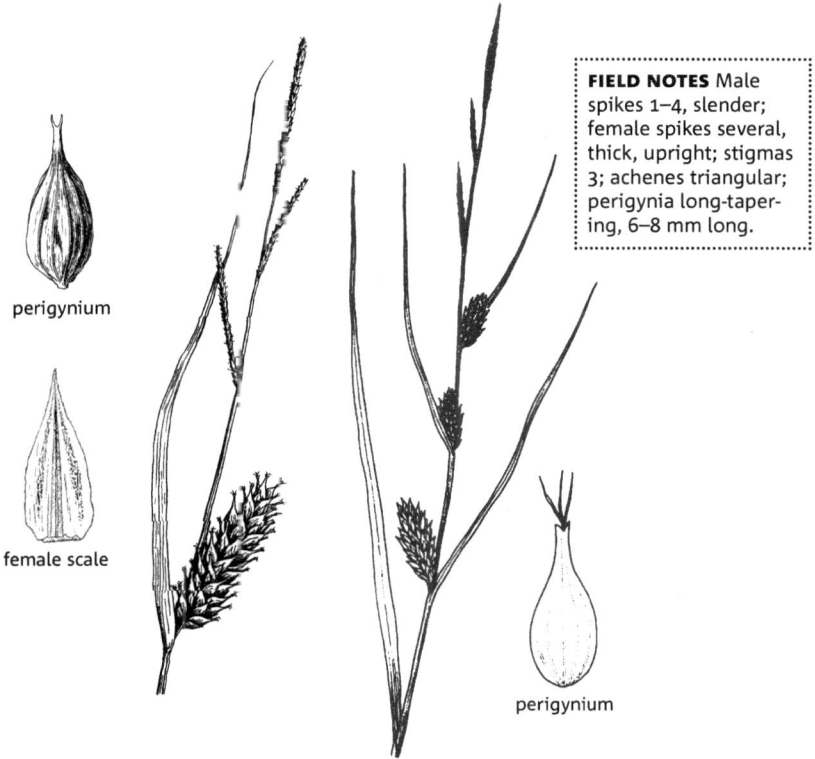

perigynium

female scale

perigynium

> **FIELD NOTES** Male spikes 1–4, slender; female spikes several, thick, upright; stigmas 3; achenes triangular; perigynia long-tapering, 6–8 mm long.

Carex vesicaria L.
INFLATED SEDGE

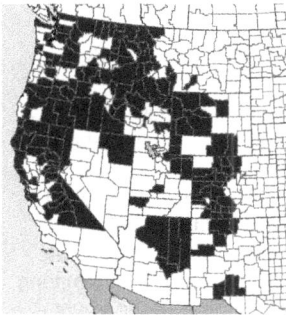

FAMILY Cyperaceae (Sedge)
FLOWERING May–August
HABITAT Marshes, wet meadows, swamps, bogs, edge of lakes and ponds, along streams.
HABIT Native perennial sedge with short, much branched rhizomes.
STEMS Upright, triangular, to 1 m tall, without hairs but sometimes rough to the touch.
LEAVES Alternate, elongated, narrow, to 8 mm wide, flat, usually septate, smooth.
FLOWERS Male and female borne in separate spikes; male spikes 1–4 in number, upright, to 8 cm long, to 3 mm thick, usually not subtended by bracts; female spikes 2–5 in number, upright, to 8 cm long, to 20 mm thick, each usually subtended by a leaf-like bract. **Scales** straw-colored to chestnut-colored, shorter and narrower than the perigynia, pointed at the tip but not awned. **Stamens** 3. Pistils enclosed in a perigynium; **perigynia** crowded, all ascending, pale green to straw-colored to chestnut-colored, lanceolate to lance-ovate, long-tapering to the tip, to 8 mm long, smooth, with 10–20 nerves; **stigmas** 3.
FRUIT Achenes triangular, to 3 mm long, with the terminal style persistent and bent.
NOTE The achenes are eaten by waterfowl.
WETLAND STATUS AW OBL | GP OBL | WMV OBL

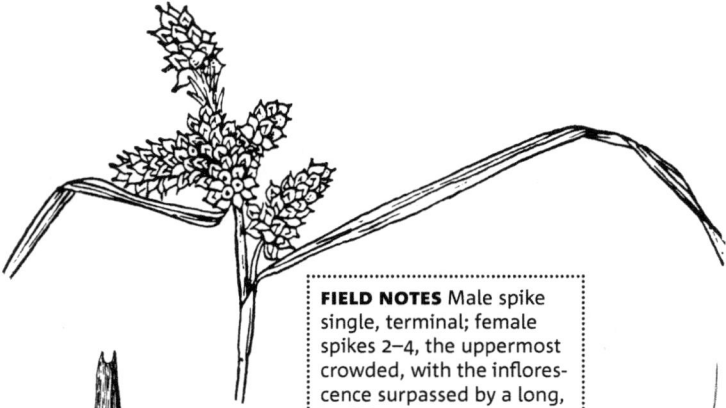

FIELD NOTES Male spike single, terminal; female spikes 2–4, the uppermost crowded, with the inflorescence surpassed by a long, leaf-like bract. Perigynium obovoid, prominently short-beaked, pale brown to straw-colored.

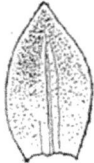

perigynium

female scale

Carex viridula Michx.
LITTLE GREEN SEDGE

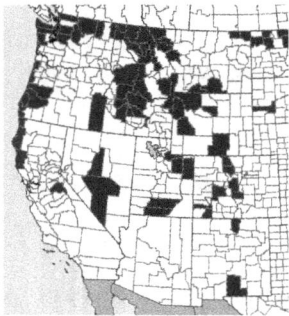

FAMILY Cyperaceae (Sedge)

FLOWERING June–September

HABITAT Wet meadows, marshes, around ponds and lakes, even in sandy and salty areas.

HABIT Native perennial sedge with fibrous roots.

STEMS Upright, unbranched, to 120 cm tall, without hairs.

LEAVES Elongated, mostly near the base of the plant, flat or channelled, to 3 mm wide, without hairs.

FLOWERS Male and female flowers borne in separate spikes, the male spike solitary, terminal, 8–25 mm long, the female spikes 2–4, 6–13 mm long; all the upper spikes sessile or nearly so, the lower spikes sometimes on short, slender stalks; at least one bract subtending the inflorescence leaf-like, much longer than the inflorescence; scales longer than the perigynia, pale brown with a green midvein and transparent margins. Stamens 3. Pistils enclosed in a perigynium, each perigynium obovoid. to 4 mm long, tapering to a slender beak, prominently nerved, pale brown to straw-colored; stigmas 3.

FRUIT Achenes triangular, to 2 mm long, smooth.

NOTE The achenes are eaten by waterfowl.

WETLAND STATUS AW OBL | GP OBL | WMV OBL

spikelet

> **FIELD NOTES** Plants small; scales of the spikelet with slender tips which are distinctly curved outward.

Cyperus squarrosus L.
AWNED FLATSEDGE

FAMILY Cyperaceae (Sedge)

FLOWERING June–October

HABITAT Wet areas, particularly along streams and around lakes and ponds; also in temporary water-filled depressions on sandstone bluffs.

HABIT Native tufted annual with fibrous roots.

STEMS Upright, slender, to 15 cm tall, smooth.

LEAVES Crowded near the base of the plant, elongated, very narrow, to 4 mm wide, smooth.

FLOWERS Crowded into flat spikelets; spikelets several, crowded into densely rounded clusters, the central cluster sessile, the other clusters, if present, on stalks; bracts much surpassing the inflorescence; each spikelet flat, less than 13 mm long, with 8–16 flowers. **Scales** lanceolate, awn-tipped, with the tip strongly recurved, green to pale brown. **Stamens** 1. **Pistils**: Ovary superior; **style** 3-parted.

FRUIT Achenes triangular, smooth, obovoid to oblongoid, brown, about 1 mm long.

NOTE The plants have the odor of slippery elm, particularly when dried. The achenes are eaten by waterfowl.

SYNONYMS *Cyperus aristatus* Rottb.

WETLAND STATUS AW OBL | GP OBL | WMV OBL

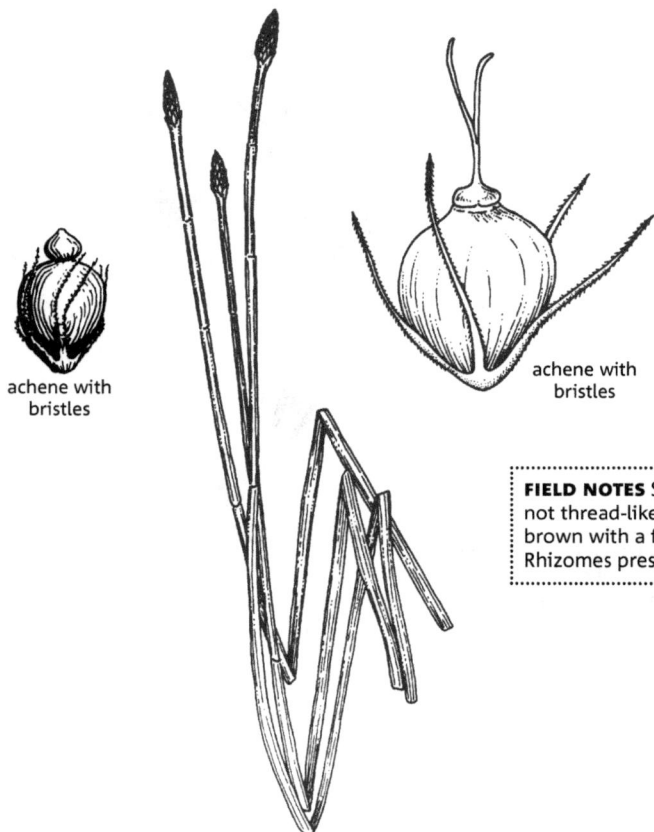

achene with
bristles

achene with
bristles

> **FIELD NOTES** Stems stiff, round,
> not thread-like; achenes yellow-
> brown with a flattened tubercle.
> Rhizomes present.

Eleocharis palustris (L.) Roemer & J. A. Schultes
CREEPING SPIKERUSH

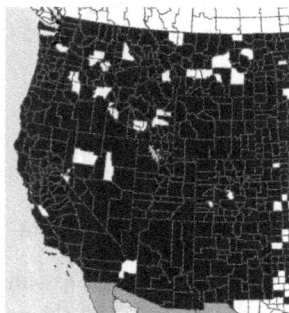

FAMILY Cyperaceae (Sedge)

FLOWERING June–August

HABITAT Wet ditches, wet meadows, around lakes and ponds.

HABIT Native perennial with branched, reddish rhizomes.

STEMS Upright, smooth, stiff, round, unbranched, to 1 m tall.

LEAVES Reduced to sheaths.

FLOWERS 1 per scale, with several scales per spikelet: spikelet one per stem, lanceoloid, to 30 mm long, with the lowest 2–3 scales empty. **Stamens** 3. **Pistils:** Ovary superior; **style** 2-cleft.

FRUIT Achenes obovoid, yellow-brown, shiny, to 2 mm long, with a flattened tubercle and subtended by as many as 8 bristles (or bristles sometimes absent).

NOTE The achenes are eaten by waterfowl.

SYNONYMS *Eleocharis macrostachya* Britt.

WETLAND STATUS AW OBL | GP OBL | WMV OBL

> **FIELD NOTES** Stems slender, nearly thread-like, bearing a small spikelet with no more than 9 flowers. Achene with distinct beak continuous with the body of the achene and about 1/4 as long.

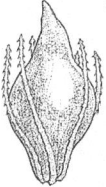

achenes with bristles

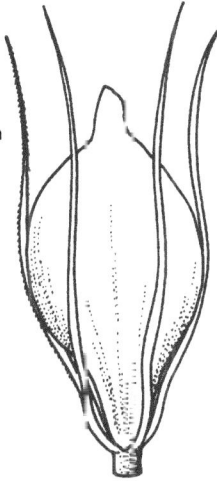

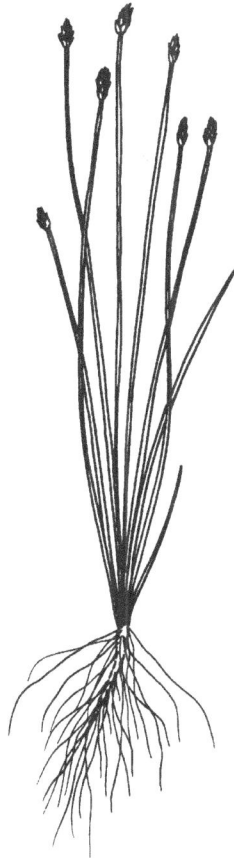

Eleocharis quinqueflora (F. X. Hartmann) Schwarz
FEW-FLOWER SPIKERUSH

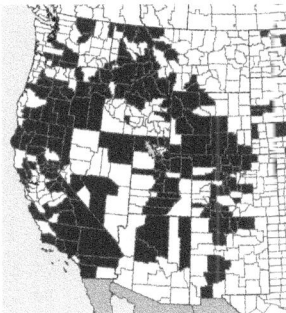

FAMILY Cyperaceae (Sedge)
FLOWERING June–August
HABITAT Bogs, fens, wet meadows, sometimes above timberline, sometimes in salty or alkaline habitats.
HABIT Native dwarf perennial herb, with short and thick as well as longer and more slender rhizomes.
STEMS Upright, very slender, angular, to 120 cm long, smooth.
LEAVES absent.
FLOWERS Borne in a solitary terminal spikelet; spikelet narrowly ovoid, pointed at the tip, to 8 mm long, with to 9 flowers. **Scales** ovate, pointed at the tip, to 6 mm long, brown with pale margins. **Stamens** 3. **Pistils**: Ovary superior; **stigmas** 3.
FRUIT Achenes triangular, to 3 mm long, abruptly tapering to a continuous short beak about 1/4 the length of the body of the achene, without hairs.
SYNONYMS *Eleocharis pauciflora* (Lightf.) Link
WETLAND STATUS AW OBL | GP OBL | WMV OBL

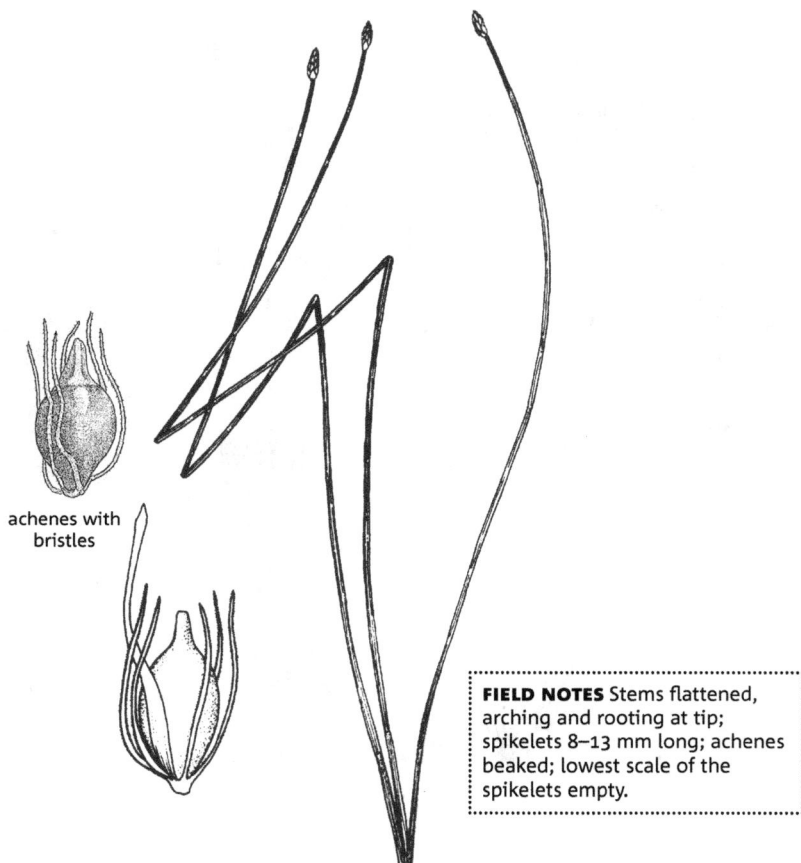

achenes with bristles

FIELD NOTES Stems flattened, arching and rooting at tip; spikelets 8–13 mm long; achenes beaked; lowest scale of the spikelets empty.

Eleocharis rostellata (Torr.) Torr.
BEAKED SPIKERUSH
FAMILY Cyperaceae (Sedge)

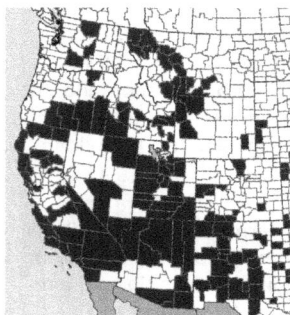

FLOWERING June–August

HABITAT Marshes, often in alkaline or salty areas.

HABIT Native perennial with stout, vertical rhizomes.

STEMS Flattened, to 1 m tall, bearing a single spikelet and no leaves. With age, the stem droops ("arches") and the tip may root where it touches a moist substrate, forming new plants.

LEAVES absent.

FLOWERS 5–25 in a spikelet, the spikelet 8–13 mm long; scales with a pale border and central area, the lowest one empty. **Stamens** 3. **Pistils:** Ovary superior; **stigmas** 3.

FRUIT Achenes more or less triangular, tapering to an elongated beak, light green to pale brown, 2–3 mm long, subtended by a few bristles usually longer than the achenes.

NOTE The achenes are eaten by waterfowl.

WETLAND STATUS AW OBL | GP OBL | WMV OBL

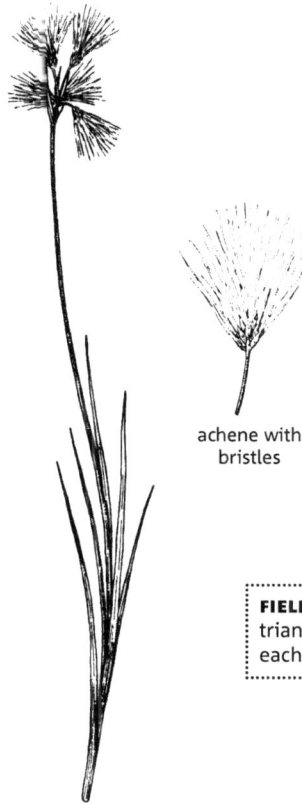

achene with
bristles

> **FIELD NOTES** Leaves channelled,
> triangular; spikelets several on
> each stem.

Eriophorum gracile W. D. J. Koch
SLENDER COTTON-GRASS

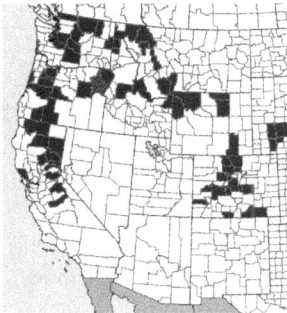

FAMILY Cyperaceae (Sedge)
FLOWERING July–August
HABITAT Bogs.
HABIT Native perennial with thickened rootstocks.
STEMS Upright or sometimes bent over, unbranched, to
50 cm tall, smooth, not triangular.
LEAVES Very narrow, elongated, alternate, triangular,
channelled, to 4 cm long, smooth, the basal leaves usually absent at flowering time.
FLOWERS Crowded together into 2–4 terminal spikelets,
each spikelet to 2.5 cm long, on a hairy stalk; all
spikelets subtended by a bract to 13 mm long; **scales**
ovate, gray to black, pointed at the tip. **Stamens** usually 3. **Pistils:** Ovary superior; **styles**
3-parted.
FRUIT Achenes triangular, obovoid to oblongoid, about 2 mm long, subtended by
bright white bristles to 2.5 cm long.
NOTE The achenes, which are wind dispersed because of the bright white bristles, are
eaten by waterfowl.
WETLAND STATUS AW OBL | GP OBL | WMV OBL

> **FIELD NOTES** Plants small, with a single spikelet at tip of stem; scales blackish green; bristles white, subtending the achene.

achene

Eriophorum scheuchzeri Hoppe
SCHEUCHZER'S COTTON-GRASS

FAMILY Cyperaceae (Sedge)
FLOWERING July–August
HABITAT Bogs, fens.
HABIT Native perennial with creeping rhizomes, often forming dense colonies.
STEMS Upright, unbranched, to 20 cm tall, smooth.
LEAVES Very few, mostly near the base of the plant, channelled and triangular, smooth, only about 1 mm wide.
FLOWERS Crowded together into a single terminal spikelet, to 2.5 cm long; **scales** narrowly lanceolate, usually blackish green, tapering to a slender point. **Stamens** usually 3. **Pistils:** Ovary superior; **styles** 3-parted.
FRUIT Achenes triangular, oblanceoloid, about 2 mm long, subtended by white bristles.
NOTE The achenes may be eaten by waterfowl.
WETLAND STATUS AW FACW | WMV FACW

FIELD NOTES Spikelets clustered, sessile, subtended by a single upright bract that appears like an extension of the stem. Stems only slightly triangular; achenes 2–3 mm long.

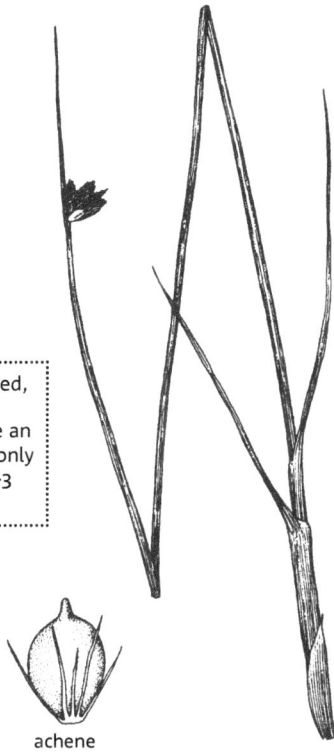

achene

Schoenoplectus americanus (Pers.) Volk. ex Schinz & R. Keller
THREE-SQUARE BULRUSH

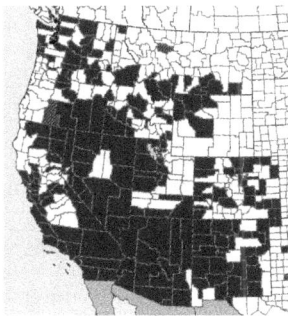

FAMILY Cyperaceae (Sedge)

FLOWERING May–August

HABITAT Wet meadows, marshes, around lakes and ponds.

HABIT Native perennial with rhizomes.

STEMS Upright, somewhat triangular, to 1 m tall, without hairs.

LEAVES Mostly arising near the base of the plant, elongated, flat or sometimes folded, to 6 mm wide, without hairs.

FLOWERS Borne in 1–6 spikelets, the spikelets sessile and subtended by an upright bract that appears like an extension of the stem; spikelets 13–25 mm long, more or less pointed at the tip; bracts to 15 cm long; **scales** brown, usually with a short awn at the notched tip. **Stamens** 3. Pistils: Ovary superior; styles 2 or 3

FRUIT Achenes lenticular or triangular, to 3 mm long, with a distinct point at the tip and with 4–6 bristles arising at the base.

NOTE This species sometimes grows in alkaline habitats. The achenes are an important food source for waterfowl.

SYNONYMS *Scirpus americanus* Pers., *Scirpus pungens* Vahl

WETLAND STATUS AW OBL | GF OBL | WMV OBL

> **FIELD NOTES** Plants stout; spikelets large, densely crowded in a sessile cluster at tip of stem.

achene

Schoenoplectus maritimus (L.) Lye
SALTMARSH BULRUSH

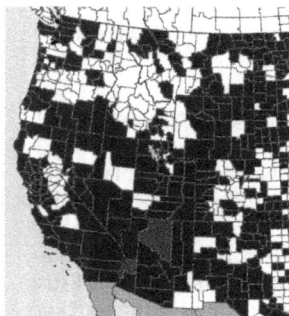

FAMILY Cyperaceae (Sedge)

FLOWERING June–September

HABITAT Wet meadows, marshes, around ponds, particularly in alkaline habitats.

HABIT Native stout perennial with rhizomes and tubers.

STEMS Upright, triangular, usually unbranched, to 1.5 m tall, without hairs.

LEAVES Alternate, elongated, flat, to 13 mm wide, without hairs.

FLOWERS Several crowded into spikelets, usually the sessile spikelets densely clustered at the tip of the stem; spikelets rounded at the tip. to 2.5 cm long, to 13 mm thick, subtended by 2–4 leaf-like bracts of different lengths; **scales** usually pale brown, minutely hairy, notched at the tip with a short awn arising from the notch. **Stamens** 3. **Pistils**: Ovary superior; **styles** 2.

FRUIT Achenes lenticular, to 4 mm long, with a minute point at the tip and with a few short bristles at the base.

NOTE The large achenes are eaten by waterfowl.

SYNONYMS *Scirpus maritimus* L.

WETLAND STATUS AW OBL | GP OBL | WMV OBL

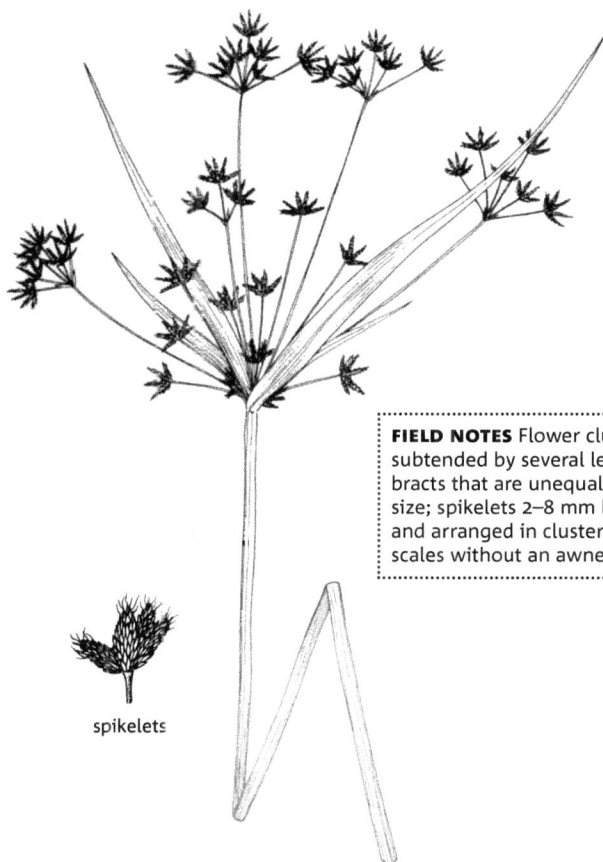

FIELD NOTES Flower clusters subtended by several leaf-like bracts that are unequal in size; spikelets 2–8 mm long and arranged in clusters; scales without an awned tip.

spikelets

Scirpus microcarpus J. & K. Presl
SMALL-FRUIT BULRUSH

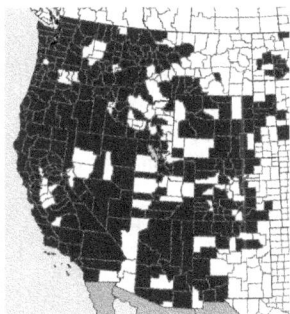

FAMILY Cyperaceae (Sedge)
FLOWERING June–August
HABITAT Along streams, wet ditches, marshes.
HABIT Native perennial with long, stout rhizomes.
STEMS Upright, triangular, stout, smooth, to 1.5 m tall.
LEAVES Several, alternate, elongated, narrow, usually rough along the edges, to 20 mm wide; sheaths tinged with red or purple.
FLOWERS Borne in spikelets, with several spikelets in clusters, some of the clusters sometimes sessile, others on long stalks, subtended by several leafy bracts of different lengths; each spikelet 2–8 mm long. **Scales** ovate, pointed but not awned at the tip, black or greenish black; bristles 4–6. **Stamens** 3. **Pistils**: Ovary superior; **style** 2-cleft.
FRUIT Achenes pale, lenticular, smooth, about 2 mm long.
NOTE The achenes are eaten by waterfowl.
WETLAND STATUS AW OBL | GP OBL | WMV OBL

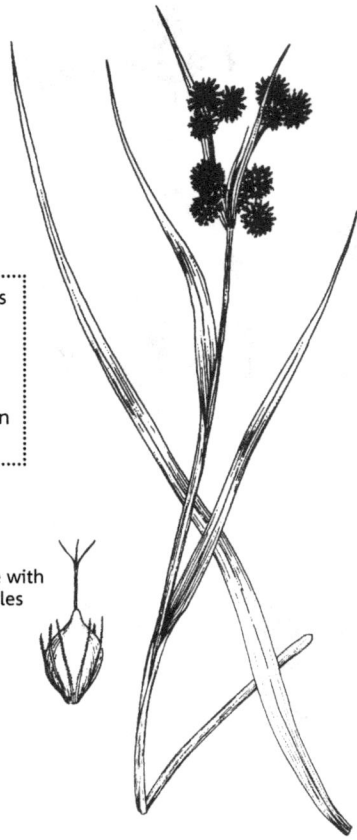

FIELD NOTES Flower clusters subtended by several leaf-like bracts that are unequal in size; spikelets less than 4 mm long and arranged in clusters; scales awn-tipped in the spikelets.

achene with bristles

Scirpus pallidus (Britton) Fernald
CLOAKED BULRUSH

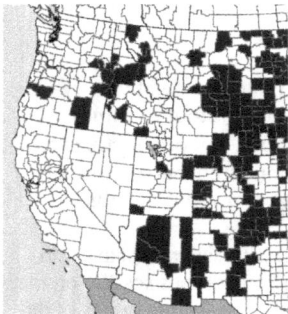

FAMILY Cyperaceae (Sedge)

FLOWERING June–August

HABITAT Along streams, wet ditches, marshes.

HABIT Native perennial with short, stout rhizomes.

STEMS Upright, triangular, smooth, to 1.5 m tall.

LEAVES Several, alternate, elongated, narrow, usually rough along the edges, to 20 mm wide; sheaths green or pale green.

FLOWERS Borne in spikelets, with several spikelets in clusters, some of the clusters sessile, others on long stalks, subtended by several leafy bracts of different lengths; each spikelet to 4 mm long. **Scales** narrow, greenish black, to 3 mm long, with a rough awn at the tip; bristles 6. **Stamens** 3. **Pistils**: Ovary superior; **styles** 3-cleft.

FRUIT Achenes white, triangular, smooth, about 1 mm long.

NOTE This species is very similar to the eastern *S. atrovirens* from which it differs by its longer scales with longer awns. The achenes are eaten by waterfowl.

WETLAND STATUS AW OBL | GP OBL | WMV OBL

> **FIELD NOTES** Plants small, with a single terminal spikelet and a single green leaf a short distance above the straw-colored scale leaves found at the plant base.

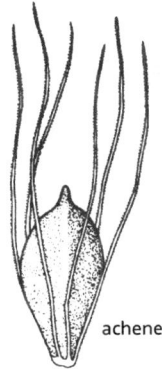

achene

Trichophorum caespitosum (L.) Hartman
TUFTED BULRUSH

FAMILY Cyperaceae (Sedge)
FLOWERING July–August
HABITAT Marshes, bogs.
HABIT Native perennial with a short rhizome and fibrous roots.
STEMS Upright, round, not triangular, pale green, to 40 cm tall, smooth.
LEAVES Mostly scale-like and straw-colored at the base of the plant; a single green leaf usually a short distance above the scale leaves, to 6 mm long, smooth.
FLOWERS Very few in a solitary, terminal spikelet; spikelet ovoid to oblongoid, to 6 mm long, subtended by a linear bract to 6 mm long; scales ovate, yellow-brown. **Stamens** 3. **Pistils**: Ovary superior, smooth; **styles** 3-cleft.
FRUIT Achenes oblongoid, pointed at the tip. triangular, brown, to 4 mm long, smooth.
SYNONYMS *Scirpus caespitosus* L.
WETLAND STATUS AW OBL | GP OBL | WMV OBL

achene

FIELD NOTES *Alisma* have 3 small white or pinkish petals, 3 sepals, and a ring of as many as 25 pistils and achenes in a single whorl. *A. gramineum* differs from others in the genus by its linear to elliptic leaves never more than 4 cm wide and a curved beak on the achene.

Alisma gramineum Lej.
NARROW-LEAF WATER-PLANTAIN

FAMILY Alismataceae (Water-plantain)

FLOWERING June–September

HABITAT Marshes, mud flats, sometimes submerged in water.

HABIT Native perennial herb with fleshy rhizomes.

STEMS Only the flower-bearing stem above ground, upright, to 45 cm tall, smooth.

LEAVES All basal, linear to elliptic to lanceolate, to 20 cm long, to 4 cm wide, pointed at the tip, smooth; leaf stalks slender, smooth.

FLOWERS Several in whorls forming a panicle, the panicle to 45 cm long; flower stalks slender, to 4 cm long, subtended by papery bracts. **Sepals** 3, free from each other, green, to 3 mm long. **Petals** 3, free from each other, pinkish, to 4 mm long. **Stamens** 6–9, about as long as the pistils. **Pistils:** Up to 20 arranged in a single whorl, each with a superior ovary.

FRUIT Achenes to 20 in a single whorl, each achene to 3 mm long, with a short, curved beak at the tip.

NOTE The leaf shape is variable in this species. The plants may be submerged in water or rooted in mud. The achenes are eaten by waterfowl.

WETLAND STATUS AW OBL | GP OBL | WMV OBL

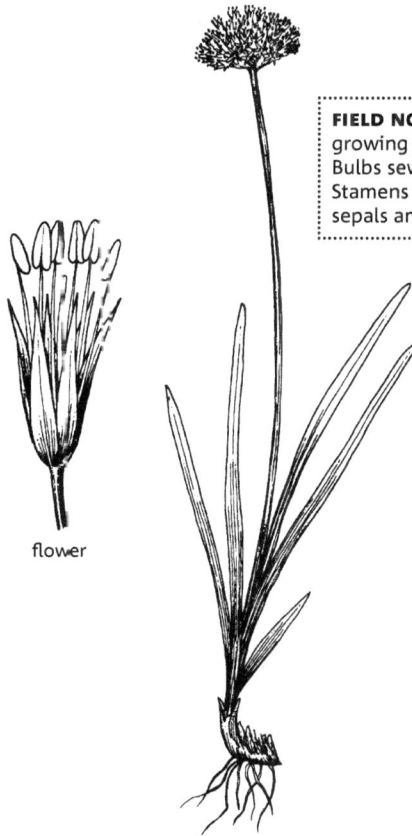

> **FIELD NOTES** Plants stout, growing to a height of 1 m. Bulbs several along a rhizome. Stamens 6, exserted above the sepals and petals.

flower

Allium validum S. Wats.
TALL SWAMP ONION

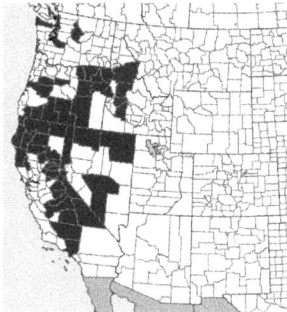

FAMILY Alliaceae (Onion)

FLOWERING June–September

HABITAT Wet meadows.

HABIT Native perennial herb, with several oblongoid to ovoid bulbs formed along a rhizome.

STEMS Upright, stout, somewhat flattened, to 1 m tall, smooth.

LEAVES Several, basal, flat, elongated, to 13 mm wide, rounded or barely pointed at the tip.

FLOWERS 15–30 in a terminal umbel, each flower on a slender stalk 13–20 mm long. **Sepals and petals** 6, similar, free from each other, pink, narrowly lanceolate, to 13 mm long, pointed at the tip. **Stamens** 6, exserted above the sepals and petals. **Pistils**: Ovary superior.

FRUIT Capsules nearly spherical, 6–8 mm in diameter, with several long, slender, not shiny seeds.

NOTE Formerly in Liliaceae; that family now divided into several new families.

WETLAND STATUS AW OBL | WMV OBL

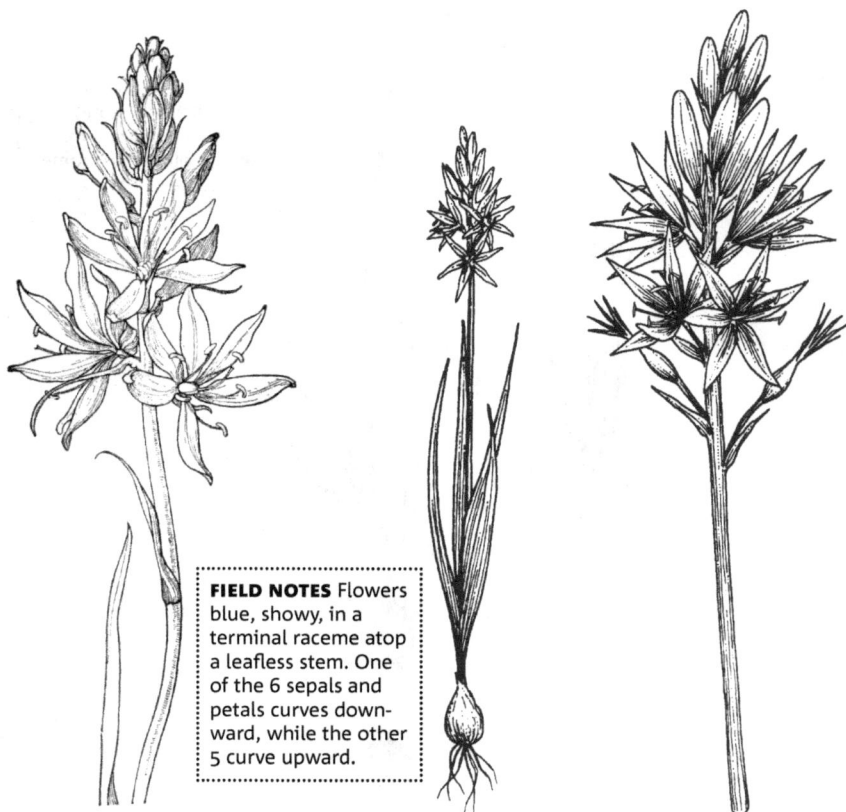

FIELD NOTES Flowers blue, showy, in a terminal raceme atop a leafless stem. One of the 6 sepals and petals curves downward, while the other 5 curve upward.

Camassia quamash (Pursh) Greene
COMMON CAMASSIA

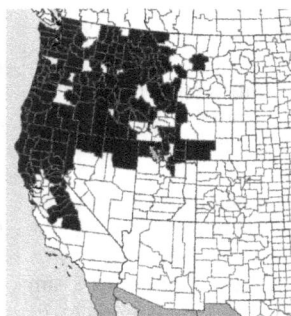

FAMILY Agavaceae (Agave)
FLOWERING April–July
HABITAT Wet meadows, damp hillsides.
HABIT Native perennial herb with a bulb.
STEMS Upright, smooth, to 60 cm tall, bearing only flowers.
LEAVES All basal, flat, elongated, to 120 cm long, to 2.5 cm wide.
FLOWERS Several in a terminal raceme to 25 cm long, each flower to 8 cm across, on stalks 8–25 mm long. **Sepals and petals** 6, all similar in size and shape, free from each other, blue, narrowly lanceolate, to 4 cm long, one of them curving downward, the other 5 curving upward. **Stamens** 6. **Pistils:** Ovary superior.
FRUIT Capsules ovoid, to 2.5 cm long, pale brown to brown.
NOTE The sepals and petals become twisted over the capsules. Formerly in Liliaceae; that family now divided into several new families.
WETLAND STATUS AW FACW | WMV FACW

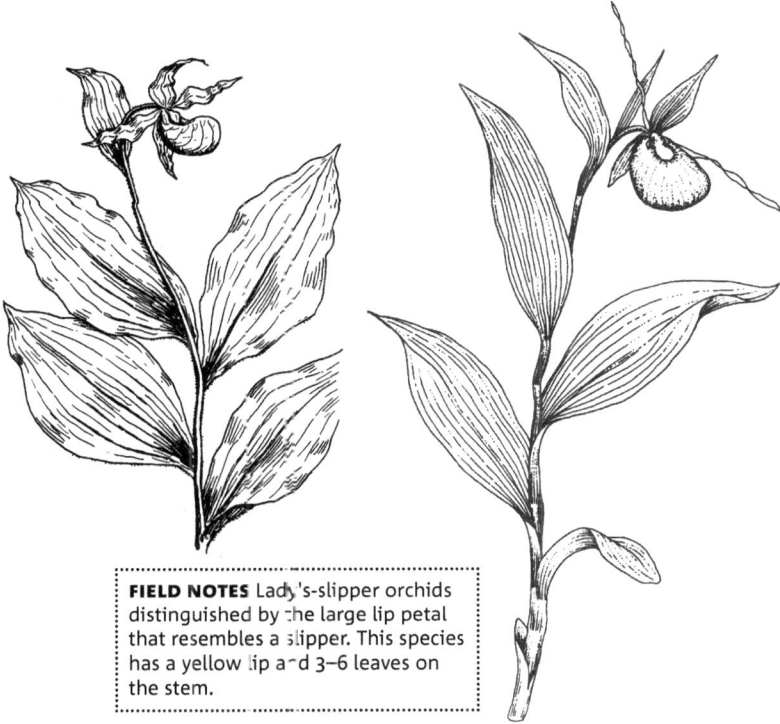

FIELD NOTES Lady's-slipper orchids distinguished by the large lip petal that resembles a slipper. This species has a yellow lip and 3–6 leaves on the stem.

Cypripedium parviflorum var. *pubescens* (Willd.) Knight
SMALL YELLOW LADY'S-SLIPPER

FAMILY Orchidaceae (Orchid)
FLOWERING April–June
HABITAT Along streams, in rich woods, in swamps.
HABIT Native perennial herb with short rhizomes and fibrous roots.
STEMS Upright, to 45 cm tall, glandular-hairy.
LEAVES Alternate, simple, 3–6 on each stem, broadly lanceolate to elliptic, to 20 cm long, to 10 cm wide, pointed at the tip, tapering to the sessile base, glandular-hairy, strongly veined.
FLOWERS Usually 1 at the tip of the stem, sometimes 2 or more, each on a stalk at least 13 mm long, subtended by a pointed bract 2–13 cm long. Sepals 3, yellow, greenish, or purple-brown, one of them narrowly ovate, pointed, and wavy along the edges, the other 2 united and situated below the lip petal. Petals 3, 2 of them yellow, greenish, or purple-brown and wavy-edged, to 10 cm long and to 13 mm wide, the other one (the lip) forming a yellow slipper to 6 cm long. Stamens 2. Pistils: Ovary inferior, ribbed.
FRUIT Capsules ellipsoid, to 5 cm long, to 13 mm wide.
NOTE Throughout its range there is much variation in size of the yellow lip and color of the other petals and sepals.
SYNONYMS *Cypripedium calceolus* L.
WETLAND STATUS AW FACW | GP FACW | WMV FACW

FIELD NOTES Flowers with 3 fringed, white petals with a yellow blotch near their base. Leaves elliptic to lanceolate, on long stalks; margins without teeth.

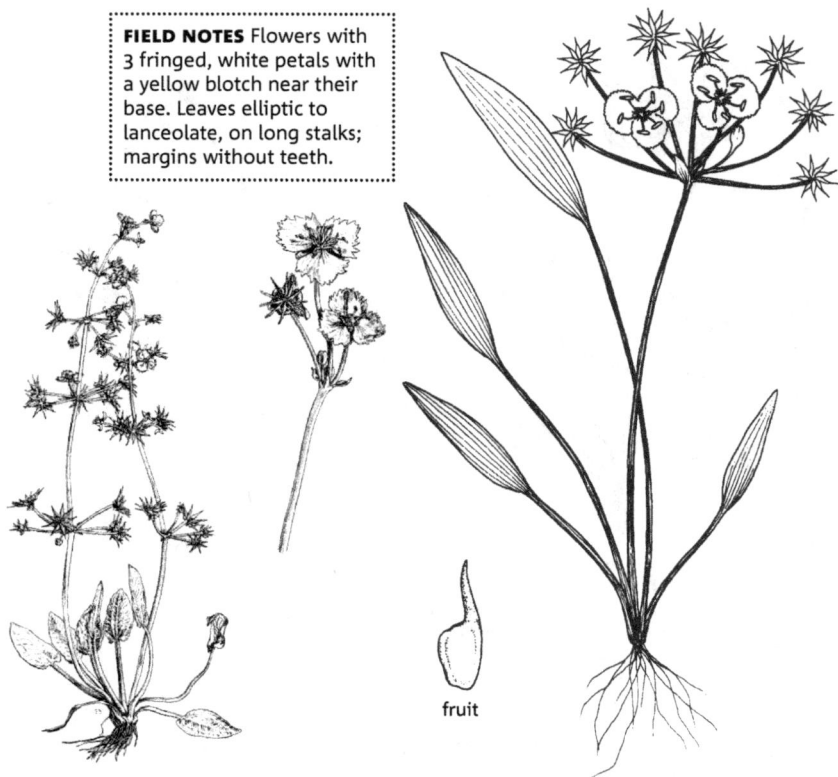

fruit

Damasonium californicum Torr.
SMALL FRINGED WATER-PLANTAIN

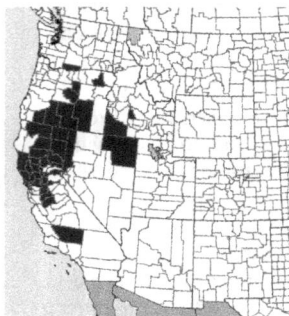

FAMILY Alismataceae (Water-plantain)

FLOWERING April–September

HABITAT Vernal pools, along streams, in mud flats.

HABIT Native perennial herb with short, fleshy rhizomes.

STEMS Only the flower-bearing stem present above ground, to 45 cm tall, smooth.

LEAVES All basal, elliptic to lanceolate, to 10 cm long, to 5 cm wide, tapering to the tip, rounded or tapering to the base, smooth, without teeth; stalks very long.

FLOWERS Several borne in whorls on a leafless stem, with the whorls arranged in a panicle; each whorl of flowers subtended by broadly lanceolate to ovate, greenish bracts. **Sepals** 3, green, free from each other, about 4 mm long, smooth. **Petals** 3, white with a yellow blotch near the base, toothed or with a fringe along the edges, nearly round, 8–13 mm across. **Stamens** 6. **Pistils**: 6–15 in a single whorl, the ovaries superior.

FRUIT 6–15 achenes spreading in a single whorl, each achene 4–13 mm long, with a short, stout beak on one side.

SYNONYMS *Machaerocarpus californicus* (Torr.) Small

WETLAND STATUS AW OBL | WMV OBL

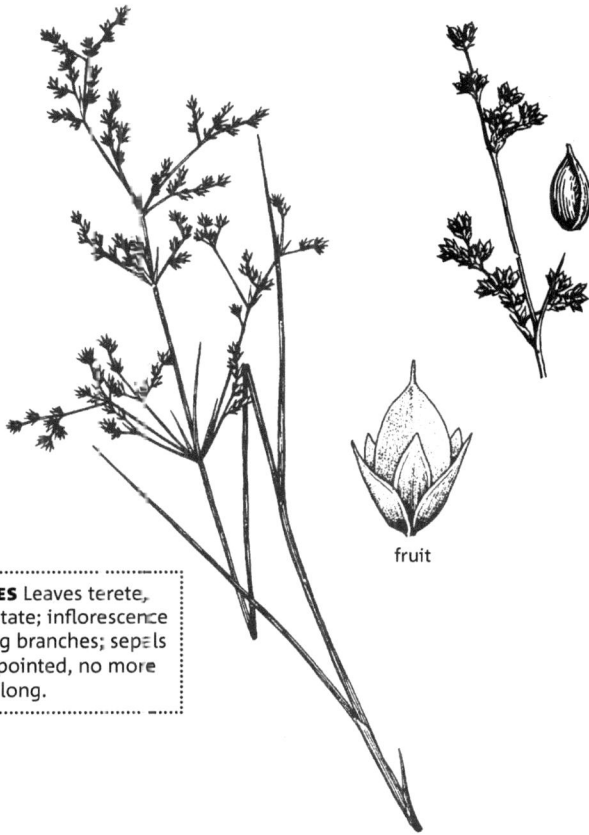

fruit

> **FIELD NOTES** Leaves terete, hollow, septate; inflorescence of spreading branches; sepals and petals pointed, no more than 3 mm long.

Juncus articulatus L.
JOINTED RUSH
FAMILY Juncaceae (Rush)

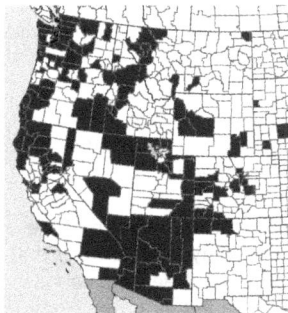

FLOWERING June–September
HABITAT Along streams, around ponds, in marshes.
HABIT Native perennial rush from a rather stout rhizome.
STEMS Upright, unbranched, to 60 cm tall, smooth.
LEAVES Terete, hollow, septate, to 15 cm long, to 2 mm wide, smooth; ligule at tip of sheath to 2 mm long.
FLOWERS 3–10 in heads, with several heads in a branched inflorescence. **Sepals** 3, green, brown or purplish, free from each other, to 3 mm long, pointed at the tip. **Petals** 3, green, brown or purplish, free from each other, to 3 mm long, pointed at the tip. **Stamens** 6.
Pistils: Ovary superior, smooth.
FRUIT Capsules ovoid, more or less triangular, pointed at the tip, to 4 mm long, longer than the sepals and petals, shiny, dark brown.
NOTE The seeds are eaten by waterfowl.
WETLAND STATUS AW OBL | GP OBL | WMV OBL

two flowered head

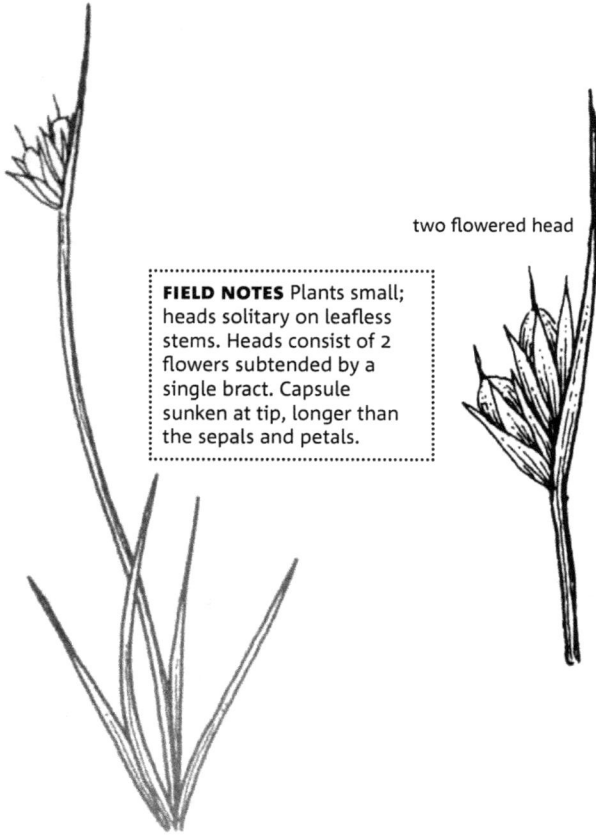

> **FIELD NOTES** Plants small; heads solitary on leafless stems. Heads consist of 2 flowers subtended by a single bract. Capsule sunken at tip, longer than the sepals and petals.

Juncus biglumis L.
TWO-FLOWER RUSH

FAMILY Juncaceae (Rush)

FLOWERING July–September

HABITAT Wet gravels, sphagnum beds, around lakes and ponds, sometimes in shallow water.

HABIT Native perennial rush with a short rhizome.

STEMS Upright, unbranched. bearing only a single head, smooth, to 30 cm tall, usually much shorter.

LEAVES All basal or nearly so. inrolled into a very slender, hollow tube.

FLOWERS 2–4 borne in a solitary head at the tip of the stem, the inflorescence subtended by a single bract that usually surpasses the head. **Sepals** 3, free from each other, purple-brown, 4–6 mm long, pointed or rounded at the tip. **Petals** 3, free from each other, purple-brown. 4–6 mm long, pointed or rounded at the tip. **Stamens** 6. **Pistils:** Ovary superior, smooth.

FRUIT Capsules obovoid, sunken at the tip, to 6 mm long, longer than the sepals and petals; seeds with a short appendage at either end.

WETLAND STATUS AW OBL | WMV OBL

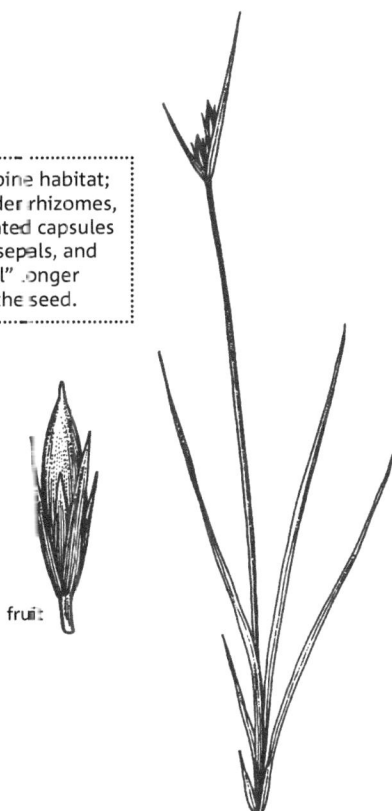

> **FIELD NOTES** Alpine habitat; plants with slender rhizomes, slender and pointed capsules longer than the sepals, and seeds with a "tail" longer than the rest of the seed.

fruit

Juncus castaneus J. E. Smith
CHESTNUT RUSH

FAMILY Juncaceae (Rush)
FLOWERING July–August
HABITAT Along streams, around and in ponds, fens, usually at or near timberline.
HABIT Native perennial rush with slender rhizomes.
STEMS Upright, smooth, to 120 cm long.
LEAVES Alternate, elongate, hollow, septate, rounded into a tube or folded, to 2 mm wide.
FLOWERS Several crowded into 1–4 heads, each head surpassed by a subtending bract; flowers to 15 per head. **Sepals** 3, purple-brown, narrowly lanceolate, pointed at the tip, to 8 mm long. **Petals** 3, purple-brown, narrowly lanceolate, pointed at the tip, to 8 mm long. **Stamens** 6. **Pistils**: Ovary superior.
FRUIT Capsules lanceoloid. narrow, pointed at the tip, to nearly 13 mm long, longer than the sepals and petals; seeds very narrow, with a tail longer than the body of the seed.
NOTE The seeds are eaten by small birds and small mammals.
WETLAND STATUS AW FACW | GP OBL | WMV FACW

> **FIELD NOTES** Flowers with 2 small bracts at their base; capsules shallowly notched; leaves nearly thread-like.

fruit

Juncus confusus Coville
COLORADO RUSH

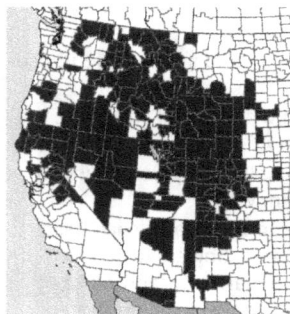

FAMILY Juncaceae (Rush)

FLOWERING June–August

HABITAT Wet meadows, along streams, in moist woods.

HABIT Native perennial rush with fibrous roots.

STEMS Upright, slender, unbranched, to 45 cm tall, smooth.

LEAVES Mostly near the base of the plant, thread-like to narrowly linear, flat or rolled up into a hollow tube, to 1 mm wide, smooth.

FLOWERS Several in small, crowded clusters to 2.5 cm long, subtended by 1–3 leafy bracts to 10 cm long, each flower subtended by a pair of small bracts. **Sepals** 3, free from each other, brown with a green mid-vein, to 6 mm long. **Petals** 3, free from each other, brown with a green mid-vien, to 6 mm long. **Stamens** 6. **Pistils:** Ovary superior, smooth.

FRUIT Capsules oblongoid, more or less tirangular, to 4 mm long, a little shorter than the sepals and petals, shallowly notched at the tip.

NOTE The seeds are eaten by waterfowl.

WETLAND STATUS AW FAC | GP FACW | WMV FAC

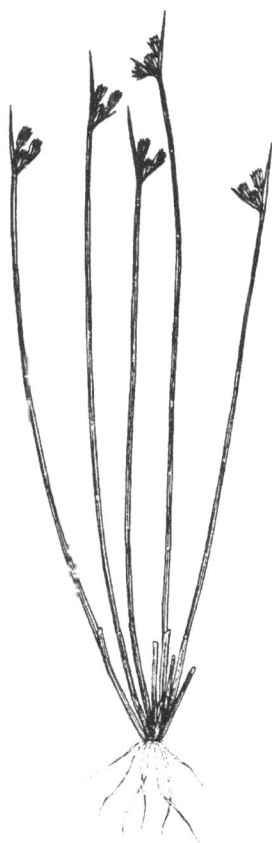

FIELD NOTES Flowers 1–4 near tip of stem, subtended by a short bract that appears to be a continuation of the stem. Capsules usually have a shallow notch at tip; seeds with a long "tail" at each end.

capsule

Juncus drummondii E. Meyer
DRUMMOND'S RUSH

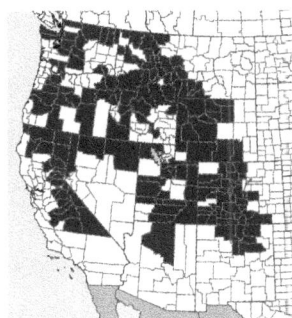

FAMILY Juncaceae (Rush)

FLOWERING July–August

HABITAT Wet meadows, moist woods, along streams, sometimes above timberline.

HABIT Native, tufted perennial rush with fibrous roots.

STEMS Upright, unbranched, terete, to 120 cm tall, smooth.

LEAVES Reduced to sheaths, or only with a bristle-like blade to 6 cm long.

FLOWERS 1–4 near the tip of the stem, subtended by a bract that appears to be a continuation of the stem; bract terete, smooth, to 5 cm long. **Sepals** 3, lanceolate, 6–8 mm long, pointed at the tip. **Petals** 3, lanceolate, 6–8 mm long but usually slightly shorter than the sepals, pointed at the tip. **Stamens** 6. **Pistils:** Ovary superior, smooth.

FRUIT Capsules ellipsoid, shallowly notched at the tip, 6–8 mm long; seeds to 2 mm long, whitish, with a slender "tail" at each end.

NOTE The seeds are eaten by waterfowl.

WETLAND STATUS AW FACW | GP FACW | WMV FACW

> **FIELD NOTES** Stems narrowly winged, flattened; bracts more than half the length of the inflorescence.

capsule

Juncus ensifolius Wikst.
DAGGER-LEAF RUSH

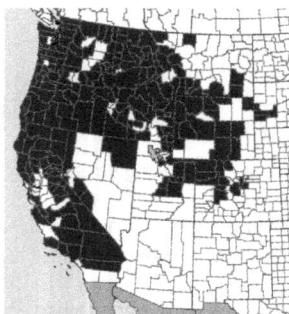

FAMILY Juncaceae (Rush)
FLOWERING July–August
HABITAT Along streams, in marshes.
HABIT Native perennial rush with creeping rhizomes.
STEMS Upright, flattened, narrowly winged, unbranched, to 60 cm tall, smooth.
LEAVES Most of the leaves near the base of the plant, flattened and folded along the mid-vein, to 40 cm long, to 8 mm wide, partially septate, smooth.
FLOWERS 4–25 borne in heads, the heads arranged in a panicle; each head spherical to hemispherical, to 13 mm across, pale brown to dark brown to purple-black. Sepals 3, free from each other, to 4 mm long, pointed at the tip. **Petals** 3, free from each other, to 4 mm long, slightly shorter than the sepals. pointed at the tip. **Stamens** 3 or 6. **Pistils:** Ovary superior, smooth.
FRUIT Capsules oblongoid, to 4 mm long, about as long as the sepals and petals, rounded at the tip or with a short point, smooth.
NOTE This is an extremely variable species. Some plants have 3 stamens per flower, others have 6; some have few-flowered heads, others have many flowers; some have dark brown sepals and petals, others have light brown. The seeds are eaten by waterfowl.
WETLAND STATUS AW FACW | GP FACW | WMV FACW

capsule

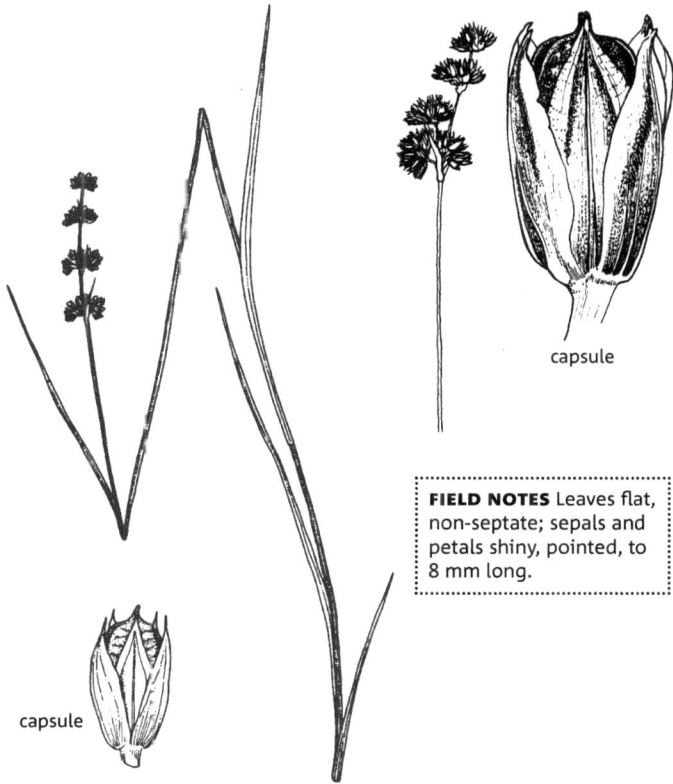

> **FIELD NOTES** Leaves flat, non-septate; sepals and petals shiny, pointed, to 8 mm long.

capsule

Juncus longistylis Torr.
LONG-STYLE RUSH

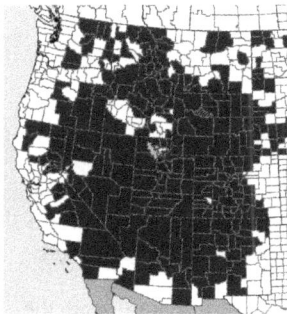

FAMILY Juncaceae (Rush)

FLOWERING June–August

HABITAT Along streams, around ponds, in fens, from valleys to the mountains.

HABIT Native perennial rush with creeping rhizomes.

STEMS Upright, to 50 cm tall, smooth.

LEAVES Alternate, sometimes crowded at the base, elongated, narrow, to 4 mm wide, smooth, flat, not septate.

FLOWERS 3–10 crowded into heads, with 2–8 heads usually somewhat separated from each other at the tip of the stem, with bracts not surpassing the inflorescence. **Sepals** 3, green with a pale margin, shiny, broadly lanceolate, to 8 mm long. **Petals** 3, green with a pale margin, shiny, broadly lanceolate, pointed at the tip, to 8 mm long, about as long as or only slightly shorter than the sepals. **Stamens** 6. **Pistils**: Ovary superior.

FRUIT Capsules angular, rounded at the tip, about as long as or slightly shorter than the sepals and petals.

NOTE The seeds are eaten by waterfowl.

WETLAND STATUS AW FACW | GP FACW | WMV FACW

> **FIELD NOTES** Heads solitary, spherical; flowers with purple-black sepals and petals; capsules round-tipped, shorter than the sepals and petals.

capsule

Juncus mertensianus Bong.
MERTEN'S RUSH

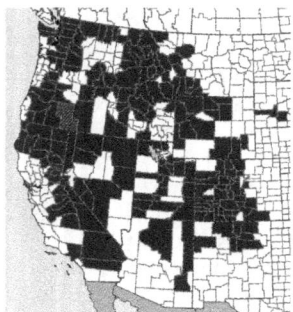

FAMILY Juncaceae (Rush)

FLOWERING June–August

HABITAT Wet meadows, along streams, in fens, at many elevations in the mountains.

HABIT Native perennial rush from stout rhizomes.

STEMS Upright, to 25 cm tall, smooth.

LEAVES Alternate, elongated, rolled into a hollow tube, septate, to 2 mm wide.

FLOWERS Numerous, crowded into a solitary head, the head spherical, to 13 mm in diameter, subtended by a bract that surpasses the inflorescence. **Sepals** 3, lanceolate, pointed at the tip, purple-black, to 4 mm long. **Petals** 3, lanceolate, pointed at the tip, purple-black, to 4 mm long, shorter than the sepals. **Stamens** 6. **Pistils:** Ovary superior.

FRUIT Capsules mostly triangular, rounded at the tip, to 6 mm long, shorter than the sepals and petals.

WETLAND STATUS AW OBL | GP OBL | WMV OBL

> **FIELD NOTES** Bract subtending inflorescence appearing to be an extension of the stem, the flowers appearing to be lateral.

capsule

Juncus mexicanus Willd. ex J. A. & J. H. Schultes
MEXICAN RUSH

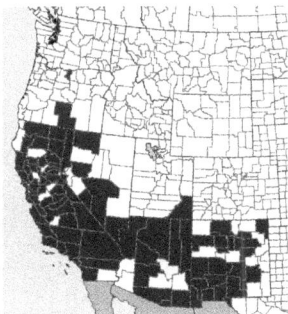

FAMILY Juncaceae (Rush)

FLOWERING May–August

HABITAT Moist, usually alkaline, areas.

HABIT Native perennial rush, with stout, creeping rhizomes.

STEMS Upright, unbranched. slender, to 60 cm tall, smooth, sometimes flat and twisted.

LEAVES Elongated, to 20 cm long, smooth.

FLOWERS Several in a loose cluster near the top of the stem, the cluster to 8 cm long; bract, which appears to be an upward extension of the stem, upright, to 15 cm long, smooth. **Sepals** 3, free from each other, lanceolate, greenish or straw-colored, to 6 mm long. **Petals** 3, free from each other, lanceolate, greenish or straw-colored, to 6 mm long. **Stamens** 6. **Pistils:** Ovary superior, smooth.

FRUIT Capsules ovoid, brown, pointed at the tip. to 6 mm long: seeds oblongoid.

NOTE Similar to *Juncus balticus* and is sometimes merged with it.

WETLAND STATUS AW FACW | GP FACW | WMV FACW

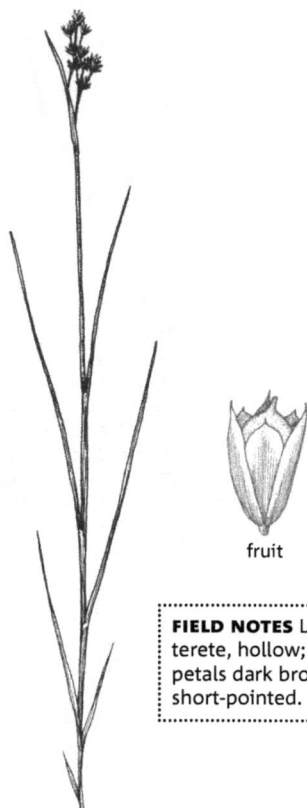

fruit

> **FIELD NOTES** Leaves septate, terete, hollow; sepals and petals dark brown; capsules short-pointed.

Juncus nevadensis S. Wats.
SIERRA RUSH

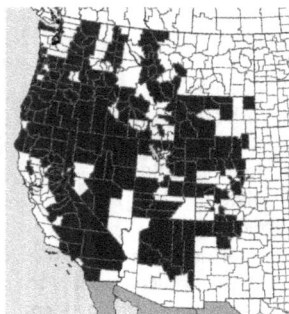

FAMILY Juncaceae (Rush)

FLOWERING June–August

HABITAT Along streams, in marshes, in wet meadows.

HABIT Native perennial rush with several stems arising from an elongated rhizome.

STEMS Upright, unbranched, very slender, to 60 cm tall, smooth.

LEAVES Elongated, terete, hollow, septate, to 25 cm long, to 2 mm wide, smooth.

FLOWERS Several crowded into heads, with to 30 heads in an inflorescence; heads hemispherical, 6–13 mm across. **Sepals** 3, free from each other, dark brown, lanceolate, pointed at the tip, to 6 mm long. **Petals** 3, free from each other, dark brown, lanceolate, pointed at the tip, a little shorter than the sepals. **Stamens** 6. **Pistils:** Ovary superior, smooth.

FRUIT Capsules oblongoid, with a short point at the rounded tip. dark brown, nearly as long as the petals.

NOTE The seeds are eaten by waterfowl.

WETLAND STATUS AW FACW | GP FACW | WMV FACW

fruit

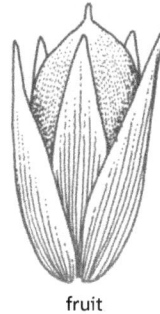

> **FIELD NOTES** Plants densely tufted; heads solitary; flowers subtended by an awn-tipped bract shorter than or barely as long as the head.

Juncus triglumis L.
THREE-FLOWER RUSH

FAMILY Juncaceae (Rush)

FLOWERING July–August

HABITAT Fens, open wet areas, at or near timberline in the mountains.

HABIT Native, densely tufted perennial rush with fibrous roots.

STEMS Upright, very slender, to 20 cm tall, smooth.

LEAVES Several crowded at the base of the plant, rolled up into a very slender hollow tube, about 1 mm wide, smooth but septate.

FLOWERS 2–several crowded into a solitary head, the head subtended by an awn-tipped bract shorter than or barely longer than the head. **Sepals** 3, whitish or brown, lanceolate, pointed at the tip, to 6 mm long. **Petals** 3, whitish or brown, lanceolate, pointed at the tip, slightly shorter than the sepals. **Stamens** 6. **Pistils:** Ovary superior.

FRUIT Capsules brown, rounded at the tip, to nearly 6 mm long, usually slightly shorter than to barely as long as the sepals.

WETLAND STATUS AW FACW | WMV FACW

> **FIELD NOTES** *Luzula* differs from *Juncus* in having hairs at least at the base of the leaves. Flowers borne singly in a much branched inflorescence.

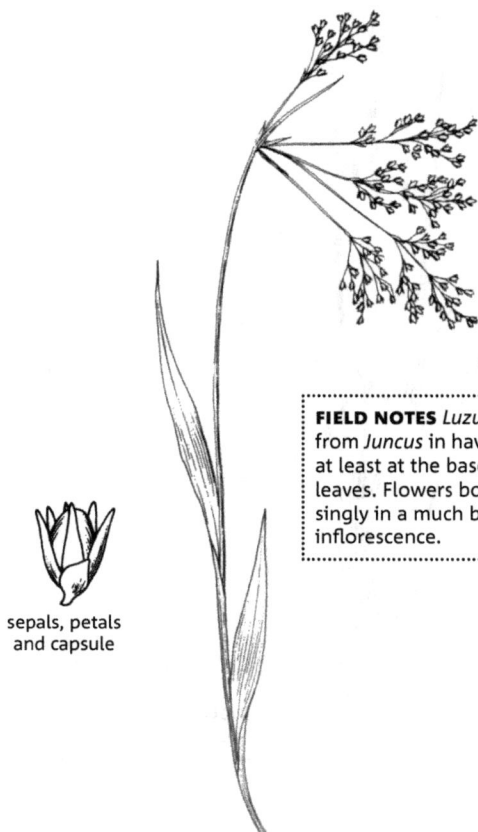

sepals, petals
and capsule

Luzula parviflora (Ehrh.) Desv.
SMALL-FLOWER WOODRUSH

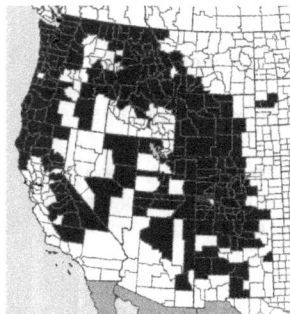

FAMILY Juncaceae (Rush)

FLOWERING June–August

HABITAT Moist woods, rarely in open habitats.

HABIT Native perennial with scaly stolons.

STEMS Upright, to 60 cm tall, smooth, slender.

LEAVES Elongated, narrow, to 20 cm long, to 13 mm wide smooth except for long hairs at the top of the sheath, the tips of the leaves rolled into a hardened point.

FLOWERS Borne singly at the tip of each branch, the branches numerous, the entire inflorescence to 15 cm long, each flower subtended by 1 or more bracts. **Sepals** 3, usually straw-colored. **Petals** 3, usually straw-colored. **Stamens** 6. **Pistils**: Ovary superior; stigmas 3.

FRUIT Capsules rusty-colored to blackish, ovoid, with a persistent style; seeds elliptic, shiny, about 1 mm long.

NOTE The much branched inflorescence is unlike that found in other woodrushes.

WETLAND STATUS AW FAC | GP UPL | WMV FAC

> **FIELD NOTES** Plant of alpine habitats; flowers borne in heads; leaves narrow, to 4 mm wide.

fruit with bract

Luzula spicata (L.) DC.
SPIKED WOODRUSH

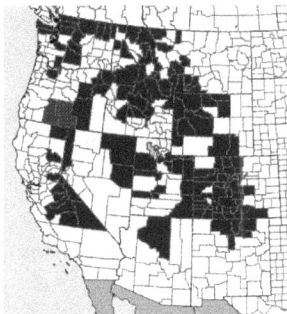

FAMILY Juncaceae (Rush)
FLOWERING June–August
HABITAT Along streams, wet meadows, often above timberline.
HABIT Native tufted perennial with fibrous roots.
STEMS Upright, unbranched, to 40 cm tall.
LEAVES Mostly near the base of the plant, elongated, narrow, flat, to 15 cm long, with long hairs at the base.
FLOWERS Crowded into a spike, the spikes sometimes interrupted, to 4 cm long, sometimes nodding, subtended by fringed bracts. **Sepals** 3, free from each other, to 3 mm long, brown with pale margins, with slender pointed tips. **Petals** 3, free from each other, to 3 mm long, brown with pale margins, with slender pointed tips. **Stamens** 6. **Pistils:** Ovary superior, smooth.
FRUIT Capsules oblongoid, a little shorter than the sepals and petals, brown.
NOTE The seeds are eaten by birds and small mammals.
WETLAND STATUS AW FACU | GP FACU | WMV FACU

flower

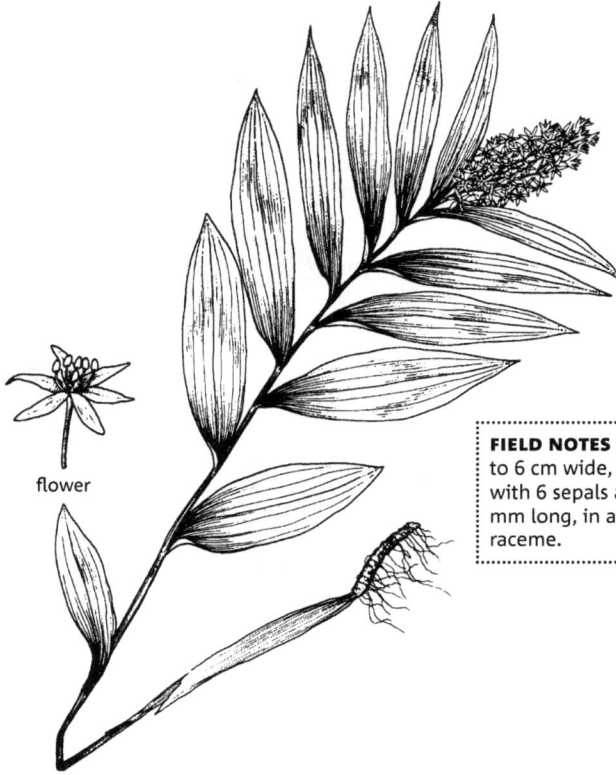

> **FIELD NOTES** Leaves several, to 6 cm wide, flowers white, with 6 sepals and petals 6–8 mm long, in a small terminal raceme.

Maianthemum stellatum (L.) Link
STARRY FALSE-SOLOMON'S-SEAL

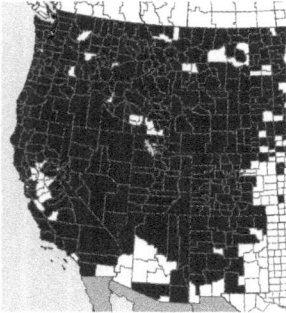

FAMILY Ruscaceae (Butcher's-broom)

FLOWERING April–June

HABITAT Moist areas.

HABIT Native perennial herb with a rhizome.

STEMS Upright, unbranched, sometimes zigzag, to 70 cm tall, smooth or minutely hairy.

LEAVES Alternate, simple, lanceolate to narrowly ovate, to 15 cm long, to 6 cm wide, pointed at the tip, rounded and sometimes clasping the stem at the base, minutely hairy on the lower surface, with several conspicuous veins.

FLOWERS Several, not crowded in a terminal raceme to 10 cm long. **Sepals and petals** 6, similar, white, free from each other, oblong. **Stamens** 6. **Pistils:** Ovary superior, smooth.

FRUIT Berries spherical, greenish yellow to reddish purple to black, 8–13 mm in diameter, with 2–3 seeds.

NOTE Several species of birds and mammals eat the berries of this species. Formerly in Liliaceae; that family now divided into several new families.

SYNONYMS *Smilacina stellata* (L.) Desf.

WETLAND STATUS AW FACU | GP FACU | WMV FAC

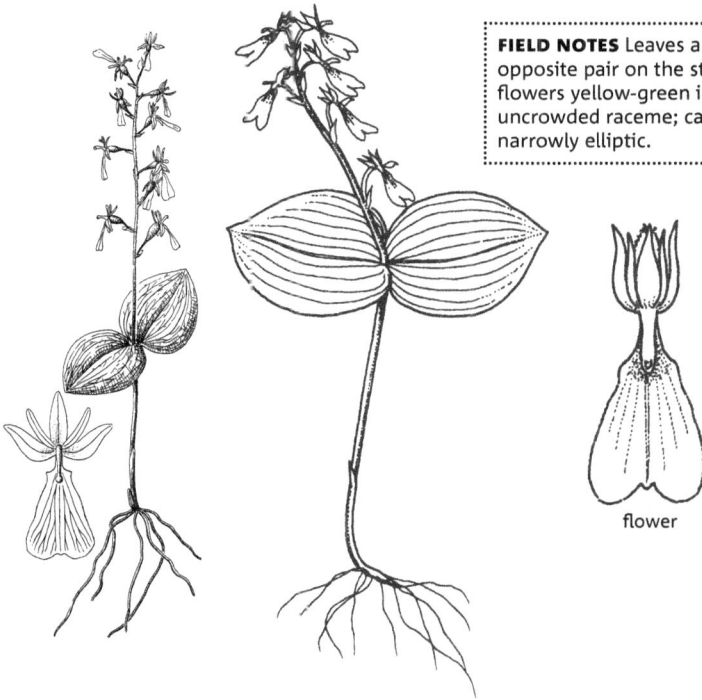

FIELD NOTES Leaves a single opposite pair on the stem; flowers yellow-green in an uncrowded raceme; capsule narrowly elliptic.

flower

Neottia convallarioides (Sw.) Rich
BROAD-LEAF TWAYBLADE

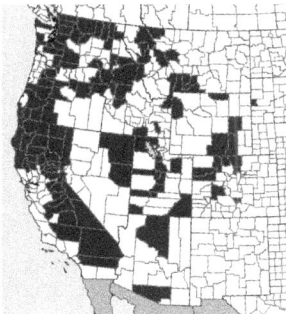

FAMILY Orchidaceae (Orchid)

FLOWERING June–July

HABITAT Along streams, damp woods at the upper elevation of mountains.

HABIT Native perennial herb with a short rhizome and fibrous roots.

STEMS Upright, unbranched, to 25 cm tall, smooth below the pair of leaves, glandular-hairy above them.

LEAVES Opposite, simple, in a single pair about halfway up the stem, nearly round, to 6 cm long, nearly as wide, rounded or with a short point at the tip, rounded at the sessile base, smooth.

FLOWERS 5–15 in an uninterrupted terminal raceme, each flower yellow-green, subtended by a bract 3–6 mm long; flower stalks slender, to 8 mm long. **Sepals** 3, free from each other, yellow-green, linear to broadly lanceolate, pointed at the tip, to 6 mm long, curved backward. **Petals** 3, 2 of them resembling the sepals, yellow-green, linear to broadly lanceolate, pointed at the tip. to 6 mm long, curved backward; lip petal 8–13 mm long, 6–8 mm wide, yellow-green, notched at the tip. abruptly tapering to a very narrow base. **Stamens** 10, associated with the pistil to form a column. **Pistils:** Ovary inferior, the pistil associated with the stamen to form a column.

FRUIT Capsules narrowly ellipsoid, to 8 mm long, smooth.

SYNONYMS *Listera convallarioides* (Sw.) Nutt. ex Ell.

WETLAND STATUS AW FAC | GF FACW | WMV FAC

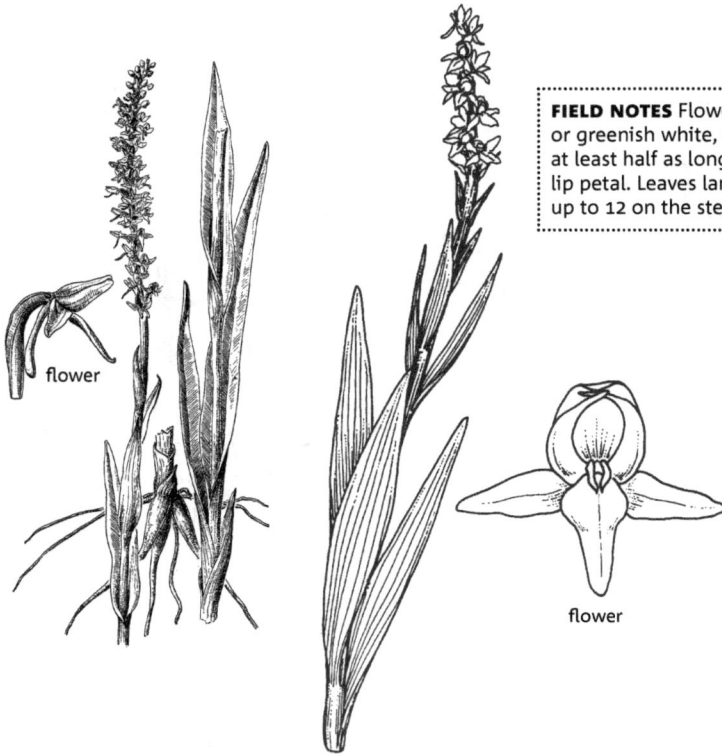

FIELD NOTES Flowers white or greenish white, the spur at least half as long as the lip petal. Leaves lanceolate, up to 12 on the stem.

flower

flower

Piperia dilatata (Pursh) Szlach. & Rutk.
LEAFY WHITE ORCHID

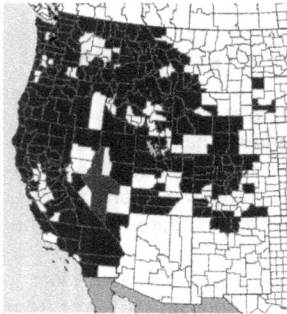

FAMILY Orchidaceae (Orchid)
FLOWERING June–September
HABITAT Bogs, along streams, in marshes, wet meadows.
HABIT Native perennial herb with slender, fleshy roots.
STEMS Upright, slender to stout, smooth, to 1 m tall.
LEAVES Alternate, simple, to 12 per stem, narrowly lanceolate to lanceolate, rounded to pointed at the tip, tapering to the base, smooth, to 30 cm long, to 6 cm wide.
FLOWERS Several in racemes, fragrant, white to greenish white. **Sepals** 3, one of them forming a hood with the petals, to 8 mm long, the other 2 elliptic to lanceolate, to nearly 13 mm long. **Petals** 3, narrowly to broadly lanceolate, to 8 mm long, the lip petal to 13 mm long, to 6 mm wide; spur more than half as long to as long as the lip petal. **Stamens** 1. **Pistils:** Ovary inferior.
FRUIT Capsules ellipsoid, 8–13 mm long, to 4 mm wide.
NOTE There is variation in the color of the flower and the length of the spur.
SYNONYMS *Habenaria dilatata* (Pursh) Hook., *Platanthera dilatata* (Pursh) Lindl. ex Beck
WETLAND STATUS AW FACW | GP FACW | WMV FACW

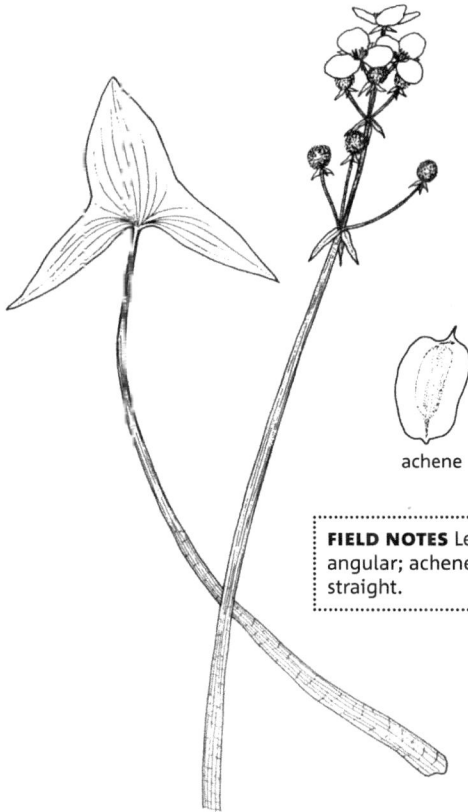

achene

FIELD NOTES Leaf stalks
angular; achene beak
straight.

Sagittaria cuneata Sheldon
NORTHERN ARROW-HEAD

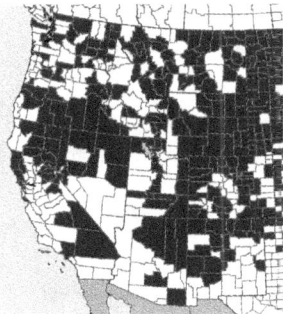

FAMILY Alismataceae (Water-plantain)
FLOWERING June–September
HABITAT Around and in lakes and ponds, along streams.
HABIT Native emersed or submersed perennial herb
with rhizomes.
STEMS Only the flowering stem above ground, to 1 m
tall, smooth.
LEAVES Basal, arrow-head shaped (sagittate), to 20 cm
long, the basal lobes usually shorter than the terminal
lobe, smooth; leaf stalk angular.
FLOWERS Male and female flowers usually borne on the
same plant in whorls, the male flowers being upper-
most; bracts lanceolate, pointed at the tip. Sepals 3, free from each other, green, ovate,
to 10 mm long. Petals 3, free from each other, white, ovate, to 20 mm long. Stamens
numerous. Pistils: Several, each with a superior ovary.
FRUIT Achenes crowded together in a spherical head, each achene to 4 mm long, with
a tiny straight beak.
NOTE The achenes are eaten by waterfowl.
WETLAND STATUS AW OBL | GP OBL | WMV OBL

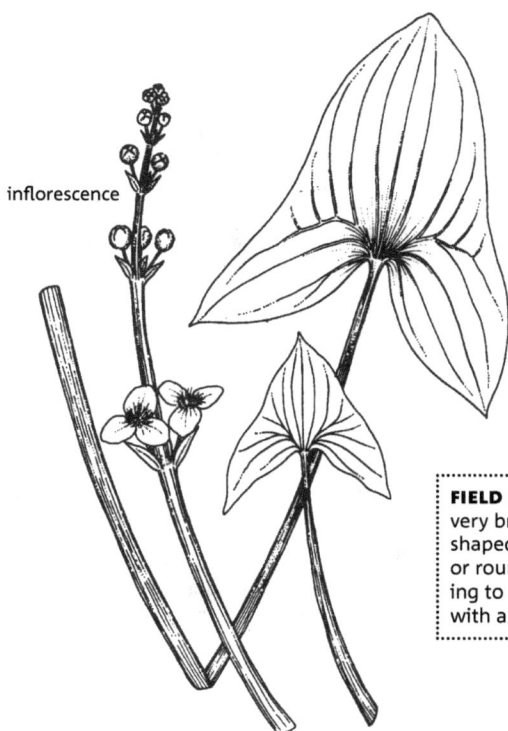

inflorescence

> **FIELD NOTES** Leaves usually very broad and arrowhead-shaped; bracts either pointed or round-tipped, but not tapering to a long point. Achenes with a horizontal beak.

Sagittaria latifolia Willd.
BROAD-LEAF ARROW-HEAD

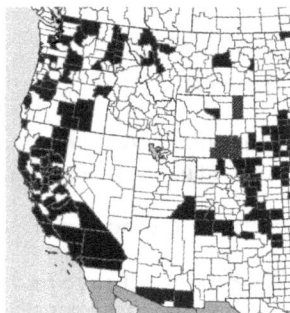

FAMILY Alismataceae (Water-plantain)

FLOWERING June–September

HABITAT Wet ditches, edges of lakes and ponds, often in shallow water.

HABIT Native (adventive in California) perennial herb with rhizomes.

STEMS The only stems are the rhizomes that are below the surface of the ground in terrestrial forms.

LEAVES Variable, but most of them arrowhead-shaped, to 45 cm long, the basal triangular lobes either longer or shorter than the terminal lobe, smooth; leaf stalks angular, not round.

FLOWERS Male and female flowers usually borne separately but on the same plant, the uppermost flowers usually male, the lowermost flowers usually female, each flower subtended by boat-shaped, pointed or round-tipped bracts to 20 mm long. **Sepals** 3, free from each other, green, to 13 mm long, eventually pointing downward. **Petals** 3, free from each other, white, to 2.5 cm long. **Stamens** numerous. **Pistils:** Ovaries superior, numerous, very crowded.

FRUIT Many achenes crowded into a rounded head, the head to 4 cm long, the achenes obovoid, to 5 mm long, winged, with a horizontal beak.

NOTE The achenes are eaten by waterfowl.

WETLAND STATUS AW OBL | GP OBL | WMV OBL

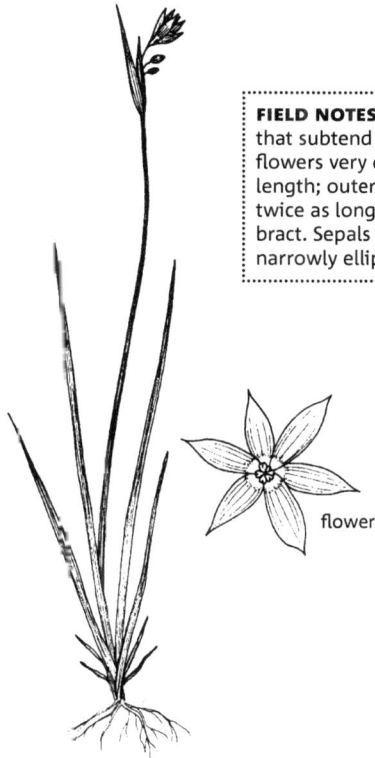

FIELD NOTES Two bracts that subtend each group of flowers very different in length; outer bract never twice as long as the inner bract. Sepals and petals narrowly elliptic.

flower

Sisyrinchium idahoense Bickn.
IDAHO BLUE-EYED-GRASS

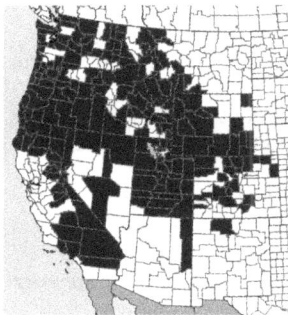

FAMILY Iridaceae (Iris)
FLOWERING May–August
HABITAT Wet meadows, along streams.
HABIT Native perennial herb with fibrous roots.
STEMS Upright, unbranched, to 50 cm tall, not conspicuously winged, smooth, with or without small teeth along the edge.
LEAVES All or nearly all basal, elongated, flattened, shorter than the flowering stems, to 4 mm wide, with or without small teeth along the edge.
FLOWERS 1–few, purple or blue with a yellow center, in an umbel at the tip of the stem, the umbel subtended and partially enclosed by a pair of leafy bracts (spathes); outer bract to 6 cm long; inner bract to 4 cm long. **Sepals and petals** 6, free from each other, blue or purple with a yellow base, narrowly elliptic, to 20 mm long, tapering to an abrupt point at the tip. **Stamens** 3. **Pistils:** Ovary inferior, glandular-hairy.
FRUIT Capsules obovoid to nearly spherical, to 6 mm long; seeds spherical, to 2 mm in diameter, minutely pitted.
NOTE There is considerable variation in this species, and it is sometimes difficult to distinguish from *S. montanum* of the Rocky Mountains.
WETLAND STATUS AW FACW | GP OBL | WMV FACW

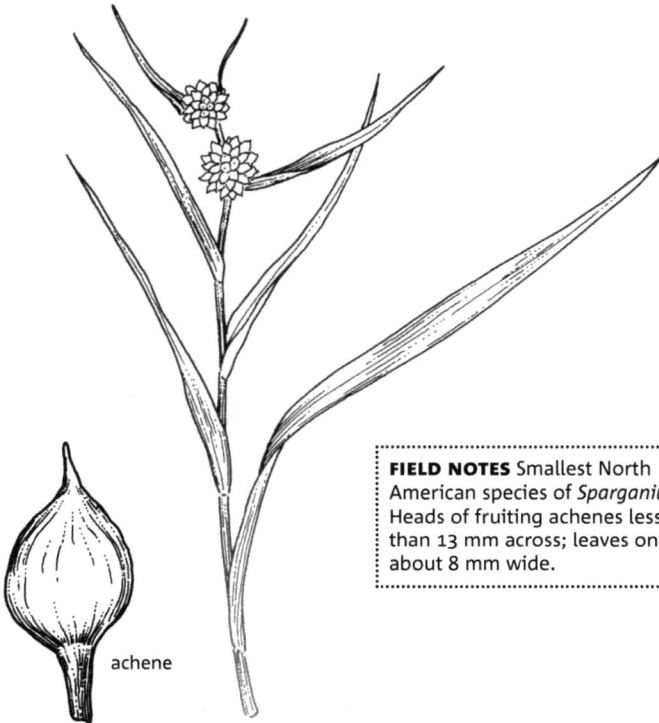

> **FIELD NOTES** Smallest North American species of *Sparganium*. Heads of fruiting achenes less than 13 mm across; leaves only about 8 mm wide.

achene

Sparganium natans L.
SMALL BURREED

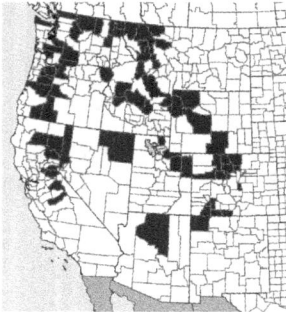

FAMILY Typhaceae (Cat-tail)

FLOWERING June–August

HABITAT Bogs, ponds, often in shallow standing water.

HABIT Native floating or emergent perennial with slender, creeping rhizomes.

STEMS Slender, floating or upright if emergent, smooth, to 60 cm long but usually shorter, particularly if emergent.

LEAVES Flat, dark green, alternate, sheathing at the base, tapering to a point at the tip, usually to 15 cm long, to 8 mm wide, smooth.

FLOWERS Male and female borne separately in spherical heads on the same plant, the male head solitary, terminal, 6–8 mm in diameter, the female heads 1–3, to nearly 13 mm in diameter. **Stamens** several, crowded together into a dense, spherical head, with whitish filaments. **Pistils** several, crowded together into a dense, spherical head.

FRUIT Achenes several, crowded together into a spherical head to nearly 13 mm across, each achene smooth, to 8 mm long, with a very short beak and very short stalk.

NOTE The achenes are eaten by waterfowl. Formerly placed in its own family, the Sparganiaceae.

SYNONYMS *Sparganium minimum* (Hartm.) Wallr.

WETLAND STATUS AW OBL | GP OBL | WMV OBL

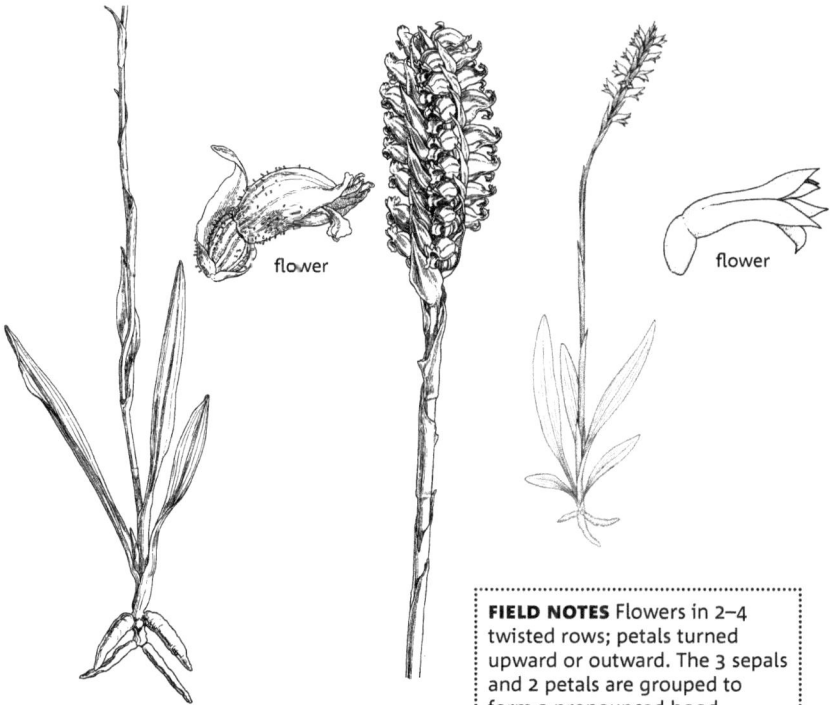

flower

flower

> **FIELD NOTES** Flowers in 2–4 twisted rows; petals turned upward or outward. The 3 sepals and 2 petals are grouped to form a pronounced hood.

Spiranthes romanzoffiana Cham.
HOODED LADIES'-TRESSES

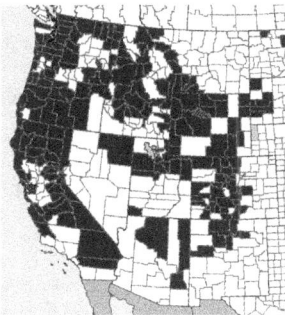

FAMILY Orchidaceae (Orchid)

FLOWERING July–September

HABITAT Along streams, in bogs, mountain meadows.

HABIT Native perennial herb with a cluster of thickened, fleshy roots.

STEMS Upright, unbranched, to 30 cm tall, smooth or glandular-hairy.

LEAVES Alternate, simple, linear-lanceolate to oblanceolate, to 10 cm long, to 13 mm wide, without teeth, smooth or glandular-hairy.

FLOWERS Many crowded into a twisted, terminal spike, the spike to 10 cm long: bracts lanceolate, longer than the flowers. **Sepals** 3, white, cream, or greenish white, grouped with 2 of the 3 petals to form a hood 8–13 mm long; lateral sepals turned downward at the tip. **Petals** 3, white, cream, or greenish white, 2 of them grouped with the 3 sepals to form a hood, the other forming an oblong lip that turns down abruptly. **Stamens** 3. **Pistils**: Ovary inferior, smooth.

FRUIT Capsules ellipsoid, smooth, to 6 mm long.

WETLAND STATUS AW FACW | GP OBL | WMV FACW

> **FIELD NOTES** Basal leaves long, narrow, with a conspicuous ligule at base. Flowers greenish yellow, 3-parted, in a slender, spike-like raceme; pistils 6 in each flower.

fruit

Triglochin maritima L.
SEASIDE ARROW-GRASS

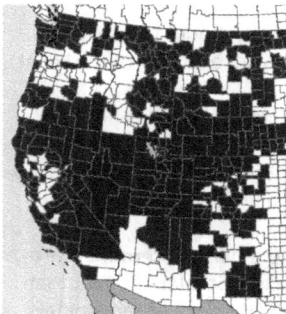

FAMILY Juncaginaceae (Arrow-grass)

FLOWERING May–August

HABITAT Marshes, wet meadows, particularly in alkaline areas.

HABIT Native perennial with a stout rhizome.

STEMS Aerial stem leafless, to 60 cm long, bearing only a terminal spike-like raceme of flowers, smooth.

LEAVES All basal, elongated, linear, to 30 cm long, usually about half as long as the plant, to 4 mm wide, pointed at the tip, smooth; ligule to 6 mm long.

FLOWERS Numerous in a terminal, spike-like raceme, the raceme to 25 cm long; bracts absent; flower stalks slender, to 6 mm long. Sepals 3, free from each other, greenish yellow, to 2 mm long. Petals 3, free from each other, greenish yellow, to 2 mm long. Stamens 6. Pistils: 6 per flower, united by their inner faces, each with a superior ovary.

FRUIT Follicles 6. united by their inner faces, oblongoid, to 4 mm long.

NOTE This species, if eaten, is poisonous to livestock.

WETLAND STATUS AW OBL | GP OBL | WMV OBL

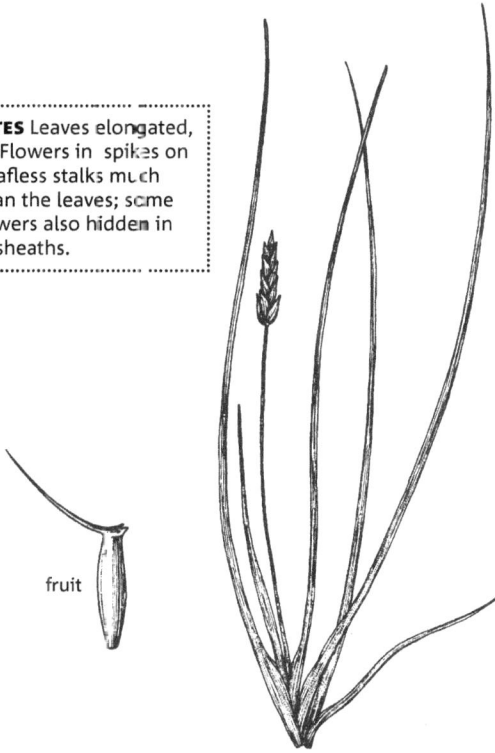

FIELD NOTES Leaves elongated, grass-like. Flowers in spikes on slender, leafless stalks much shorter than the leaves; some female flowers also hidden in basal leaf sheaths.

fruit

Triglochin scilloides (Poir.) Mering & Kadereit
FLOWERING QUILLWORT

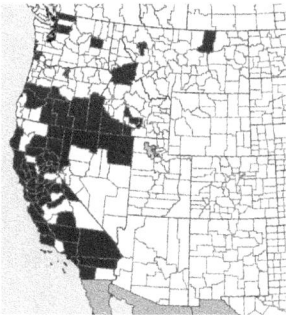

FAMILY Juncaginaceae (Arrow-grass)

FLOWERING March–October

HABITAT Around lakes and ponds in mud, in slow streams.

HABIT Native annual herb with fibrous roots.

STEMS No leafy stems present, only stems to 20 cm tall with spikes of flowers present.

LEAVES All elongated, terete, fleshy, to 30 cm long, to 5 mm wide, smooth.

FLOWERS Male and female flowers embedded in the fleshy axis of a spike, with some female flowers only borne singly and hidden by basal leaf sheaths; spikes 2–8 mm long; all flowers usually subtended by a bract. **Sepals** 1, oblong. **Petals** absent. **Stamens** 1. **Pistils**: Ovary superior; style of basal flowers to 20 cm long.

FRUIT Those hidden in the sheaths narrowly oblongoid, flattened, light brown, 6–8 mm long, often with 1 or more slender horns at the tip; those in the spikes broadly lanceoloid, green, narrowly winged.

NOTE The flowering spike usually is borne at water level. Sometimes some of the flowers are not subtended by bracts.

SYNONYMS *Lilaea scilloides* (Poir.) Haum.

WETLAND STATUS AW OBL | GP OBL | WMV OBL

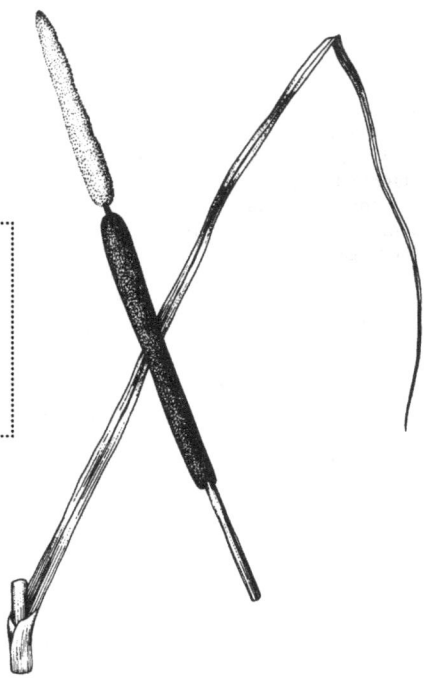

> **FIELD NOTES** Leaves up to 8 per stem, usually less than 13 mm wide. Upper male spike often separated from lower female spike; female spike usually less than 2.5 cm thick.

Typha angustifolia L.
NARROW-LEAF CAT-TAIL

FAMILY Typhaceae (Cat-tail)

FLOWERING June–August

HABITAT Marshes, in and along streams, around lakes and ponds.

HABIT Introduced, rather slender perennial herb with slender branching rhizomes.

STEMS Upright, to 1.5 m tall, smooth.

LEAVES Alternate, narrow, elongated, to 8 per stem, usually less than 13 mm wide.

FLOWERS Male and female borne separately in spikes on the same plant; male spike formed directly above female spike but usually separated from the female spike by a short interval, much narrower than the female spike, and falling away after pollen is shed, pale brown; female spike to 20 cm long, usually less than 2.5 cm thick, dark brown. **Sepals** absent. **Petals** absent. **Stamens** 2–5 per flower. **Pistils** 1 per flower, the ovary superior.

FRUIT Achenes ellipsoid, about 1 mm long, subtended by fine hairs.

NOTE The male spike is often separated from the female spike by a short interval; however, this is not always true. The achenes are eaten by waterfowl.

WETLAND STATUS AW OBL | GP OBL | WMV OBL

RELATED SPECIES *Typha domingensis* Pers. occurs in brackish or nutrient-rich wetlands of the American Southwest. Distinguished from *Typha angustifolia* in the following key:

Upper surface of leaf blade gland-dotted near base; pistillate spike cinnamon to medium brown . . . . **T. domingensis**
Glands absent on upper surface of leaf blade; pistillate spike dark brown . . . . **T. angustifolia**

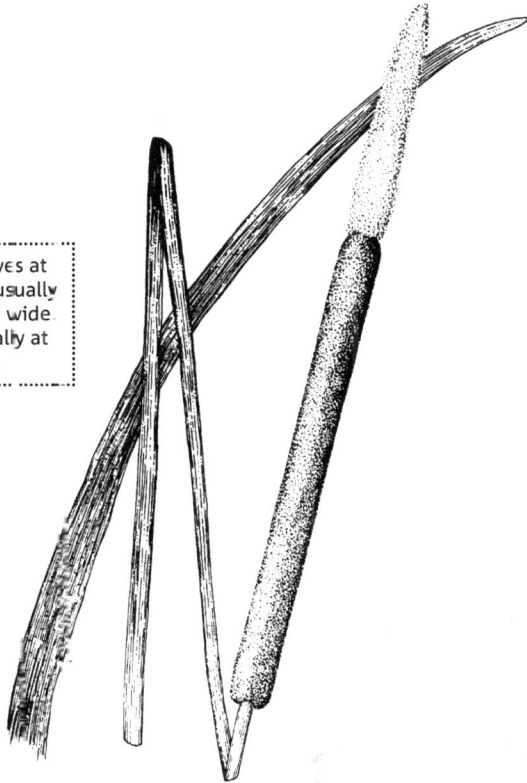

FIELD NOTES Leaves at least 8 per stem, usually more than 13 mm wide. Female spike usually at least 2.5 cm thick.

Typha latifolia L.
BROAD-LEAF CAT-TAIL

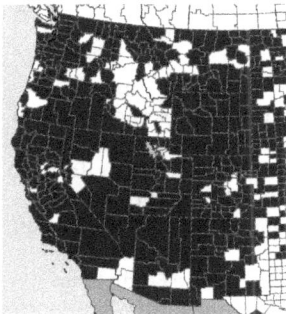

FAMILY Typhaceae (Cat-tail)

FLOWERING June–August

HABITAT Marshes, in and along streams, around lakes and ponds.

HABIT Native, coarse perennial herb with stout, branching rhizomes.

STEMS Upright, to 3 m tall, smooth.

LEAVES Alternate, elongated, at least 8 per stem, some of them usually at least 2.5 cm wide, always at least 13 mm wide.

FLOWERS Male and female borne separately in spikes on the same plant; male spike formed directly above the female spike and falling away after pollen is shed, pale brown; female spike thick, to 25 cm long, to 4 cm thick, dark brown. **Sepals** absent. **Petals** absent. **Stamens** 2–5 per flower. **Pistils** 1 per flower, the ovary superior.

FRUIT Achenes ellipsoid, about 1 mm long, subtended by fine hairs.

NOTE There is considerable intergradation and probable hybridization between cattails so that intermediate specimens are often encountered. The achenes are eaten by waterfowl. The leaves are used by muskrats for building nests.

WETLAND STATUS AW OBL | GP OBL | WMV OBL

> **FIELD NOTES** Plants coarse; leaves large, coarsely veined; flowers white or greenish white, in a terminal, erect panicle.

inflorescence

flower

Veratrum californicum Dur.
CALIFORNIA FALSE HELLEBORE

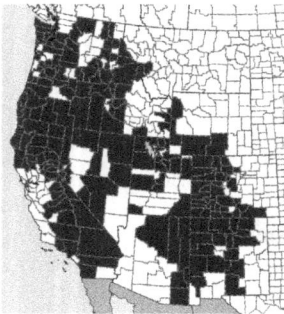

FAMILY Melanthiaceae (False Hellebore)

FLOWERING June–August

HABITAT Along streams, around lakes, in wet meadows.

HABIT Native, coarse perennial herb with short, thick rhizomes.

STEMS Upright, stout, unbranched except in the inflorescence, smooth or hairy in the inflorescence, to 2.5 m tall.

LEAVES Alternate, simple, broadly elliptic to ovate, to 40 cm long, to 20 cm wide, hairy, with conspicuous veins, pointed or rounded at the tip. rounded and sometimes clasping the stem at the base.

FLOWERS Many crowded into an erect, terminal panicle, the panicle to 60 cm long; flower stalks to 6 mm long. **Sepals and petals** 6, similar, white or greenish white, lanceolate to elliptic, to 20 mm long, united below. **Stamens** 6. **Pistils:** Ovary superior; styles 3.

FRUIT Capsules ovoid, to 4 cm long, smooth, containing many pale, flat seeds to 20 mm long.

NOTE Some of the flowers may only have stamens, while others may have both stamens and pistils. The stems and leaves are sometimes browsed by larger mammals. Formerly in Liliaceae; that family now divided into several new families.

WETLAND STATUS AW FACW | GP OBL | WMV FAC

lower leaf

> **FIELD NOTES** Individual leaflets lanceolate to ovate; fruits not or only slightly-winged.

Angelica arguta Nutt.
LYALL'S ANGELICA

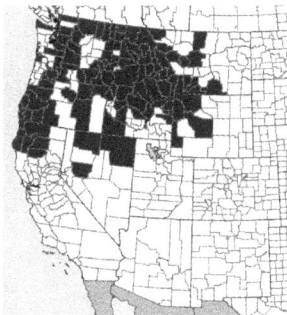

FAMILY Apiaceae (Carrot)

FLOWERING July–August

HABITAT Wet woods, bogs.

HABIT Native perennial herb, with a stout taproot.

STEMS Stout, usually branched, to 2 m tall, smooth or slightly rough to the touch.

LEAVES Alternate, 2- or 3-pinnately compound, the leaflets lanceolate to ovate, to 15 cm long, to 5 cm wide, more or less pointed at the tip, rounded at the base, toothed, smooth or slightly rough to the touch; leaf stalks to 30 cm long.

FLOWERS Many borne in several compound umbels, the rays of each umbel to 10 cm long, the stalks of each flower to 13 mm long, with a conspicuous web at the base. **Sepals** minute or absent. **Petals** 5, white, free from each other, to 3 mm long. **Stamens** 5. **Pistils:** Ovary inferior, smooth.

FRUIT Oval to obovate, to 8 mm long, to 6 mm broad, narrowly winged.

NOTE The fruits are eaten by birds.

WETLAND STATUS AW FACW | WMV FACW

FIELD NOTES Stipules united; flowers large, 2–2.5 cm long, crowded into ovoid heads.

pod

Astragalus agrestis Dougl. ex G. Don
FIELD MILKVETCH

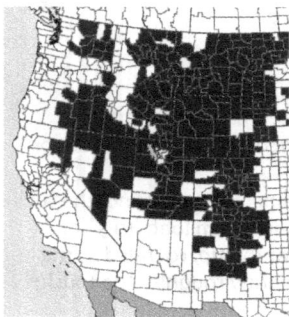

FAMILY Fabaceae (Pea)
FLOWERING May–August
HABITAT Pastures, hillsides, prairies, along roads.
HABIT Native perennial herb with a taproot.
STEMS Upright, to 40 cm tall, gray-hairy.
LEAVES Alternate, pinnately compound with 9–25 leaflets, each leaflet oblong to oblong-ovate, to 4 cm long, to 8 mm wide, pointed or rounded at the tip, tapering or rounded at the base, with appressed hairs; stipules united, to 20 mm long.
FLOWERS Up to 50 crowded into ovoid heads arising from the leaf axils, the heads to 15 cm long; flower stalks about 1 mm long. **Sepals** 5, united below into a tube, to 13 mm long, with black and white hairs. **Petals** 5, arranged in the form of a sweet pea flower, purple to pink to blue to even whitish, to 2.5 cm long. **Stamens** 10. **Pistils:** Ovary superior, minutely hairy.
FRUIT Pods to 13 mm long, to 4 mm wide, with appressed hairs, sessile; seeds smooth, brown, to 3 mm long.
NOTE The hairiness of this species is extremely variable, as is the flower color.
WETLAND STATUS AW FAC | GP FACU | WMV FACW

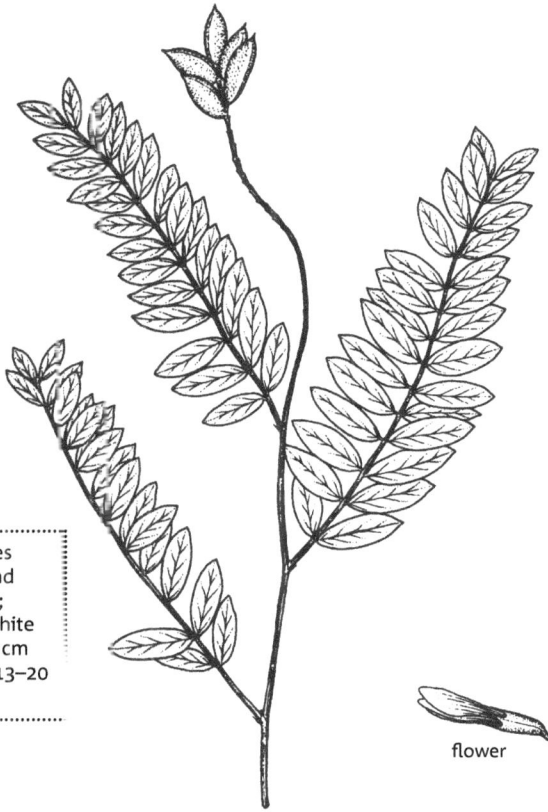

FIELD NOTES Leaves pinnately compound with 15–35 leaflets; flowers greenish white to yellowish, 2–2.5 cm long; pods terete, 13–20 mm long.

flower

Astragalus canadensis L.
CANADA MILKVETCH

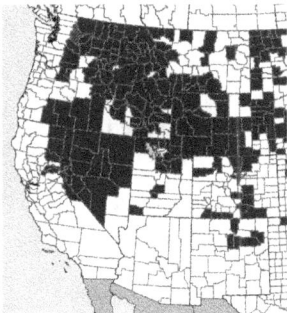

FAMILY Fabaceae (Pea)

FLOWERING May–August

HABITAT Along rivers and streams, moist prairies, open woods; also on dry bluffs.

HABIT Native perennial herb with rhizomes.

STEMS Upright, usually branched, to 1 m tall, with appressed hairs.

LEAVES Alternate, pinnately compound, with 15–35 leaflets; leaflets lanceolate to ovate to elliptic, to 5 cm long, pointed at the tip, tapering to the base, without teeth. appressed-hairy.

FLOWERS Several borne in dense racemes, the racemes to 20 cm long, the teeth 1–4 mm long, smooth or sparsely hairy. **Sepals** 5, green, united below to form a tube, the tube 2–8 mm long, the teeth 1–4 mm long, smooth or sparsely hairy. **Petals** 5, arranged to form a sweetpea-shaped flower, greenish white to yellowish, 2–2.5 cm long. **Stamens** 10. **Pistils:** Ovary superior, usually smooth.

FRUIT Pods terete, 13–20 mm long, usually smooth, with a slender beak at the tip to 4 mm long; seeds smooth, brown.

WETLAND STATUS AW FAC | GF FAC | WMV FACW

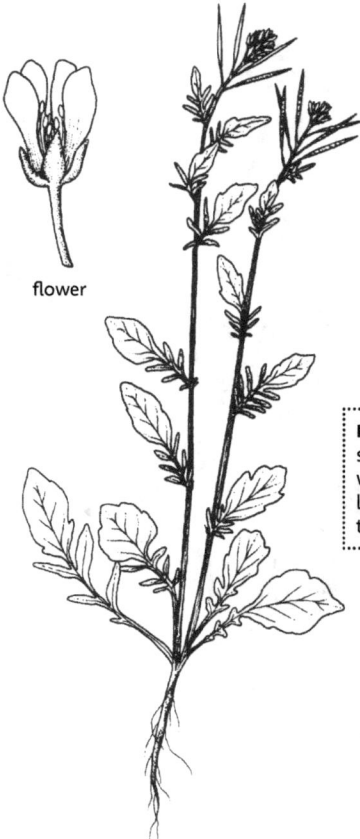

flower

> **FIELD NOTES** Petals pale yellow; some leaves pinnately divided with a very large terminal leaflet; pods slender, ascending, to 5 cm long.

Barbarea orthoceras Ledeb.
AMERICAN WINTER-CRESS

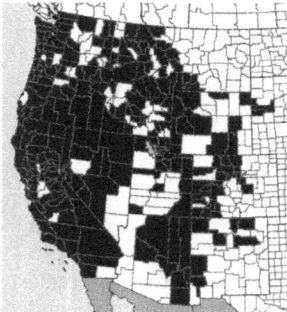

FAMILY Brassicaceae (Mustard)

FLOWERING March–October

HABITAT Along streams, wet meadows, around springs.

HABIT Native perennial herb with a thickened root.

STEMS Upright, branched, stout, to 45 cm tall, smooth, sometimes purple-tinged.

LEAVES Basal and alternate, smooth, some of the basal leaves sometimes simple and undivided, the rest of the leaves pinnately divided, the terminal leaflet often several times larger than the other leaflets: uppermost leaves merely pinnately lobed.

FLOWERS Several in a terminal raceme; flower stalks rather thick, 3–6 mm long. **Sepals** 4, yellow-green, free from each other, to 3 mm long. **Petals** 4, pale yellow, free from each other, to 6 mm long. **Stamens** 6. **Pistils**: Ovary superior, smooth.

FRUIT Pods elongated, ascending, to 5 cm long, to 2 mm wide, smooth, with a short beak; seeds brown, about 1 mm long.

WETLAND STATUS AW FACW | GP OBL | WMV FACW

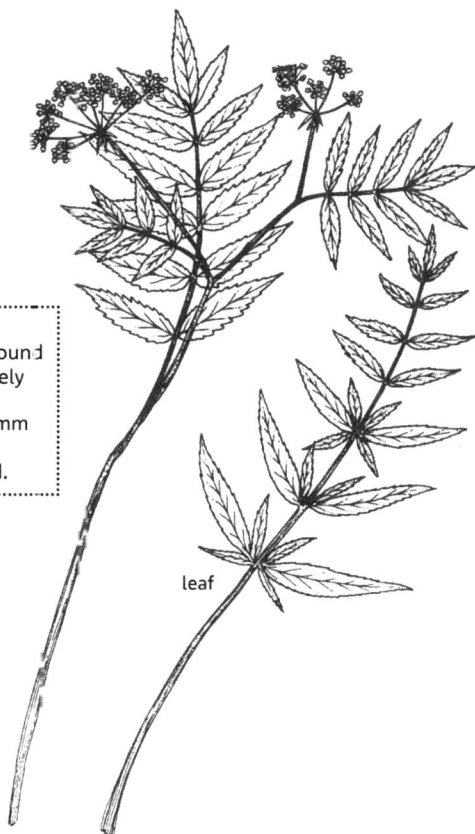

> **FIELD NOTES** Flowers white, borne in compound umbels; leaves pinnately compound with 9–23 oblong leaflets to 20 mm wide, the uppermost leaflets coarsely lobed.

leaf

Berula erecta (Huds.) Coville
CUT-LEAF WATER PARSNIP

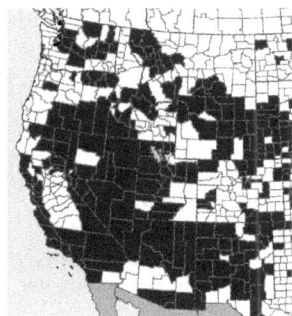

FAMILY Apiaceae (Carrot)

FLOWERING May–October

HABITAT Swamps, springs, bogs, sometimes in shallow water.

HABIT Native upright or reclining perennial herb, with clusters of thickened roots.

STEMS Upright or reclining, branched, to 1 m long, smooth.

LEAVES Alternate, pinnately compound with 9–23 leaflets, each leaflet oblong, to 20 mm wide, with or without teeth, but those of the upper leaves deeply jagged lobed, smooth.

FLOWERS Many borne in compound umbels to 8 cm across, the umbels with as many as 20 rays; bracts 4–8, linear; flower stalks to 6 mm long. **Sepals** 5, green, minute. **Petals** 5, free from each other, white, to 2 mm long. **Stamens** 5. **Pistils:** Ovary inferior.

FRUIT Ovoid, flattened laterally, to 2 mm long, smooth.

SYNONYMS *Berula pusilla* Fern.

WETLAND STATUS AW OBL | GP OBL | WMV OBL

> **FIELD NOTES** Lower leaves simple; upper leaves pinnately compound.

Cardamine breweri S. Wats.
BREWER'S BITTER-CRESS

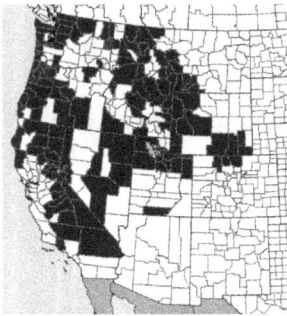

FAMILY Brassicaceae (Mustard)

FLOWERING May–July

HABITAT Along streams in the mountains.

HABIT Native perennial herb, with slender, creeping rootstocks.

STEMS Spreading and rooting at the nodes or upright, usually unbranched, to 60 cm tall, smooth or sparsely hairy.

LEAVES Basal leaves usually simple, ovate, pointed at the tip. heart-shaped at the base, smooth or nearly so. to 5 cm long, usually without teeth, on a long stalk: upper leaves alternate, pinnately compound with 3–5 leaflets, the leaflets ovate, shallowly lobed or wavy-toothed, the lateral leaflets smaller than the terminal one. smooth or nearly so.

FLOWERS Several in an uncrowded raceme, each flower on a stalk 6–20 mm long. **Sepals** 4, green, free from each other, about 2 mm long. **Petals** 4, white, free from each other, to 6 mm long. **Stamens** 6. **Pistils:** Ovary superior, smooth.

FRUIT Pods elongated, ascending to erect, smooth, to 2.5 cm long, less than 2 mm wide, with 10–20 nearly spherical seeds.

WETLAND STATUS AW FACW | GP FACW | WMV FACW

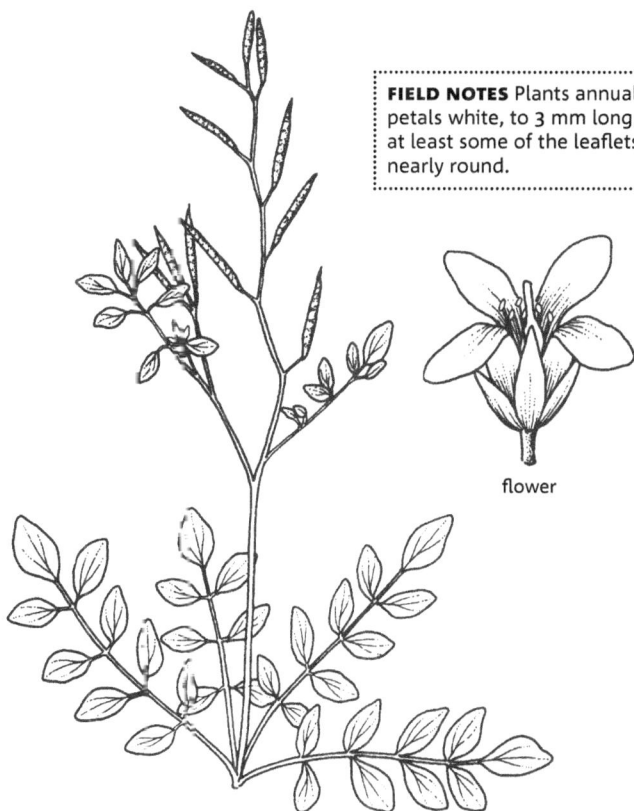

> **FIELD NOTES** Plants annual; petals white, to 3 mm long; at least some of the leaflets nearly round.

flower

Cardamine oligosperma Nutt.
FEW-SEED BITTER-CRESS

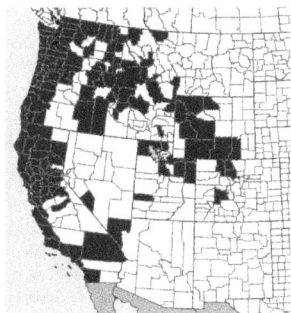

FAMILY Brassicaceae (Mustard)

FLOWERING March–July

HABITAT Moist woods.

HABIT Native annual (sometimes biennial) herb with a taproot.

STEMS Upright, usually unbranched. to 50 cm tall, with short, spreading hairs or smooth.

LEAVES Mostly basal in a rosette, with a few alternate leaves on the stem, pinnately compound with 5–11 leaflets, the leaflets oval to nearly round, to 8 mm long, with or without lobes or teeth, short-hairy or smooth, on distinct stalks; leaflets of stem leaves much narrower.

FLOWERS 2–10 in a terminal raceme, without bracts. **Sepals** 4, green, free from each other, to 1 mm long. **Petals** 4, white, free from each other, spatulate, to 3 mm long. **Stamens** 6. **Pistils:** Ovary superior.

FRUIT Pods very upright, elongated, very narrow, to 2.5 cm long; seeds winged.

NOTE The leaves are sometimes browsed by deer.

WETLAND STATUS AW FAC | GP FAC | WMV FAC

fruit

> **FIELD NOTES** Leaves large, much divided, alternate; flowers white, in large umbels. Fruit smooth, ovoid to nearly spherical.

Cicuta douglasii (DC.) Coult. & Rose
WESTERN WATER-HEMLOCK

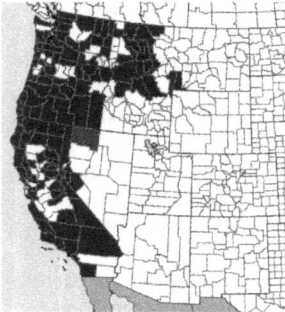

FAMILY Apiaceae (Carrot)

FLOWERING June–September

HABITAT Swamps, wet roadside ditches.

HABIT Native, stout perennial herb with an enlarged rootstock.

STEMS Upright, stout, branched, to 2 m tall, smooth.

LEAVES Alternate, much divided, sometimes at least 3-pinnate, the leaflets narrowly lanceolate, to 10 cm long, toothed or even shallowly lobed, smooth.

FLOWERS Many borne in large umbels, the umbels to 10 cm across; each umbel sometimes subtended by 1 or more bracts. **Sepals** 5, green, very tiny. **Petals** 5, white, free from each other, about 2 mm long. **Stamens** 5. **Pistils:** Ovary inferior.

FRUIT Ovoid or nearly spherical, to 4 mm long, smooth but with low ribs.

NOTE All parts of this plant are poisonous if eaten.

WETLAND STATUS AW OBL | GP OBL | WMV OBL

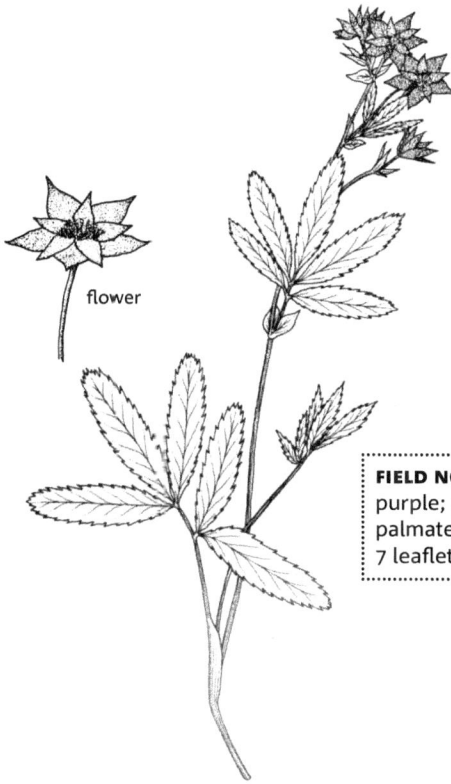

flower

> **FIELD NOTES** Flowers reddish purple; petals pointed; leaves palmately compound with 5–7 leaflets.

Comarum palustre L.
MARSH CINQUEFOIL

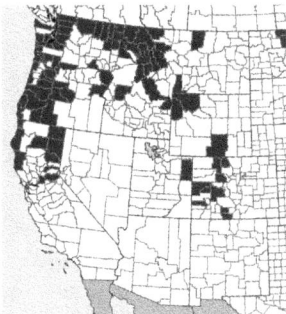

FAMILY Rosaceae (Rose)

FLOWERING May–August

HABITAT Swamps, bogs, often in shallow water.

HABIT Native perennial herb with creeping rhizomes.

STEMS Upright or ascending, stout, to 45 cm tall, glandular-hairy in the upper half, more or less smooth in the lower half.

LEAVES Alternate, palmately compound, with 5–7 leaflets; leaflets lanceolate to oblong to oblanceolate, to 6 cm long, rounded at the tip, tapering to the base, toothed, smooth on the upper surface, pale and smooth or hairy on the lower surface.

FLOWERS Few borne in cymes; flowers subtended by bracts to 8 mm long. **Sepals** 5, united below to form a short floral tube, ovate, to 20 mm long, pointed at the tip, purplish. **Petals** 5, free from each other, ovate-lanceolate, pointed at the tip, to 8 mm long, reddish purple. **Stamens** numerous. **Pistils:** Many, each with a superior ovary.

FRUIT Achenes many, ellipsoid.

NOTE The achenes are eaten by waterfowl.

SYNONYMS *Potentilla palustris* (L.) Scop.

WETLAND STATUS AW OBL | GP OBL | WMV OBL

inflorescence

leaf

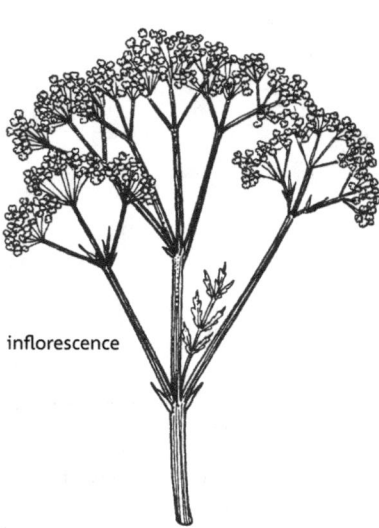

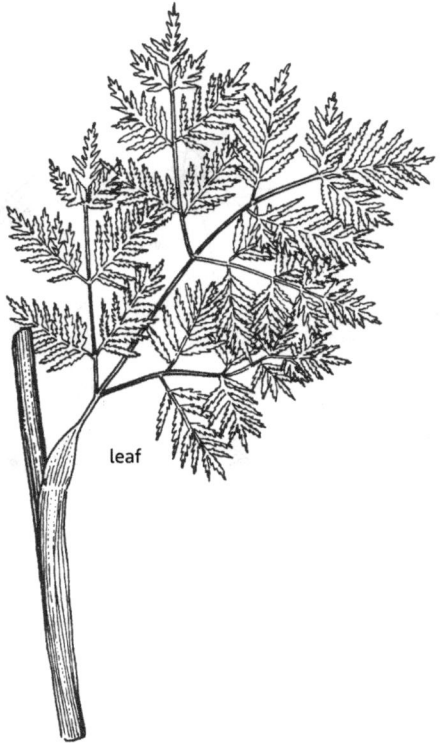

> **FIELD NOTES** Plants large,
> to 3 m tall, its of flowers
> white, in umbels; stems
> purple-spotted; leaflets
> much-divided, less than
> 13 mm wide.

Conium maculatum L.
POISON-HEMLOCK

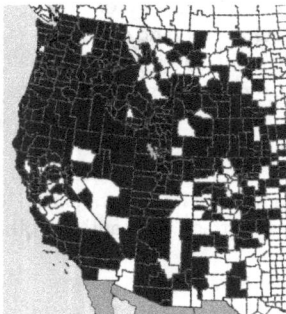

FAMILY Apiaceae (Carrot)

FLOWERING April–July

HABITAT Wet disturbed areas, particularly in roadside ditches.

HABIT Introduced, robust biennial herb with a taproot.

STEMS Upright, stout, branched, to 3 m tall, purple-spotted, smooth.

LEAVES Alternate, 3- to 4-pinnately compound, to 120 cm long, finely divided into leaflets to 13 mm wide, each leaflet deeply toothed or lobed, smooth.

FLOWERS Borne in umbels, with many umbels per plant, each umbel to 8 cm across, subtended by small lanceolate bracts. **Sepals** absent. **Petals** 5, free from each other, white. **Stamens** 5. **Pistils:** Ovary inferior, smooth.

FRUIT Ovoid, compressed laterally, to 3 mm long, smooth, with pale brown ribs.

NOTE All parts of this plant are extremely poisonous when eaten. It is native to Europe and Asia.

WETLAND STATUS AW FACW | GP FACW | WMV FAC

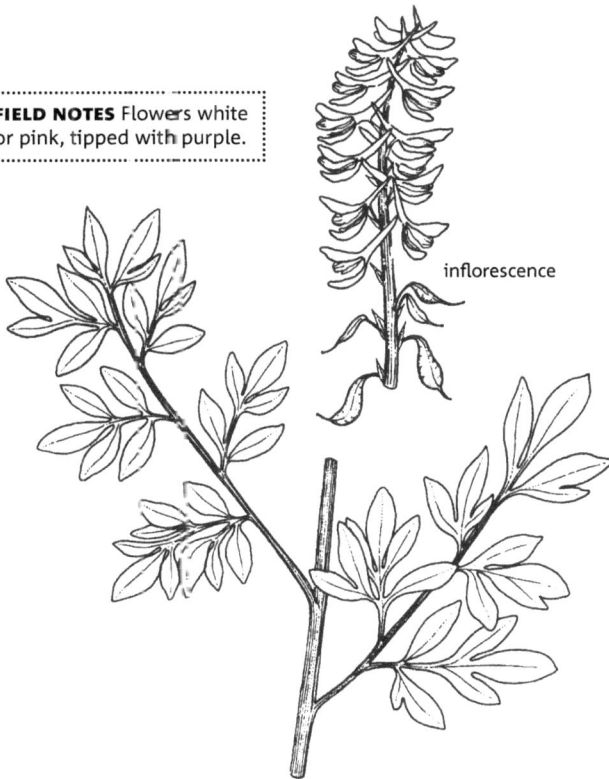

FIELD NOTES Flowers white
or pink, tipped with purple.

inflorescence

Corydalis caseana Gray
SIERRA CORYDALIS

FAMILY Fumariaceae (Fumitory)

FLOWERING June–August

HABITAT Along streams, in shaded woods.

HABIT Native perennial herb with thickened roots.

STEMS Upright, stout and rather succulent, to 30 cm long, each leaflet elliptic to lanceolate to ovate, pointed at the tip, tapering to the base, smooth, to 2.5 cm long, without teeth.

LEAVES Alternate, doubly pinnately compound, to 30 cm long, each leaflet elliptic to lanceolate to ovate, pointed at the tip, tapering to the base, smooth, to 2.5 cm long, without teeth.

FLOWERS Many, in racemes to 13 cm long, each flower on a short stalk. **Sepals** 2, green, free from each other, falling away early. **Petals** 4, white or pink with purple tips, to 15 mm long, one of the petals spurred. **Stamens** 6. **Pistils**: Ovary superior; style persistent on the fruit.

FRUIT Capsule oblongoid, 13–20 mm long, smooth, with numerous black, shiny, minutely warty seeds.

NOTE Formerly placed in Papaveraceae (Poppy Family).

WETLAND STATUS AW FACW | GP FACW | WMV FACW

> **FIELD NOTES** Stems and leaves glandular-hairy; leaves pinnately compound with 5–9 leaflets; petals yellowish, not more than twice as long as the sepals.

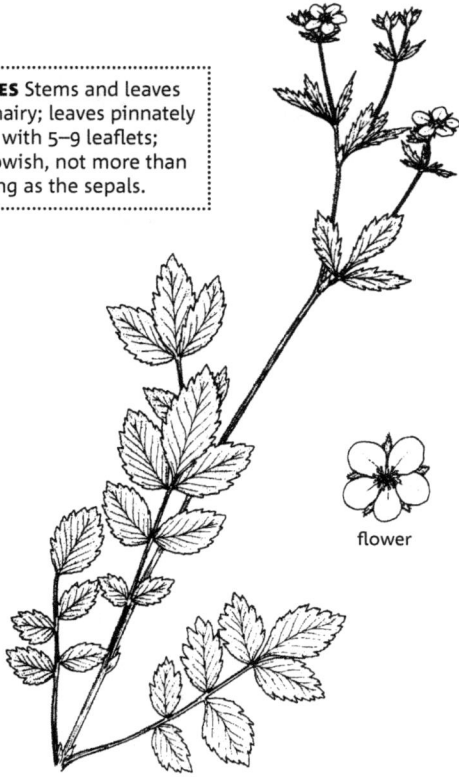

flower

Drymocallis glandulosa (Lindl.) Rydb.
GLAND CINQUEFOIL

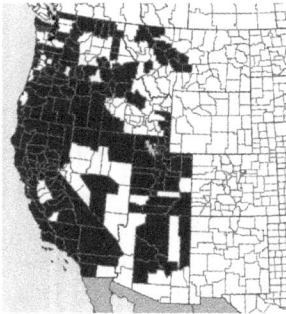

FAMILY Rosaceae (Rose)

FLOWERING June–August

HABITAT Open woods.

HABIT Native perennial herb with a thickened rootstock and short rhizomes.

STEMS Upright, usually unbranched, to 45 cm tall, glandular-hairy.

LEAVES Alternate and basal, pinnately compound with 5–9 leaflets; leaflets obovate, to 4 cm long, glandular-hairy, sharply toothed; stipules ovate-lanceolate, to 8 mm long.

FLOWERS Few in a cyme; bractlets subtending each flower often nearly as long as the sepals, hairy. **Sepals** 5, green, lanceolate to ovate, to 13 mm long, hairy. **Petals** 5, yellowish, free from each other, ovate, 6–20 mm long, shorter or longer than the sepals. **Stamens** About 25. **Pistils:** Numerous, each with a superior ovary.

FRUIT Achenes yellow-brown, smooth, to 2 mm long.

NOTE Sometimes the hairs on the plant are not glandular.

SYNONYMS *Potentilla glandulosa* Lindl.

WETLAND STATUS AW FAC | GP FAC | WMV FAC

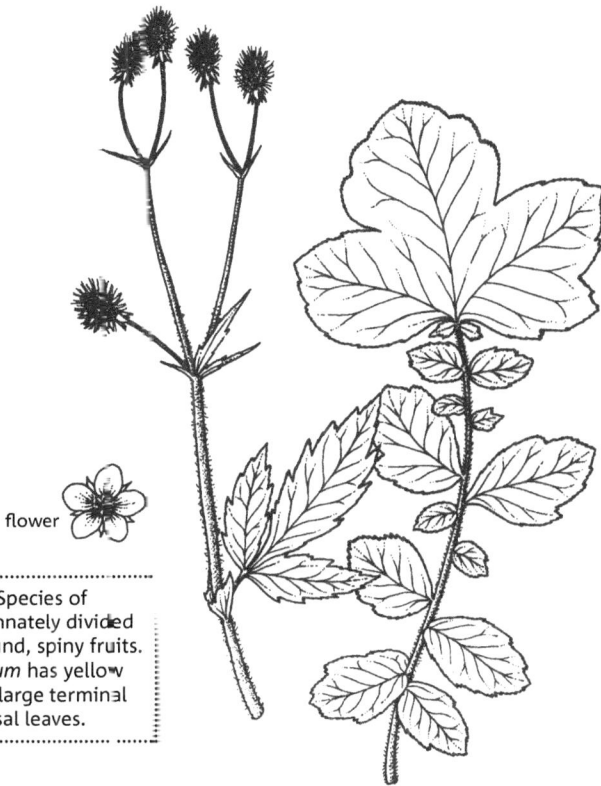

flower

> **FIELD NOTES** Species of *Geum* have pinnately divided leaves and round, spiny fruits. *G. macrophyllum* has yellow flowers and a large terminal part of the basal leaves.

Geum macrophyllum Wild.
LARGE-LEAF AVENS

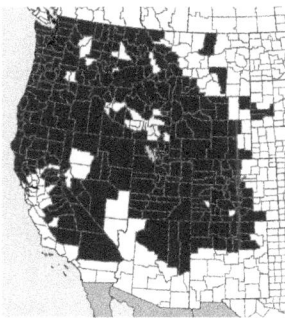

FAMILY Rosaceae (Rose)

FLOWERING June–August

HABITAT Along streams, in moist woods.

HABIT Native perennial herb with a short, thick rootstock and narrow stolons.

STEMS Upright, to 60 cm tall, usually unbranched, with spreading hairs, some of which may be gland-tipped.

LEAVES Basal leaves pinnately divided, to 30 cm long, the terminal segment much larger than the lateral segments, with spreading hairs; stem leaves deeply 3-lobed or divided into 3 leaflets, smaller than the basal leaves, with spreading hairs.

FLOWERS Several in cymes, each flower on a sometimes glandular stalk. **Sepals** 5, green, to 6 mm long, turned downward, hairy, sometimes glandular. **Petals** 5, free from each other, yellow, to 8 mm long. **Stamens** numerous. **Pistils**: Several, free from each other.

FRUIT Spherical, to nearly 2.5 cm in diameter, consisting of several achenes, each achene elliptic, flattened, to 4 mm long, with the persistent style forming a terminal spine, the entire head appearing prickly.

WETLAND STATUS AW FACW | GP FACW | WMV FAC

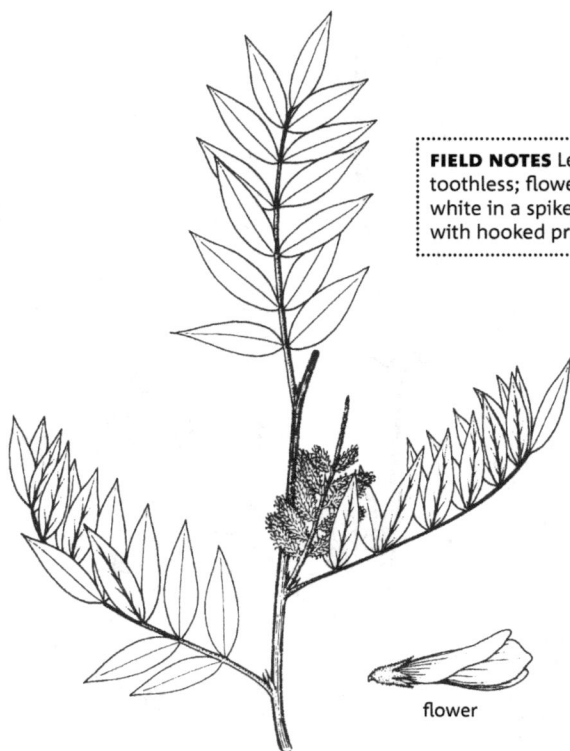

> **FIELD NOTES** Leaflets 11–19 toothless; flowers yellowish white in a spike; pods covered with hooked prickles.

flower

Glycyrrhiza lepidota Pursh
AMERICAN LICORICE

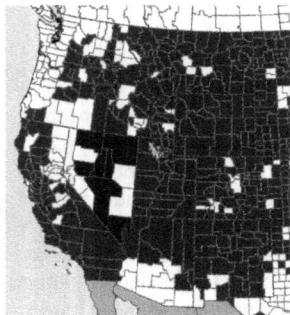

FAMILY Fabaceae (Pea)

FLOWERING May–August

HABITAT Low ground, particularly in disturbed areas.

HABIT Native perennial herb with thick, licorice-scented roots.

STEMS Upright, branched, usually somewhat sticky-hairy, less commonly smooth, to 1 m tall.

LEAVES Alternate, pinnately compound with 11–19 leaflets, the leaflets lanceolate to oblong, to 4 cm long, pointed at the tip, tapering to the base, without teeth, smooth or minutely hairy; stipules linear-lanceolate, 2–6 mm long.

FLOWERS Many in a spike, the spike oblongoid, to 6 cm long, shorter than the leaves. **Sepals** 5, green, attached to each other, the upper 2 shorter than the lowest 3, to 6 mm long, glandular. **Petals** 5, arranged in the form of a sweetpea flower, yellowish white, 8–13 mm long. **Stamens** 10. **Pistils:** Ovary superior.

FRUIT Pods oblongoid, 13–20 mm long, covered with short hooked prickles; seeds flat, brown, to 3 mm long.

NOTE The prickly pods resemble the fruit of a cocklebur. They become entangled in the fur of mammals and are dispersed in that manner.

WETLAND STATUS AW FAC | GP FACU | WMV FAC

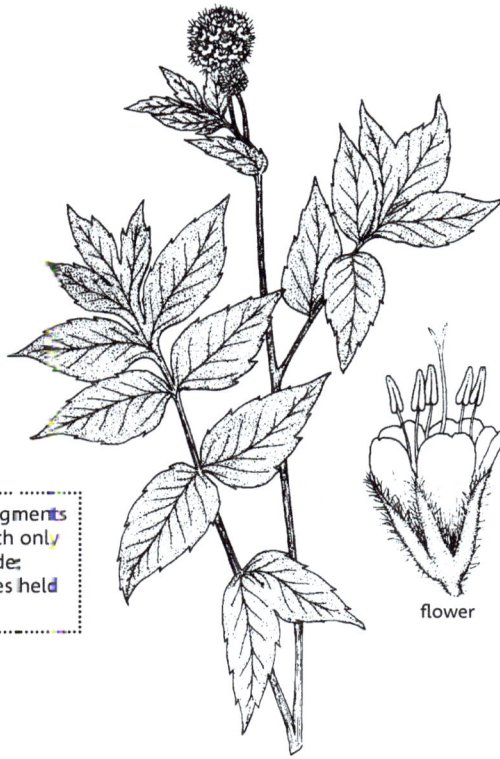

> **FIELD NOTES** Leaf segments coarsely toothed, with only 2–4 teeth on each side; flowers blue, in cymes held above the leaves.

flower

Hydrophyllum occidentale (S.Wats.) Gray
CALIFORNIA WATER-LEAF

FAMILY Boraginaceae (Borage Family)

FLOWERING April–June

HABITAT Moist woods and thickets, from foothills to mid-mountain elevations.

HABIT Native perennial herb with rhizomes and fleshy roots.

STEMS Upright, often branched near the base, to 45 cm tall, with spreading hairs.

LEAVES Basal and alternate, pinnately divided into 7–15 segments; each segment to 4 cm wide, pointed at the tip, with 2–4 coarse teeth on each side, with spreading hairs.

FLOWERS Borne in crowded, rounded cymes elevated above the leaves, the inflorescence on hairy stalks to 15 cm long; individual flower stalks to 6 mm long. **Sepals** 5, green, united below, 3–4 mm long, hairy and ciliate. **Petals** 5, blue-violet or even white, united to form a bell, 6–13 mm long, usually with a very shallow notch at the tip. **Stamens** 5. attached to the tube of the petals. **Pistils:** Ovary superior; style 2-cleft.

FRUIT Capsules ovoid, about 4 mm wide, containing 1–3 seeds.

NOTE There is variation in the degree of hairiness in this species. Formerly placed in Hydrophyllaceae (Water-leaf Family).

WETLAND STATUS AW FACW | WMV FACW

inflorescence

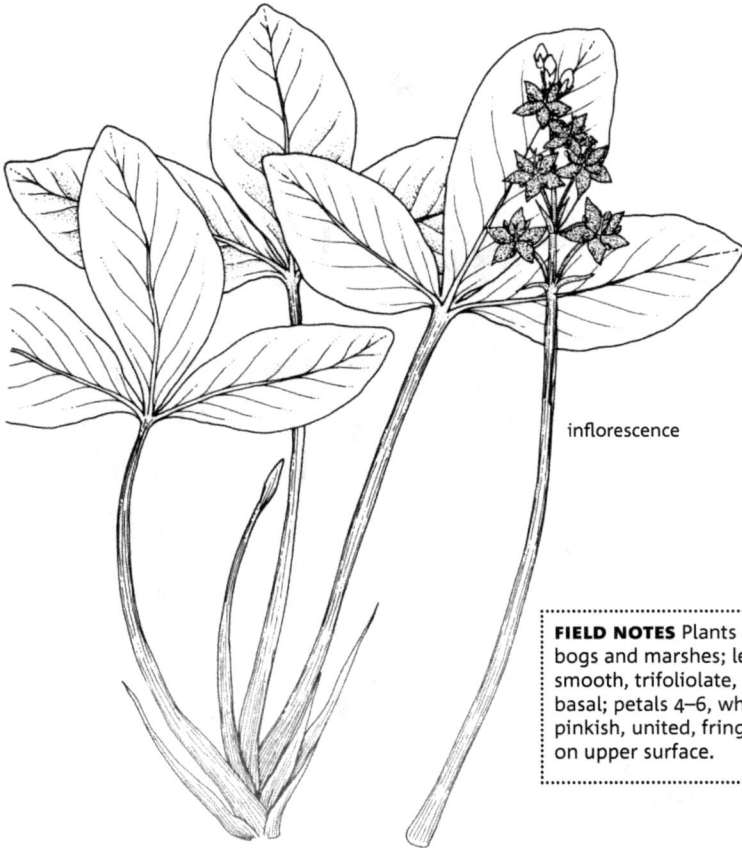

> **FIELD NOTES** Plants of bogs and marshes; leaves smooth, trifoliolate, all basal; petals 4–6, white or pinkish, united, fringed on upper surface.

Menyanthes trifoliata L.
BUCKBEAN

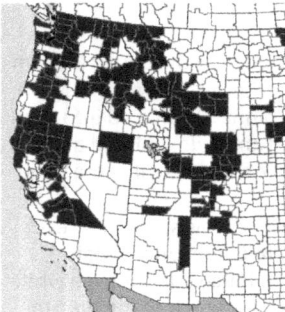

FAMILY Menyanthaceae (Buckbean)
FLOWERING May–August
HABITAT Marshes, bogs, swamps, edges of lakes.
HABIT Native perennial herb with creeping rootstocks.
STEMS All underground or underwater as rhizomes.
LEAVES Basal, divided into 3 leaflets; leaflets oblong to ovate, pointed or nearly rounded at the tip, tapering to the sessile base, to 8 cm long, to 5 cm wide, with or without teeth; leaf stalks to 25 cm long, smooth.
FLOWERS 7–20 borne on a separate leafless stalk; each flower stalk to 2.5 cm long, with a bract at its base. **Sepals** usually 5, united at the base, green, oblong to ovate, to 4 mm long. **Petals** 4–6, usually 5, united to form a short funnel, white to pinkish, the tube and lobes to 13 mm long. **Stamens** usually 5, borne on the tube of the petals. **Pistils:** Ovary superior or partly inferior; stigma 2-lobed.
FRUIT Capsules ovoid or spherical, to 13 mm in diameter; seeds yellow-brown, shiny.
WETLAND STATUS AW OBL | GP OBL | WMV OBL

> **FIELD NOTES** Leaves pinnately compound with 3–11 leaflets; petals 4, white, about twice as long as the sepals; pods long, slender, straight or curving, with seeds in two rows.

Nasturtium officinale Ait f.
WATERCRESS

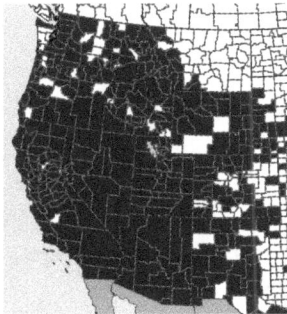

FAMILY Brassicaceae (Mustard)

FLOWERING April–October

HABITAT Springs, brooks, cool water, usually in water.

HABIT Introduced perennial herb with fibrous roots.

STEMS Creeping or floating, rather fleshy, smooth, rooting at many of the nodes.

LEAVES Alternate, pinnately compound, with 3–11 leaflets; leaflets oblong to oval to nearly spherical, rounded at the tip, smooth, without teeth or with wavy-toothed margins.

FLOWERS Borne in terminal racemes and in racemes from the axils of the uppermost leaves, each flower 4–6 mm across, not subtended by bracts. **Sepals** 4, green, free from each other, elliptic to narrowly oblong, smooth, to 3 mm long. **Petals** 4, white, free from each other, broadly oblong to nearly spherical, rounded at the tip, up to 6 mm long. **Stamens** 6. **Pistils:** Ovary superior, smooth.

FRUIT Pods elongated, linear, cylindric, straight or curved, on slender stalks, smooth, to 30 mm long, up to 3 mm thick, with or without a minute beak at the tip.

NOTE This species is collected for use in salads or as a garnish.

SYNONYMS *Rorippa nasturtium-aquaticum* (L.) Hayek

WETLAND STATUS AW OBL | GP OBL | WMV OBL

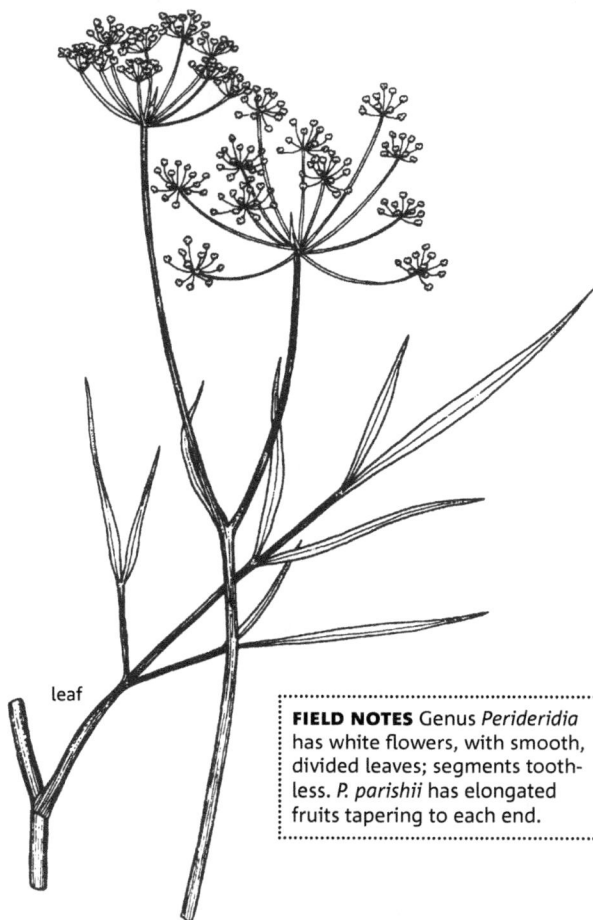

leaf

> **FIELD NOTES** Genus *Perideridia* has white flowers, with smooth, divided leaves; segments toothless. *P. parishii* has elongated fruits tapering to each end.

Perideridia parishii (Coult. & Rose) A. Nels. & J. F. Macbr.
PARRISH'S YAMPAH

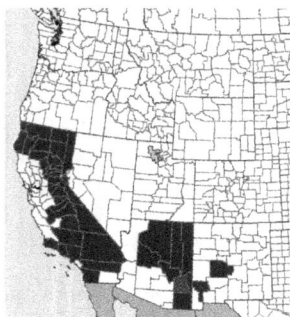

FAMILY Apiaceae (Carrot)

FLOWERING July–September

HABITAT Wet meadows.

HABIT Native perennial herb with one or more tubers.

STEMS Upright, slender, branched, to 75 cm tall, smooth.

LEAVES Alternate, sometimes simple but usually divided into very narrow, toothless segments to 10 cm long, smooth.

FLOWERS Borne in umbels, with usually 8–15 umbels in an inflorescence; bractlets linear to obovate. **Sepals** 5, green, very tiny. **Petals** 5, white, free from each other, to 3 mm long. **Stamens** 5. **Pistils:** Ovary inferior, smooth.

FRUIT Oblongoid to ovoid, to 4 mm long, longer than broad, tapering to each end.

WETLAND STATUS AW FAC | GP FAC | WMV FAC

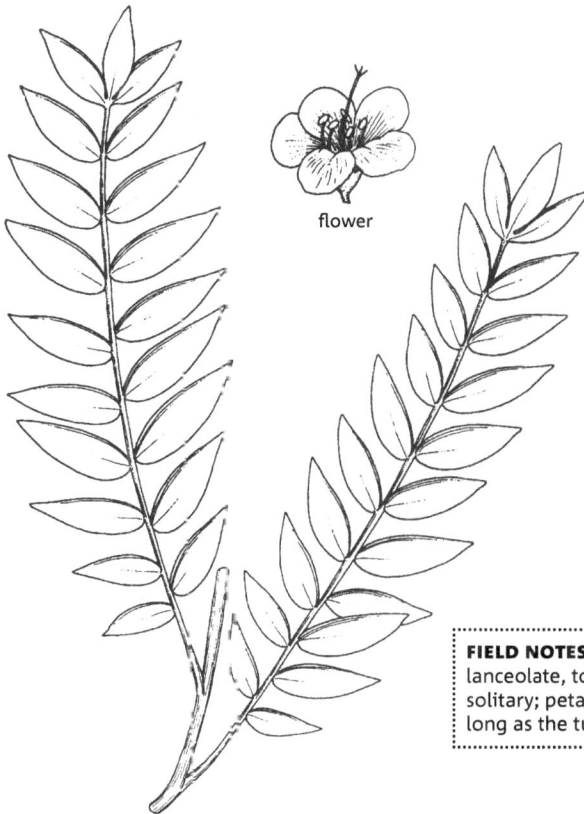

flower

> **FIELD NOTES** Leaflets 11–27, lanceolate, toothless; stems solitary; petals about twice as long as the tube of the petals.

Polemonium occidentale Greene
WESTERN JACOB'S-LADDER

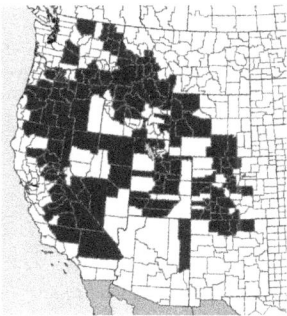

FAMILY Polemoniaceae (Phlox)

FLOWERING June–August

HABITAT Along streams, swamps, wet places in the mountains.

HABIT Native perennial herb from rhizomes.

STEMS Solitary, upright, smooth or glandular-hairy, to 1 m tall.

LEAVES Alternate, pinnately compound with 11–27 leaflets; each leaflet lanceolate, pointed at the tip, tapering to the base, to 45 mm long, to 13 mm wide, without teeth, smooth or glandular-hairy.

FLOWERS Several crowded into a branched but narrow inflorescence, subtended by lobed or toothless bracts. **Sepals** 5, green, united to form a cup, glandular-hairy, to 13 mm long, the lobes about as long as the teeth. **Petals** 5, blue, united about halfway to form a tube, 13–20 mm long, the lobes longer than the tube. **Stamens** 5, attached to the tube of the petals. **Pistils:** Ovary superior; styles exserted beyond the petals.

FRUIT Capsules ovoid, about 3 mm long, with several elongated seeds.

WETLAND STATUS AW FACW | GP FACW | WMV FACW

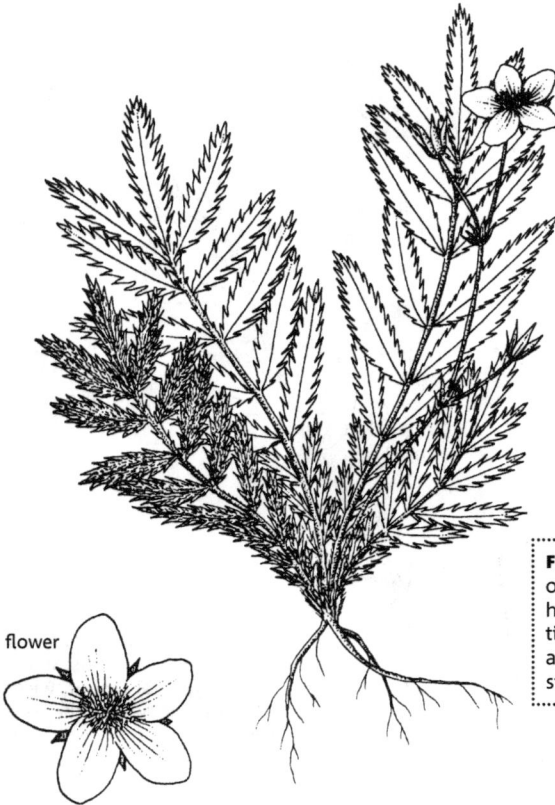

FIELD NOTES Lower surface of leaflets with white, silky hairs. Flowers solitary at tips of leafless stalks that arise from the creeping stolons.

flower

Potentilla anserina L.
SILVERWEED

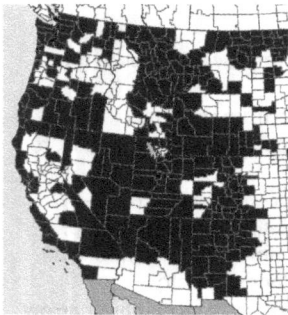

FAMILY Rosaceae (Rose)

FLOWERING May–October

HABITAT Along streams, moist soils, particularly in alkaline areas.

HABIT Native perennial herb with stolons.

STEMS Only flower-bearing upright stems present, to 15 cm long, hairy.

LEAVES Nearly all basal, pinnately compound, with 7–31 leaflets; each leaflet obovate to oblong, interspersed with smaller leaflets, to 4 cm long, pointed or rounded at the tip, tapering or rounded at the base, silvery-silky, particularly on the lower surface, toothed.

FLOWERS Solitary at the tip of often leafless stalks; bractlets usually a little longer than the sepals. **Sepals** 5, green, united to form a cup, the lobes 2–6 mm long, pointed, hairy, spreading or turned downward at flowering time. **Petals** 5, yellow, free from each other, 6–13 mm long, rounded at the tip. **Stamens** 20–25. **Pistils**: Several, free from each other; ovary superior.

FRUIT Achenes ovoid, to 3 mm long, corky, grooved, yellow-brown.

WETLAND STATUS AW OBL | GP FACW | WMV OBL

> **FIELD NOTES** Basal leaves palmately divided into 5 or 7 sharply toothed leaflets; flowers yellow 2.5 cm across with about 20 stamens.

Potentilla glaucophylla Lehm.
MOUNTAIN-MEADOW CINQUEFOIL

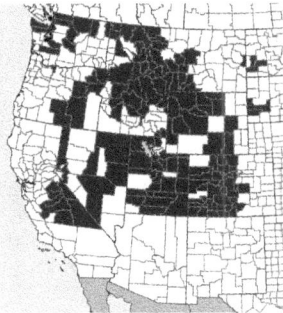

FAMILY Rosaceae (Rose)

FLOWERING July–August

HABITAT Along streams, in rocky woods, wet meadows.

HABIT Native perennial herb with a thickened rootstock.

STEMS Upright, rather slender, to 1 m tall, smooth or with appressed hairs.

LEAVES Mostly palmately divided into 5 or 7 leaflets, rarely pinnately divided, the leaflets obovate to oblanceolate, to 5 cm long, pointed or rounded at the tip, tapering to the base, sharply toothed, usually somewhat hairy.

FLOWERS Several in cymes; flowers subtended by small, lanceolate bracts to 6 mm long. **Sepals** 5, green, united below to form a tube (hypanthium), the lobes ovate to lanceolate, to 6 mm long. **Petals** 5, yellow, free from each other, to 13 mm long. **Stamens** about 20. **Pistils** many, each with a superior ovary.

FRUIT Achenes several in a cluster, smooth, to 2 mm long, pale brown.

NOTE variable in the shape and number of leaflets.

SYNONYMS *Potentilla diversifolia* Lehm.

WETLAND STATUS AW FACU | GP FACW | WMV FACU

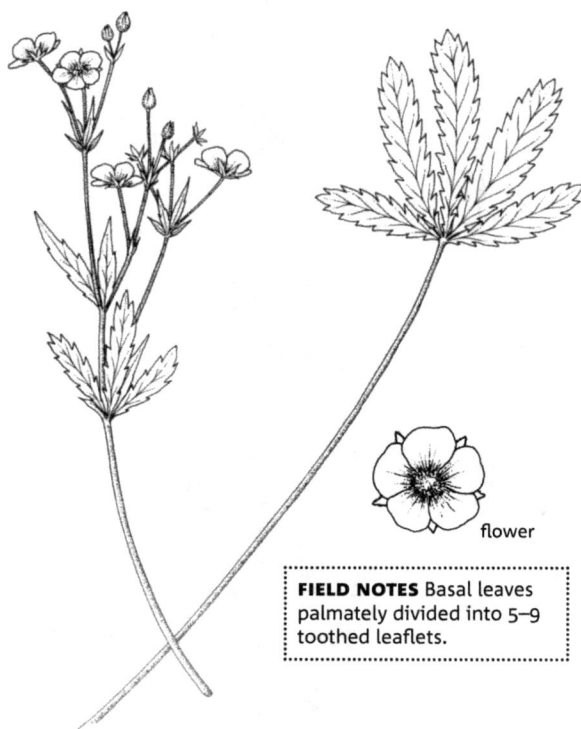

FIELD NOTES Basal leaves palmately divided into 5–9 toothed leaflets.

flower

Potentilla gracilis Dougl. ex Hook.
NORTHWEST CINQUEFOIL

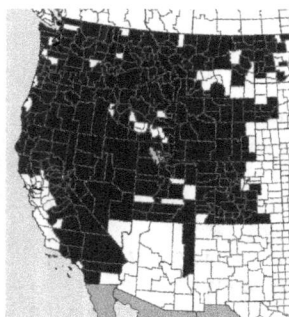

FAMILY Rosaceae (Rose)

FLOWERING May–August

HABITAT Wet meadows, damp forests, common in the mountains.

HABIT Native perennial herb from a short, stout rootstock.

STEMS Upright, to 60 cm tall, usually with some sort of hairiness.

LEAVES Palmately divided into 5–9 leaflets; each leaflet oblanceolate, to 9 cm long, green or gray, variously hairy or sometimes nearly smooth on the lower surface, toothed or divided only to about the middle; stipules toothed or toothless, to 2.5 cm long.

FLOWERS Many in an open cyme, each flower subtended by a bract. **Sepals** 5, green, alternating with 5 small bractlets on the outside, lanceolate to ovate, 6–13 mm long, usually silky-hairy **Petals** 5, yellow, free from each other, obovate, usually with a rather shallow notch at the tip, 6–20 mm long. **Stamens** usually 20. **Pistils** several to many, located in a cup (hypanthium); ovary superior.

FRUIT Achenes smooth, greenish yellow, about 2 mm long.

NOTE Throughout its range, this is an extremely variable species, particularly in the coloration, hairiness, and division of the leaves.

WETLAND STATUS AW FAC | GP FAC | WMV FAC

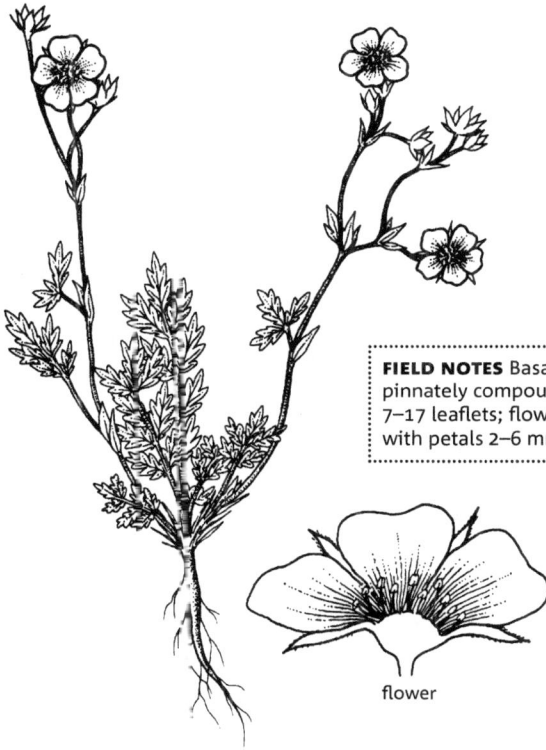

> **FIELD NOTES** Basal leaves pinnately compound with 7–17 leaflets; flowers yellow with petals 2–6 mm long.

flower

Potentilla plattensis Nutt.
PLATTE CINQUEFOIL

FAMILY Rosaceae (Rose)

FLOWERING June–August

HABITAT Wet meadows, moist or dry prairies.

HABIT Native perennial herb with a thickened rootstock.

STEMS Upright or lying flat, usually branched, to 25 cm long, smooth or with appressed hairs, sometimes reddish.

LEAVES Basal and alternate, pinnately divided into 7–17 leaflets; leaflets obovate to oblong, to 13 mm long, deeply lobed, smooth or with appressed hairs; stipules leafy, lanceolate, 6–13 mm long; basal leaves with stalks, stem leaves sessile.

FLOWERS Few to several in an open cyme, each flower on a slender stalk, subtended by an oblong bract to 2 mm long. Sepals 5, green, united below to form a short floral tube, the lobes triangular to lanceolate, to 4 mm long, with appressed hairs. **Petals** 5, yellow, free from each other, to 6 mm long. **Stamens** 20. **Pistils** several, each with a superior ovary.

FRUIT Achenes many, greenish to dark brown, smooth.

NOTE The achenes are eaten by birds.

WETLAND STATUS AW FACW | GP FACW | WMV FACW

WESTERN WETLAND FLORA · 187

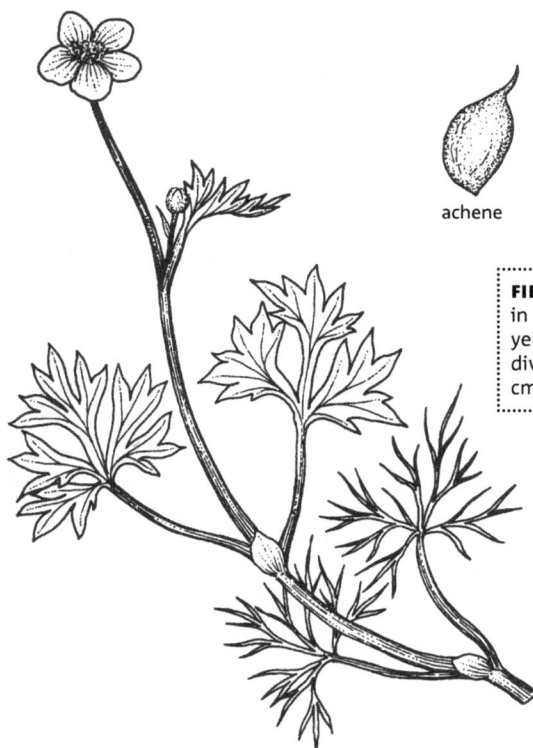

achene

> **FIELD NOTES** Plants usually in standing water; flowers yellow; floating leaves divided into segments as 2.5 cm wide, not thread-like.

Ranunculus gmelinii DC.
SMALL YELLOW WATER BUTTERCUP

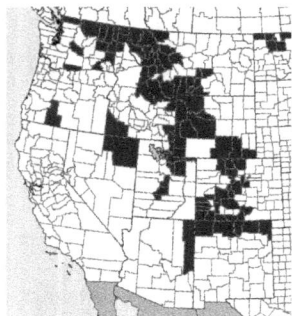

FAMILY Ranunculaceae (Buttercup)

FLOWERING May–September

HABITAT Along streams, around ponds, in marshes, in ditches, often in standing water.

HABIT Native perennial herb with slender rhizomes.

STEMS Submersed stems floating on water, terrestrial stems rooting at the nodes, sparsely branched, to 45 cm long, smooth or hairy.

LEAVES Alternate, those floating on the water surface 3-lobed or once- or twice-divided into leaflets, the leaflets linear to ovate, those submersed much more narrowly divided, to 2.5 cm long, to 2.5 cm wide, smooth or hairy.

FLOWERS 1-several. **Sepals** 5, greenish yellow, free from each other, spreading, ovate, to 6 mm long, smooth or hairy. **Petals** 5, sometimes 6, 7, or 8, yellow, free from each other, obovate to nearly spherical, to 8 mm long. **Stamens** 20–45. **Pistils** many, each with a superior ovary.

FRUIT Many achenes crowded into an ovoid or spherical head 8–13 mm across; each achene obovoid, to 2 mm long, with a flat, narrowly triangular beak.

NOTE The achenes are eaten by waterfowl.

WETLAND STATUS AW FACW | GP FACW | WMV FACW

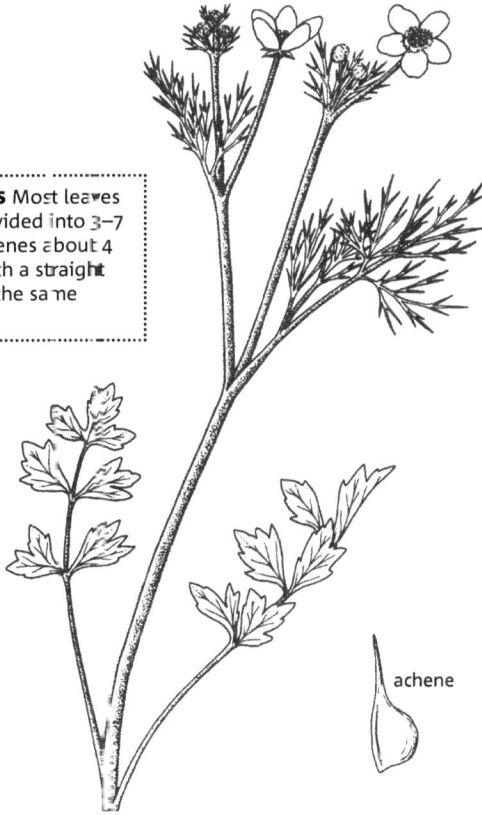

> **FIELD NOTES** Most leaves pinnately divided into 3–7 leaflets; achenes about 4 mm long with a straight beak about the same length.

achene

Ranunculus orthorhynchus Hook.
STRAIGHT-BEAK BUTTERCUP

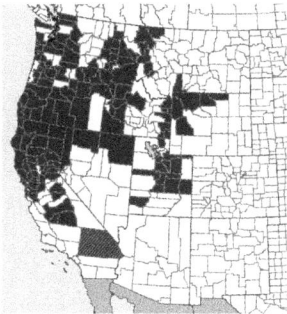

FAMILY Ranunculaceae (Buttercup)

FLOWERING May–July

HABITAT Wet meadows.

HABIT Native perennial herb with rather fleshy roots.

STEMS Upright, usually branched, hollow, to 45 cm tall, with spreading hairs.

LEAVES At least the basal leaves pinnately divided into 3–7 leaflets, with each leaflet divided again into narrow, pointed segments, hairy; stalks to 15 cm long, smooth or hairy.

FLOWERS Several in a branched cluster, to 6 cm across. Sepals 5, yellow-green, free from each other, to 8 mm long, hairy, turned downward. Petals 5, yellow, sometimes reddish on the back, free from each other, 8 mm to nearly 2.5 cm long. **Stamens** 20–40. **Pistils** many in a cluster, each with a superior ovary, smooth

FRUIT Achenes 12–20 in a cluster, ellipsoid, to 4 mm long, with a straight beak as long as the achenes or longer.

NOTE The achenes are eaten by waterfowl.

WETLAND STATUS AW FACW | GP FACW | WMV FACW

achene

basal leaf

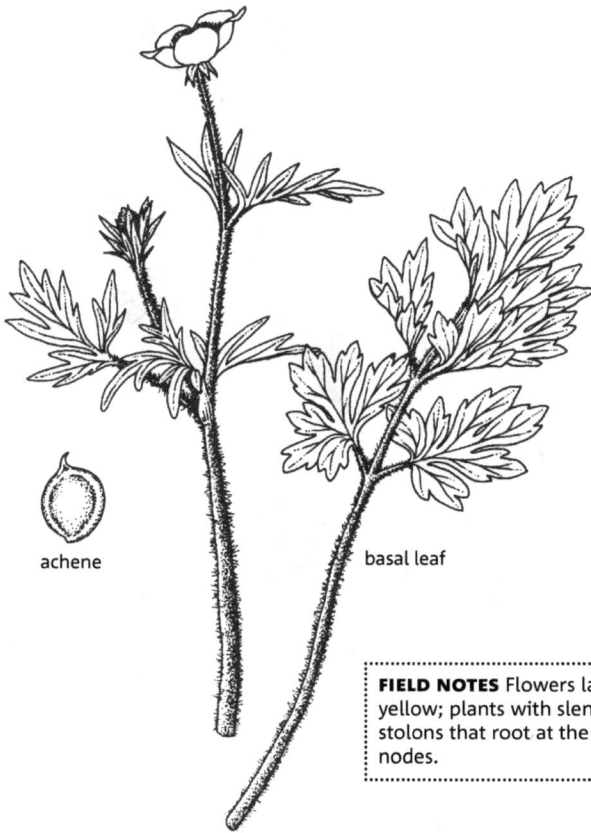

> **FIELD NOTES** Flowers large, yellow; plants with slender stolons that root at the nodes.

Ranunculus repens L.
CREEPING BUTTERCUP

FAMILY Ranunculaceae (Buttercup)

FLOWERING May–August

HABITAT Moist, disturbed soil.

HABIT Introduced perennial herb with slender stolons and with thread-like roots.

STEMS Ascending, branched, to 60 cm tall, hairy.

LEAVES Alternate, ternately compound, the leaflets both lobed and toothed, to 5 cm long, usually hairy.

FLOWERS 1–few on ascending stems, each flower to 30 mm across. **Sepals** 5, green, free from each other, to 8 mm long. **Petals** 5, yellow, free from each other, averaging about 13 mm long (or a little longer). **Stamens** many. **Pistils** many in each flower, each with a superior ovary.

FRUIT Achenes to 25 in a head, each achene obovoid, 3–4 mm long, with a hooked beak at the tip.

NOTE This buttercup is native to Europe and Asia, and sometimes planted in gardens because of its showy flowers. The achenes may be eaten by birds.

WETLAND STATUS AW FAC | GP FACW | WMV FAC

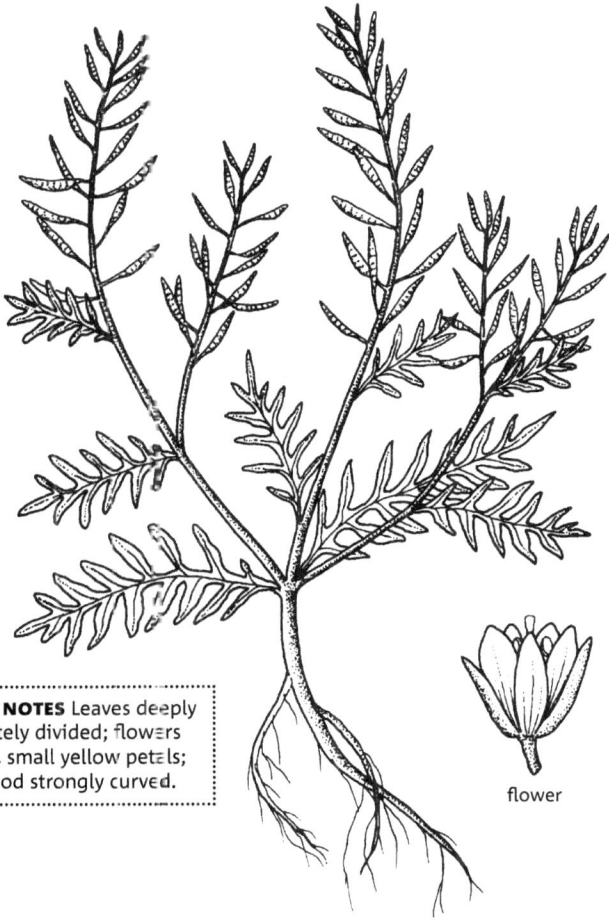

> **FIELD NOTES** Leaves deeply pinnately divided; flowers with 4 small yellow petals; fruit pod strongly curved.

flower

Rorippa curvisiliqua (Hook.) Bessey ex Britt.
CURVE-POD YELLOW-CRESS

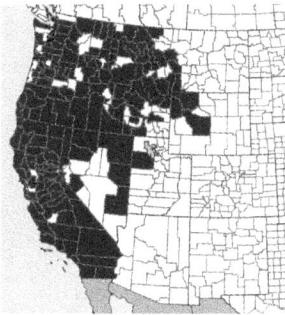

FAMILY Brassicaceae (Mustard)
FLOWERING March–November
HABITAT Damp or wet places.
HABIT Native annual or biennial herb with fibrous roots.
STEMS Upright, much branched, to 30 cm tall, smooth.
LEAVES Alternate, deeply pinnately divided, to 8 cm long, the lobes lanceolate to oblong, toothed or untoothed, smooth.
FLOWERS Several in short, axillary racemes; each flower on a stalk to 6 mm long. **Sepals** 4, green, free from each other, to 2 mm long. **Petals** 4, yellow, free from each other, to 2 mm long. **Stamens** 6. **Pistils:** Ovary superior, smooth.
FRUIT Pods linear, strongly curved, to 13 mm long, containing many brown seeds.
WETLAND STATUS AW OBL | GP OBL | WMV OBL

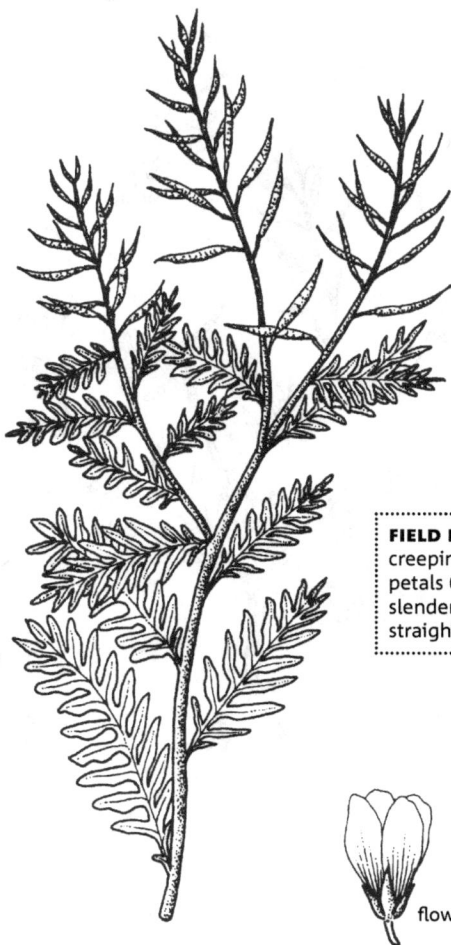

> **FIELD NOTES** Plants with creeping rootstocks, yellow petals 6–8 mm long, and slender, more or less straight, pods.

flower

Rorippa sinuata (Nutt.) A. S. Hitchc.
SPREADING YELLOW-CRESS

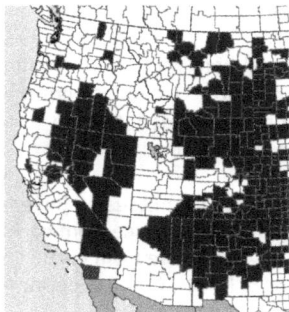

FAMILY Brassicaceae (Mustard)
FLOWERING April–September
HABITAT Moist areas.
HABIT Native perennial herb with creeping rootstocks.
STEMS Upright, branched, to 30 cm tall, smooth.
LEAVES Alternate, deeply pinnately divided into 11–19 lobes, the lobes linear to oblong, smooth.
FLOWERS Several in terminal racemes; each flower with a stalk 6–13 mm long. **Sepals** 4, green, free from each other, to 4 mm long. **Petals** 4, yellow, free from each other, 2–6 mm long. **Stamens** 6. **Pistils:** Ovary superior.
FRUIT Pods linear, more or less straight. 8–13 mm long, smooth, with small brown seeds.
WETLAND STATUS AW FACW | GP FACW | WMV FAC

> **FIELD NOTES** Heads ovoid, of white (less commonly pinkish) flowers; leaves on stalks longer than the leaflets.

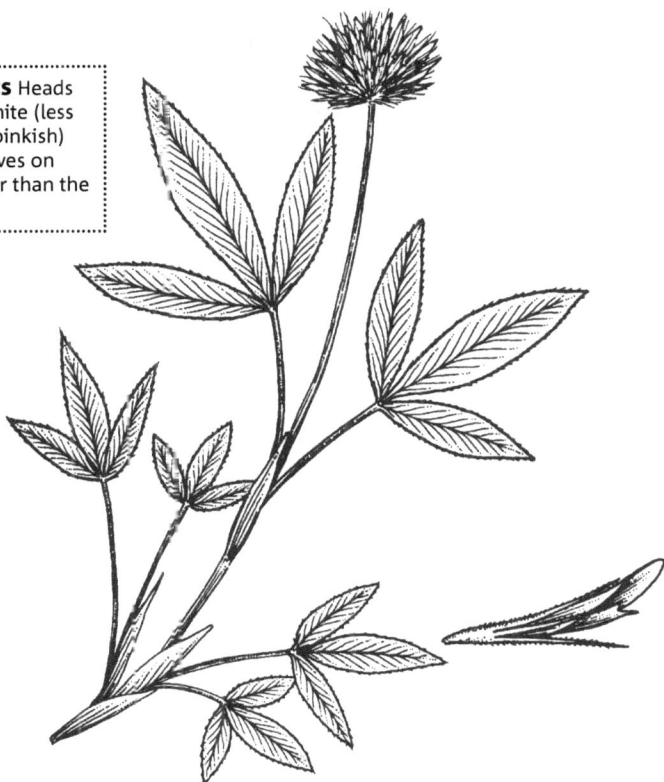

Trifolium longipes Nutt.
LONG-STALK CLOVER

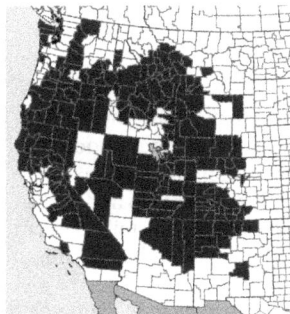

FAMILY Fabaceae (Pea)

FLOWERING June–September

HABITAT Wet meadows in the mountains.

HABIT Native perennial herb with branched rhizomes and a taproot.

STEMS Spreading to erect, to 22.5 cmes long, smooth or with some appressed hairs.

LEAVES Alternate, divided into 3 leaflets; each leaflet narrow, lanceolate to obovate, pointed or rounded at the tip, tapering to the base, to 5 cm long, finely toothed, hairy on lower surface; leaf stalks longer than the leaflets; stipules lanceolate, to 2.5 cm long.

FLOWERS Many crowded into ovoid heads to 15 mm across, on long stalks to 15 cm long. **Sepals** 5, united below into a tube, 8 to nearly 13 mm long, the slender teeth longer than the tube, hairy. **Petals** 5, white (less commonly pinkish), arranged to form a sweet pea-shaped flower 8–13 mm long. **Stamens** 10. **Pistils:** Ovary superior, hairy.

FRUIT Pods a little longer than broad hairy near the tip, containing 2–4 seeds.

NOTE The petals are sometimes tinged with purple. The fruits and leaves are eaten by some mammals.

WETLAND STATUS AW FACW | GP FACW | WMV FAC

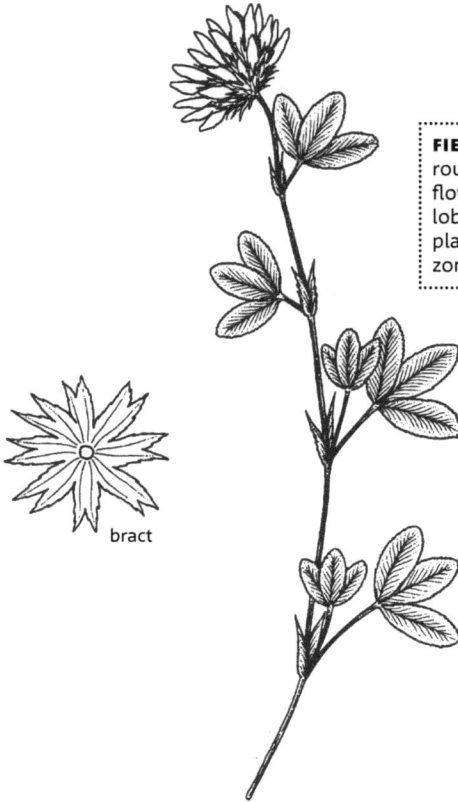

> **FIELD NOTES** Heads nearly round, of white and purple flowers subtended by a flat, lobed and toothed bract; plants with creeping rhizomes.

bract

Trifolium wormskioldii Lehm.
COW CLOVER

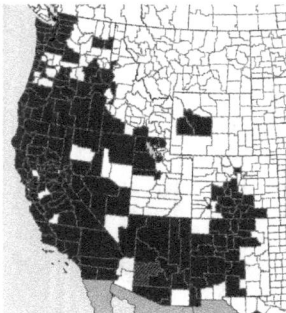

FAMILY Fabaceae (Pea)
FLOWERING May–October
HABITAT Most moist areas.
HABIT Native perennial herb with creeping rhizomes.
STEMS Spreading, branched, to 30 cm long, smooth.
LEAVES Alternate, divided into 3 leaflets; leaflets oblanceolate to obovate, rounded at the tip and at the base, to 4 cm long, toothed, smooth; leaf stalks to 8 cm long: leaflets sessile or nearly so; stipules lanceolate, deeply toothed.
FLOWERS Many in rounded heads to 4 cm in diameter, white and purple, each head on a stalk to 6 cm long and subtended by a lobed and toothed bract, the bract to 20 mm across and flat. **Sepals** 5, green, united to form a tube, to nearly 13 mm long, the 5 awn-like teeth longer than the tube, smooth. **Petals** 5, white and purple, arranged in the shape of a sweetpea flower, to 13 mm long. **Stamens** 10. **Pistils:** Ovary superior, smooth.
FRUIT Pods ovoid, smooth, with 2–6 seeds.
NOTE Deer browse on this species.
WETLAND STATUS AW FACW | GP FACW | WMV FACW

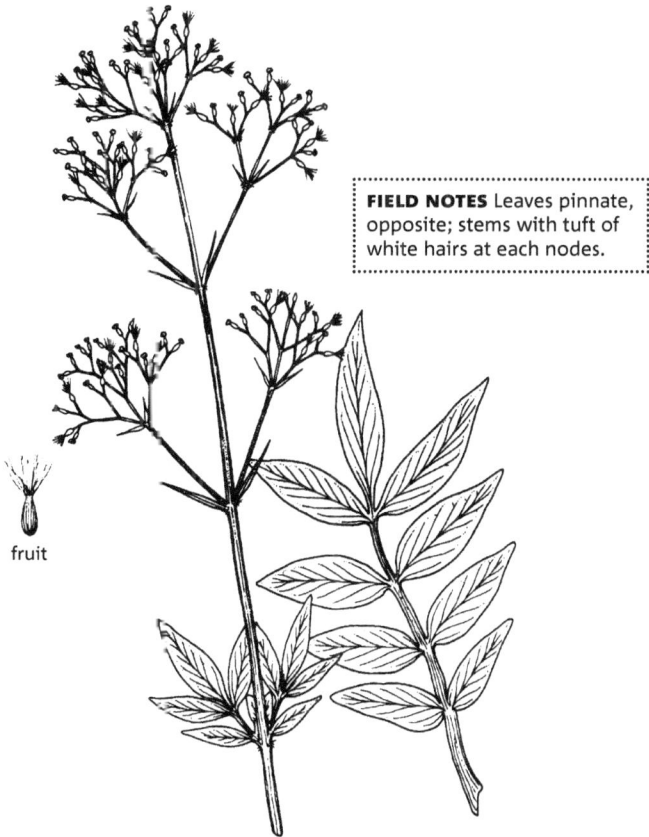

> **FIELD NOTES** Leaves pinnate, opposite; stems with tuft of white hairs at each nodes.

fruit

Valeriana occidentalis Heller
WESTERN VALERIAN

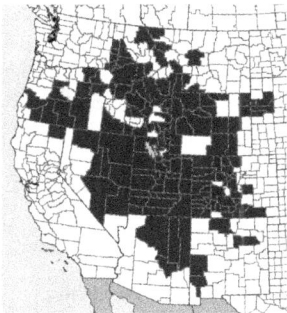

FAMILY Valerianaceae (Valerian)
FLOWERING May–September
HABITAT Wet meadows, moist soil.
HABIT Native perennial herb from rhizomes.
STEMS Upright, to 75 cm tall, smooth or nearly so except for a tuft of white hairs at the nodes.
LEAVES Basal and opposite on the stem, simple or pinnately divided into as many as 13 segments, oblong to narrowly ovate, smooth or nearly so.
FLOWERS Several crowded in a terminal cyme, the cyme to 6 cm long; bracts to 6 mm long. **Sepals** nearly absent and inconspicuous in flower, developing into as many as 16 plume-like bristles during fruiting. **Petals** 5, white, united below, 3–4 mm long. **Stamens** 3, attached to the petals. **Pistils:** Ovary inferior.
FRUIT Achenes linear to narrowly oblong, tan, to 6 mm long, with plumose bristles at the top.
NOTE The leaves and stems are browsed by deer and elk.
WETLAND STATUS AW FAC | GP FAC | WMV FAC

FIELD NOTES Leaflets with terminal tendril; flowers 2–10, 1–2.5 cm long, in racemes.

flower

Vicia americana Muhl. ex Willd.
AMERICAN PURPLE VETCH

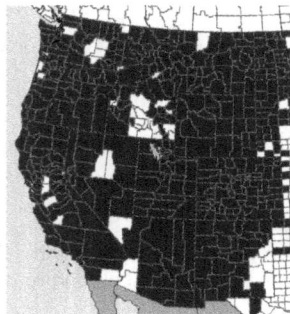

FAMILY Fabaceae (Pea)

FLOWERING May–August

HABITAT Along streams, damp thickets, prairies, road-sides, open places.

HABIT Native, sprawling perennial vine, sometimes scrambling over other vegetation.

STEMS Sprawling or climbing, to 1 m long, smooth or sometimes hairy.

LEAVES Alternate, pinnately compound, with 4–16 leaflets and a branched tendril where a terminal leaflet should be; leaflets linear to oblong to ovate, to 5 cm long, rounded but with a small projection at the tip, with or without teeth, smooth or hairy.

FLOWERS Flowers 2–9 in racemes from the axils of the leaves, the racemes on stalks shorter than the leaves. **Sepals** 5, united below, green, the tube to 6 mm long, the teeth very unequal in size and shape. **Petals** 5, purple, to 2.5 cm long, with the configuration of a sweetpea flower. **Stamens** 10. **Pistils**: Ovary superior, smooth.

FRUIT Pods to nearly 5 cm long, to nearly 13 mm wide, smooth, containing 8–14 seeds; seeds black, 2–6 mm in diameter.

NOTE There is considerable variation in leaflet size and number.

WETLAND STATUS AW FAC | GP FACU | WMV FAC

FIELD NOTES Plants mat-forming; leaves smooth, opposite, toothless; flowers small, white, in long-stalked spikes.

Alternanthera philoxeroides (Mart.) Griseb.
ALLIGATOR-WEED

FAMILY Amaranthaceae
FLOWERING April–October
HABITAT Wet to moist soils and shallow water, forming mats over still water, including water-filled ditches, sloughs, swales, lakes, ponds.
HABIT Introduced perennial herb, forming mats and potentially invasive.
STEMS Trailing, mat-forming, hollow, to 1 m long, smooth except for a few hairs where the leaves are attached; flowering stems upright, to 30 cm tall.
LEAVES Opposite, simple, elliptic to oblanceolate, pointed or somewhat rounded at the tip, tapering to the sessile base, smooth, toothless, somewhat fleshy, to 10 cm long.
FLOWERS Small, crowded into terminal or axillary spikes; spike about 12 mm long, on a stalk to 8 cm long. **Sepals** 4–5, white, free from each other, to 6 mm long, minutely toothed near the tip. **Petals** absent. **Stamens** 5, united at base. **Pistils:** Ovary superior; **styles** 2.
FRUIT Small, indehiscent, 1-seeded.
NOTE This species, native of South America, can be an aggressive weed in wetlands. The flowers have a sweetish odor. The alligator weed flea beetle (*Agasicles hygrophila*) has been introduced in the United States as a biological control.
WETLAND STATUS AW OBL | CP OBL | WMV OBL

> **FIELD NOTES** Differs from other species of *Arnica* by having 5–10 pairs of leaves on the stem and pale yellow flower heads to 5 cm across, each head subtended by bracts with a rounded tip and a tuft of hairs at tip of the bract.

bract

Arnica chamissonis Less.
LEAFY ARNICA

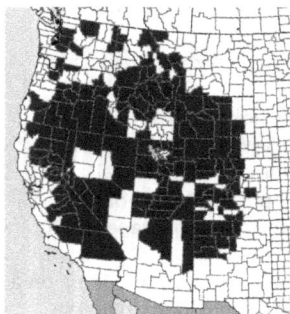

FAMILY Asteraceae (Aster)

FLOWERING June–August

HABITAT Wet meadows and other moist places in the mountains.

HABIT Native perennial herb with elongated rhizomes.

STEMS Upright, unbranched, to 75 cm tall, hairy, with the hairs sometimes glandular and sticky.

LEAVES Opposite, simple, with 5–10 pairs per stem, lanceolate to oblanceolate, to 20 cm long, to 5 cm wide, hairy, with or without teeth, all but the lowermost sessile.

FLOWERS Crowded into heads to 5 cm across, pale yellow, with 12–18 ray flowers and many disk flowers; heads subtended by several bracts, the bracts rounded at the tip, to 13 mm long, with a tuft of hairs at the tip. **Sepals** absent. **Petals** 5, pale yellow, some of them united to form a ray, others united to form a tube. **Stamens** 5. **Pistils**: Ovary inferior, somewhat hairy.

FRUIT Achenes usually somewhat hairy, narrowly ellipsoid, to 6 mm long, with a tuft of tawny bristles at the tip.

WETLAND STATUS AW FACW | GP FACW | WMV FACW

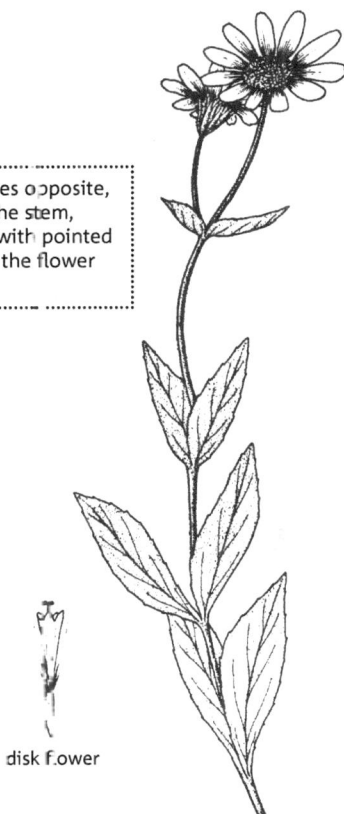

FIELD NOTES Leaves opposite, 5–12 pairs along the stem, margins toothed, with pointed bracts subtending the flower heads.

disk flower

Arnica lanceolata Nutt.
STREAMBANK ARNICA

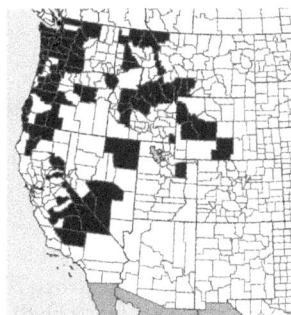

FAMILY Asteraceae (Aster)

FLOWERING July–August

HABITAT Along streams, moist woods.

HABIT Native perennial herb with much branched rhizomes.

STEMS Upright, usually unbranched, to 75 cm tall, usually glandular-hairy, sometimes more or less smooth.

LEAVES Opposite, simple, with 5–12 pairs on the stem, lanceolate to narrowly ovate, to 11 cm long, pointed or somewhat rounded at the tip, rounded at the base, toothed, glandular-hairy or nearly smooth.

FLOWERS Crowded into heads to 5 cm across, consisting of 8–14 pale yellow rays and a yellow disk; each head subtended by narrow pointed bracts 13–20 mm long. **Sepals** absent. **Petals** 5, some of them united to form rays 10–25 mm long, others united to form tubular flowers in a central disk. **Stamens** 5. **Pistils:** Ovary inferior, hairy.

FRUIT Achenes sparsely hairy, with a tuft of tawny-colored soft bristles at the tip.

SYNONYMS *Arnica amplexicaulis* Nutt.

WETLAND STATUS AW FACW | GP FAC | WMV FACW

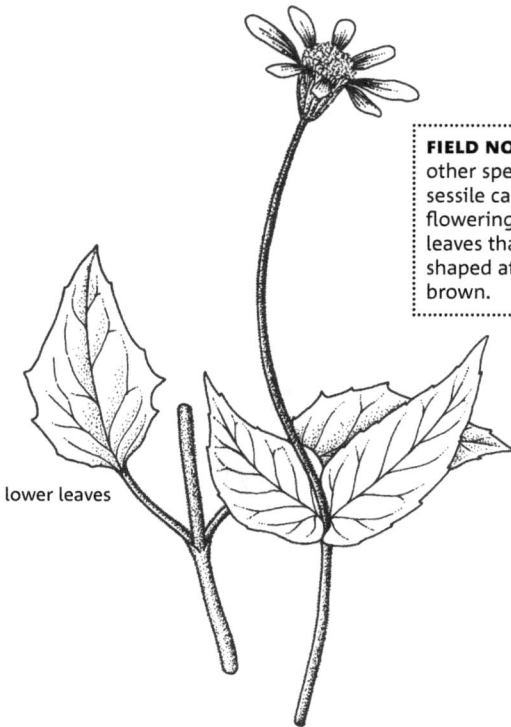

> **FIELD NOTES** Differs from other species of *Arnica* by its sessile cauline leaves, 1–3 flowering heads, and basal leaves that are not heart-shaped at the base; achenes brown.

lower leaves

Arnica latifolia Bong.
MOUNTAIN ARNICA

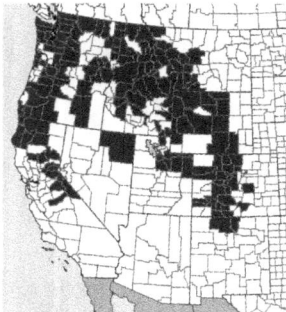

FAMILY Asteraceae (Aster)

FLOWERING June–August

HABITAT Moist forests and wet meadows in the mountains.

HABIT Native perennial herb with slender, elongated rhizomes.

STEMS Upright, to 60 cm tall, mostly smooth.

LEAVES Basal and cauline, sometimes glandular, sometimes hairy, toothed, the basal ones ovate to lanceolate, usually pointed at the tip, truncate or rounded at the base but not heart-shaped, to 25 cm long, with a distinct stalk, the cauline leaves 2–4 pairs, opposite, sessile, to 13 cm long, rounded or pointed at the tip.

FLOWERS Several crowded together into usually 1–3 heads; each head to 5 cm across, subtended by lanceolate, pointed, green, usually hairy and often glandular bracts, with 8–12 yellow rays and central yellow disk. **Sepals** absent. **Petals:** Some united to form yellow rays 1–2.5 cm long, others united to form yellow tubular flowers in a central disk. **Stamens** 5. **Pistils:** Ovary inferior.

FRUIT Achenes brown, smooth or minutely hairy, with a tuft of white hairs.

NOTE At higher elevations, a dwarf variation of this plant occurs, sometimes referred to as var. *gracilis*.

WETLAND STATUS AW FAC | GP FAC | WMV FAC

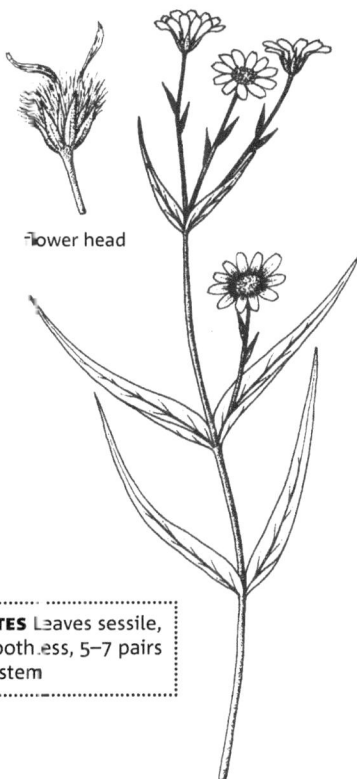

Flower head

FIELD NOTES Leaves sessile, margins toothless, 5–7 pairs along the stem

Arnica longifolia D.C. Eat.
SEEP SPRING ARNICA

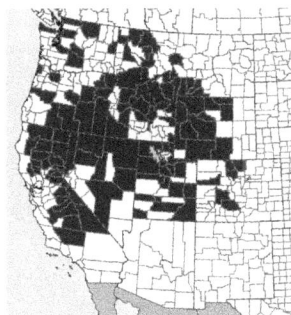

FAMILY Asteraceae (Aster)

FLOWERING July–August

HABITAT Around springs, along rivers, wet places in woods.

HABIT Native perennial herb with a thickened rootstock and often short rhizomes.

STEMS Up to 60 cm tall, sparsely hairy, some of the stems often not bearing flower heads.

LEAVES Opposite, simple. 5–7 pairs per stem, lanceolate to narrowly elliptic, to 10 cm long, pointed at the tip, tapering to the sessile or somewhat clasping base, sparsely hairy. usually without teeth.

FLOWERS Borne in heads 2–5 cm across, each head consisting of 8–13 yellow ray flowers and several yellow disk flowers; bracts subtending each head long-pointed at the tip, glandular-hairy. **Sepals** absent. **Petals** 5, some united to form yellow rays, others united to form short yellow tubes that comprise the disk. **Stamens** 5. **Pistils:** Ovary inferior, smooth or hairy.

FRUIT Achenes elongated, smooth or hairy, with a tuft of tawny hairs at the tip.

NOTE The achenes may be eaten by birds.

WETLAND STATUS AW FACW | GP FACW | WMV FACW

flower head

> **FIELD NOTES** Leaves hairy, opposite, 3–4 pairs; flowers few, yellow, in heads to 6 cm across.

Arnica mollis Hook.
HAIRY ARNICA

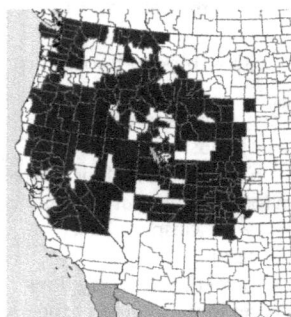

FAMILY Asteraceae (Aster)

FLOWERING June–September

HABITAT Moist areas in the mountains, including spruce-pine forests.

HABIT Native perennial herb with dark brown rhizomes.

STEMS Upright, unbranched, to 60 cm tall, with short or long hairs, some of which may be glandular.

LEAVES Basal leaves oblanceolate to spatulate, borne on stalks; leaves on the stem in 3–4 opposite pairs, lanceolate to ovate, to 8 cm long, sessile, usually rough-hairy, with or without teeth.

FLOWERS Crowded together into a head, the head to 6 cm across, some of the flowers yellow and ray-like surrounding a yellow disk; rays 12–18, 2–3 cm long; disk to 4 cm across; each head subtended by several pointed, long-hairy bracts. **Sepals** absent. **Petals:** Some united to form flat rays, others united to form short tubes. **Stamens** 5. **Pistils:** Ovary inferior, hairy.

FRUIT Achenes to 4 mm long, hairy, with a tuft of tawny-colored hairs at the tip,

WETLAND STATUS AW FAC | GP FAC | WMV FAC

> **FIELD NOTES** Leaves 3 or 4 pairs on the stem; margins irregularly toothed; flower heads with 8–15 pale yellow rays 2–2.5 cm long; achenes with tawny-colored soft bristles.

Arnica ovata Greene
STICKY-LEAF ARNICA

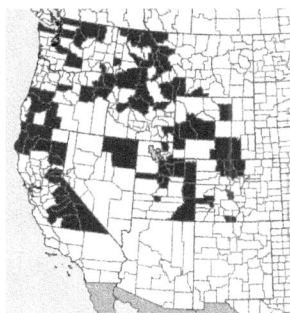

FAMILY Asteraceae (Aster)
FLOWERING July–September
HABITAT Wet, often rocky, areas.
HABIT Native perennial herb with branching rhizomes.
STEMS Upright, unbranched, to 40 cm tall, smooth or glandular-hairy.
LEAVES Leaves on the stem 3–4 pairs, opposite, elliptic to ovate, to 6 cm long, to 6 cm wide, rounded or pointed at the tip, rounded at the base, irregularly toothed, rough-hairy, sessile or on short stalks; basal leaves, if present, smaller, on stalks.
FLOWERS Crowded into heads, with few to several heads at the tip of the stem; heads to 5 cm across, bearing both ray flowers and disk flowers; heads subtended by pointed, narrow, green bracts that have some long hairs, some of which may be glandular. **Sepals** absent. **Petals:** Some united to form 8–15 pale yellow rays 2–2.5 cm long; others united into tubes that comprise the central yellow disk. **Stamens** 5. **Pistils:** Ovary inferior, hairy.
FRUIT Achenes short-hairy, with tawny soft bristles at the tip.
NOTE This is a rather variable species, similar to *Arnica mollis* but differing mostly by having the stem leaves larger than the basal leaves.
SYNONYMS *Arnica xdiversifolia* Greene
WETLAND STATUS AW FACW | GP FACW | WMV FACW

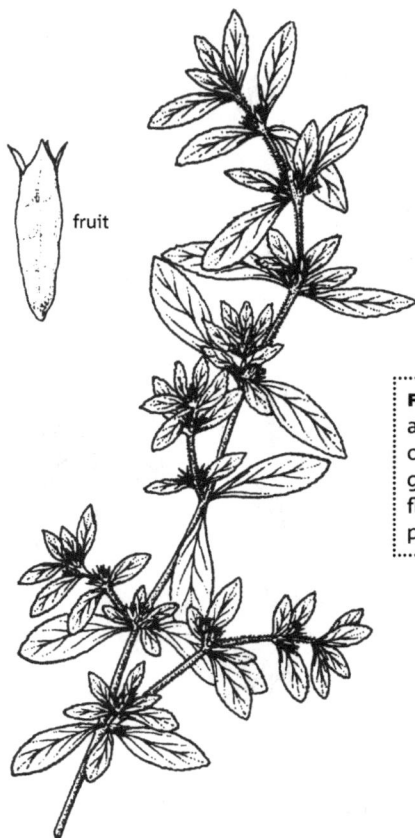

fruit

> **FIELD NOTES** Plants
> annual, much branched,
> often sprawling; leaves
> glandular-hairy, opposite;
> flowers tiny, axillary;
> petals 5, free, white.

Bergia texana (Hook.) Seub. ex Walpers
TEXAS BERGIA

FAMILY Elatinaceae (Waterwort)
FLOWERING July–October
HABITAT Mud flats, muddy shores of ponds, sandy lake beds.
HABIT Native annual herb with a slender taproot.
STEMS Spreading or ascending, much branched, to 40 cm long, glandular-hairy, usually reddish.
LEAVES Opposite, simple, elliptic to oblong, to 4 cm long, to 20 mm wide, pointed at the tip, tapering to the base, glandular-toothed, glandular-hairy.
FLOWERS Solitary in the axils of the leaves, borne on short stalks. **Sepals** 5, free from each other, persistent on the fruit, green with a whitish margin, 3–4 mm long, with a thickened vein down the middle. **Petals** 5, free from each other, about as long as the sepals. **Stamens** 5 or 10. **Pistils:** Ovary superior; styles 3 or 5.
FRUIT Capsules spherical to ovoid, to 3 mm long, containing minute curved seeds.
NOTE The seeds are eaten by waterfowl.
WETLAND STATUS AW OBL | GP OBL | WMV OBL

> **FIELD NOTES** Flowers with
> 2 sepals, 2 notched white
> petals, 2 stamens, and an
> inferior ovary. Capsule
> small, covered with hooked
> hairs.

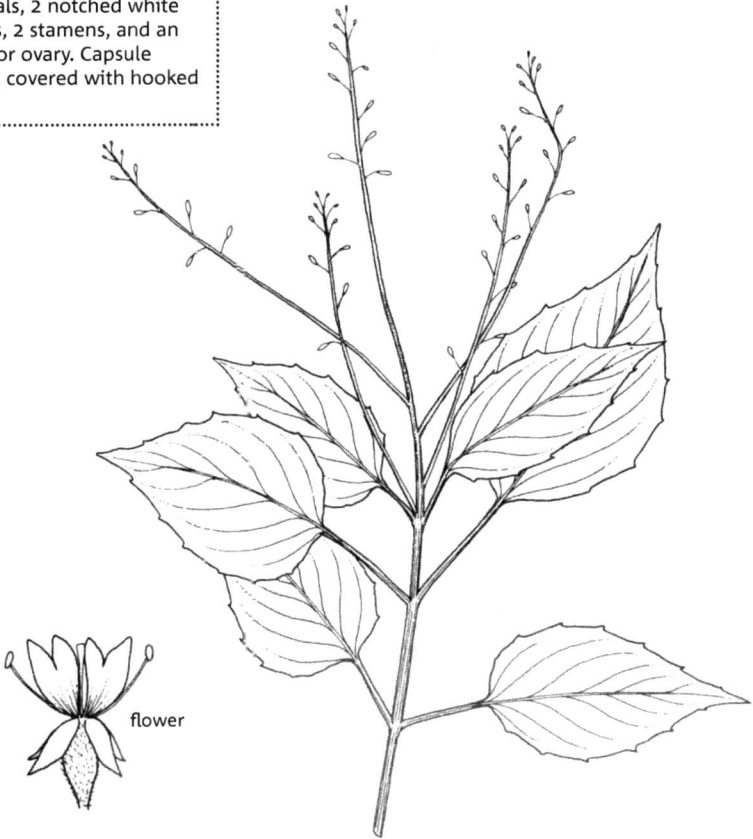

flower

Circaea alpina L.
SMALL ENCHANTER'S NIGHTSHADE

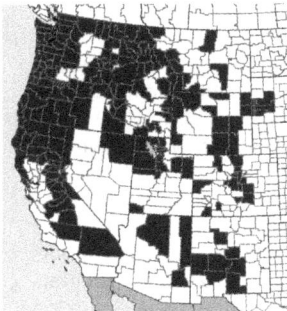

FAMILY Onagraceae (Evening-primrose)
FLOWERING June–August
HABITAT Moist woods.
HABIT Native perennial herb from a tuberous rootstock.
STEMS Upright, unbranched. to 60 cm tall, usually with a few appressed hairs.
LEAVES Opposite, simple, ovate, to 8 cm long, pointed at the tip, rounded or heart-shaped at the base, with or without teeth, smooth or sparsely hairy: leaf stalks 2–4 cm long.
FLOWERS Few in a raceme, without bracts. **Sepals** 2. green, free from each other, about 1 mm long. **Petals** 2, white, free from each other, notched. about 1 mm long. **Stamens** 2. **Pistils**: Ovary superior.
FRUIT Capsules obovoid, to 2 mm long, covered with hooked hairs.
WETLAND STATUS AW FAC | GP FACW | WMV FAC

> **FIELD NOTES** Stems and leaves glaucous; rootstocks scaly.

capsule

Epilobium glaberrimum Barbey
GLAUCOUS WILLOWHERB

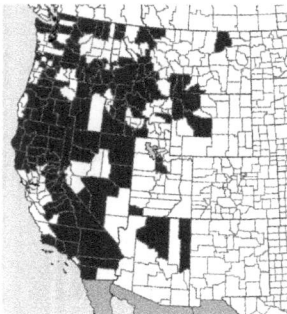

FAMILY Onagraceae (Evening-primrose)

FLOWERING July–August

HABITAT Along streams, moist woods.

HABIT Native perennial herb with scaly rootstocks.

STEMS Upright, unbranched. slender, to 60 cm tall, smooth or rarely with glandular hairs, glaucous, often purplish.

LEAVES Opposite, simple, ascending, oblong to lanceolate, to 5 cm long, more or less rounded at the tip, tapering to the sessile base, smooth, glaucous, with or without minute teeth; flower stalks very short.

FLOWERS Several in a terminal cluster, usually upright, sometimes drooping. **Sepals** 4, green, free from each other, about 2 mm long. **Petals** 4, purple to white, free from each other, usually slightly notched at the tip, 4–8 mm long. **Stamens** 8. **Pistils:** Ovary inferior.

FRUIT Capsules ellipsoid, slender, 4–8 cm long; seeds to about 1 mm long, covered with minute warts, bearing a tuft of whitish hairs.

WETLAND STATUS AW FACW | GP FACW | WMV FACW

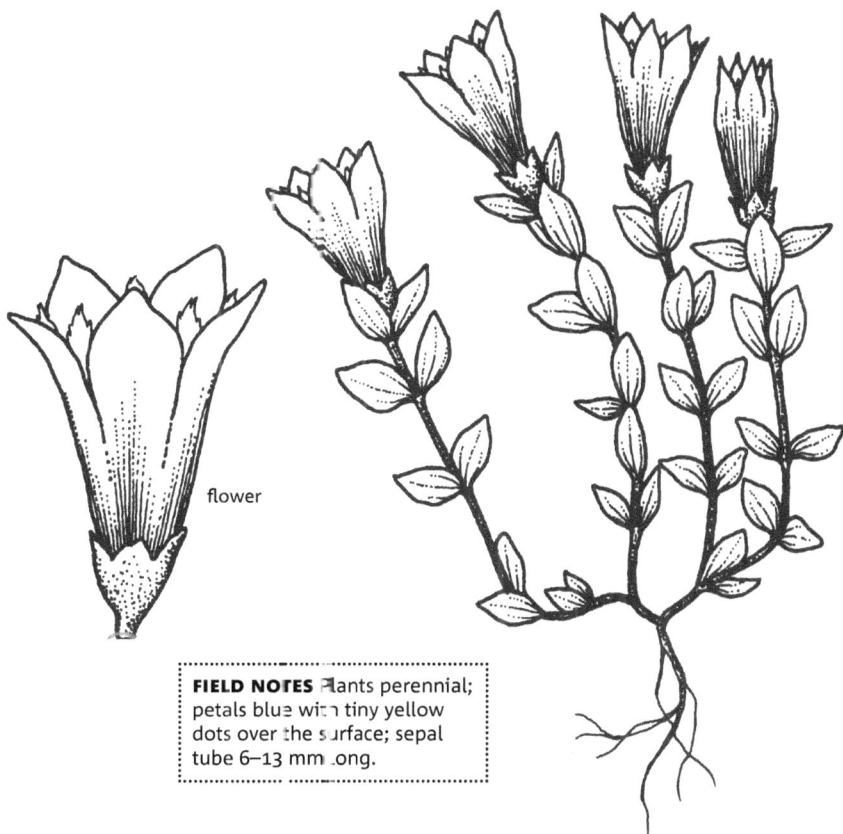

flower

FIELD NOTES Plants perennial; petals blue with tiny yellow dots over the surface; sepal tube 6–13 mm long.

Gentiana calycosa Griseb.
RAINIER PLEATED GENTIAN

FAMILY Gentianaceae (Gentian)

FLOWERING July–September

HABITAT Wet meadows, along streams, mostly in the mountains.

HABIT Native perennial herb with a thick rootstock.

STEMS Upright, unbranched, to 30 cm tall, smooth.

LEAVES Opposite, simple, ovate to suborbicular, to 5 cm long, pointed or rounded at the tip, rounded at the base, smooth, without teeth.

FLOWERS Showy, 1–3 in a cluster, subtended by lanceolate to ovate bracts. **Sepals** 5, green, united below to form a tube, the tube 6–13 mm long, the lobes 3–8 mm long. **Petals** 5, deep blue, united to form a bell-shaped tube, the tube to 4 cm long, the lobes 8–13 mm long, the surface covered with tiny yellow dots. **Stamens** 5, attached to the tube of the petals. **Pistils:** Ovary superior, smooth; stigmas 2.

FRUIT Capsules narrowly ellipsoid, 2–2.5 cm long, with a stalk at the base; seeds numerous, narrowly winged, about 2 mm long.

WETLAND STATUS AW FACW | WMW FACW

FIELD NOTES *Gentianella* differs from other genera of the Gentianaceae by lacking fringed petals, by lacking folds between the petals, and by having the petals united for more than 2/3 their length.

flower

Gentianella amarella (L.) Boerner
NORTHERN GENTIAN

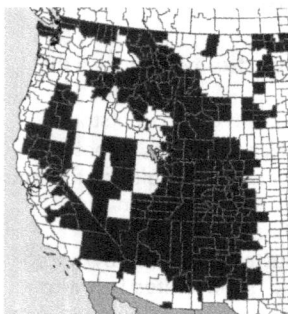

FAMILY Gentianaceae (Gentian)

FLOWERING June–September

HABITAT Wet meadows in the mountains.

HABIT Native annual herb with fibrous roots.

STEMS Upright, branched, to 45 cm tall, smooth.

LEAVES Opposite, simple, smooth, without teeth, the basal leaves spatulate, pointed or rounded at the tip, tapering to the base, to 5 cm long, the cauline leaves ovate to lanceolate, more or less pointed at the tip, to 4 cm long.

FLOWERS 2–10 in the axils of the leaves and also terminal; flower stalks slender, to 5 cm long, smooth. **Sepals** usually 5, united for more than half their length, green, to 6 mm long. **Petals** usually 4, united for more than 2/3 their length, usually blue or purple, nearly 2.5 cm long. **Stamens** 5. **Pistils:** Ovary superior; stigmas 2.

FRUIT Capsules cylindrical, to 13 mm long; seeds minute, smooth.

SYNONYMS *Gentiana amarella* L., *Gentiana acuta* Michx.

WETLAND STATUS AW FACW | GP FACW | WMV FACW

> **FIELD NOTES** Flower heads to 6 cm wide; bracts subtending flower heads may be as long as 2.5 cm.

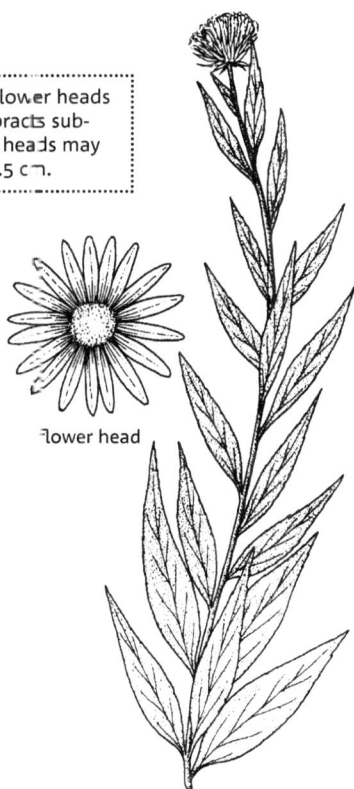

Flower head

Helianthus nuttallii Torr & Gray
NUTTALL'S SUNFLOWER

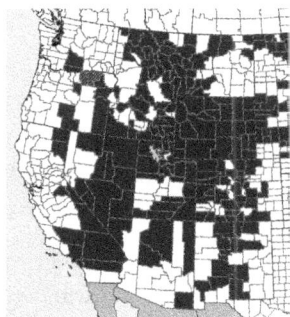

FAMILY Asteraceae (Aster)

FLOWERING July–November

HABITAT Moist or dry meadows.

HABIT Native perennial herb with thickened roots and short rhizomes.

STEMS Upright, branched or unbranched, to 4 m tall, smooth or less commonly hairy, more or less glaucous.

LEAVES All opposite, or the uppermost alternate, simple, narrowly lanceolate to lanceolate, to 15 cm long, to 4 cm wide, pointed at the tip, tapering to the sessile or short-stalked base, with or without teeth, short-hairy and rough to the touch.

FLOWERS Many crowded together into heads to 6 cm across, each head consisting of 12–20 yellow rays and a central disk of yellow-tubular flowers; bracts surrounding the heads elongated, pointed at the tip, rough-hairy, to 2.5 cm long. **Sepals** absent. **Petals** yellow, some of them united to form rays, others united to form tubular flowers that comprise the central disk. **Stamens** 5. **Pistils:** Ovary inferior.

FRUIT Achenes ellipsoid, to 4 mm long, smooth.

NOTE The achenes are eaten by birds.

WETLAND STATUS AW FACW | GP FACW | WMV FACW

FIELD NOTES Leaves toothless, undivided, and in whorls; one stamen per flower.

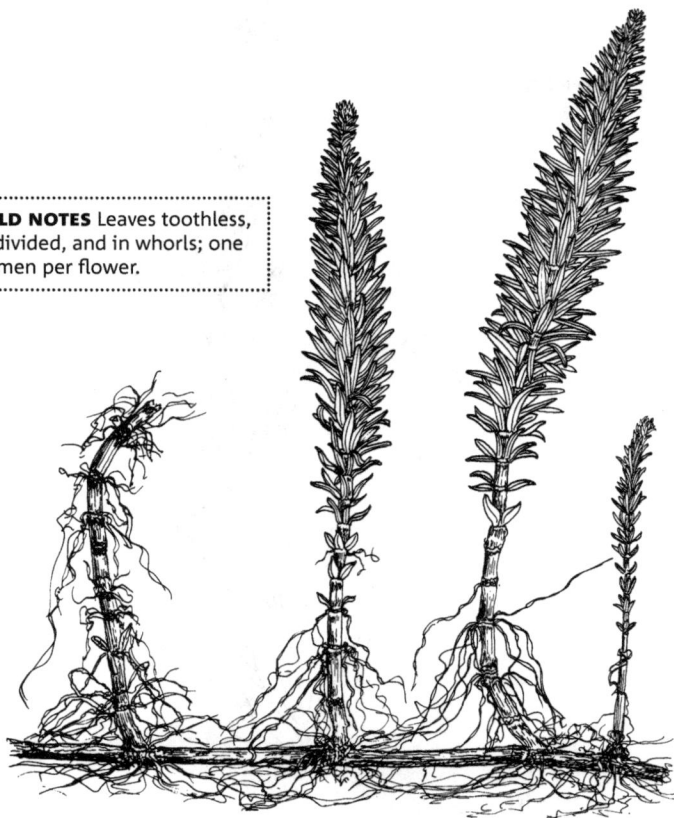

Hippuris vulgaris L.
COMMON MARE'S-TAIL

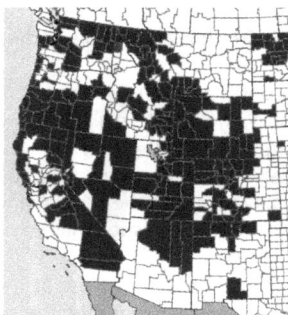

FAMILY Plantaginaceae (Plantain)

FLOWERING June–August

HABITAT In and around lakes, ditches, sloughs, ponds, and streams.

HABIT Native submerged or emersed perennial from slender, creeping rhizomes.

STEMS Upright, unbranched, to 60 cm long, smooth.

LEAVES 6–12 in whorls, the whorls 8–13 mm apart on the stem, each leaf linear, sessile, to 4 cm long, without teeth or divisions, smooth; submerged leaves sometimes reduced to short scales.

FLOWERS Solitary in the axils of emersed leaves. **Sepals** very tiny, green, united to the ovary. **Petals** absent. **Stamens** 1. **Pistils**: Ovary inferior.

FRUIT Up to 3 mm long, nut-like, ellipsoid, hard, not dehiscent, 1-seeded.

NOTE Leaves that are in the water are usually very soft and weak, resembling those of *Elodea*, waterweed, but leaves of *Elodea* are rarely in whorls of more than 3. The fruits are eaten by waterfowl. Formerly placed in own family: Hippuridaceae.

WETLAND STATUS AW OBL | GP OBL | WMV OBL

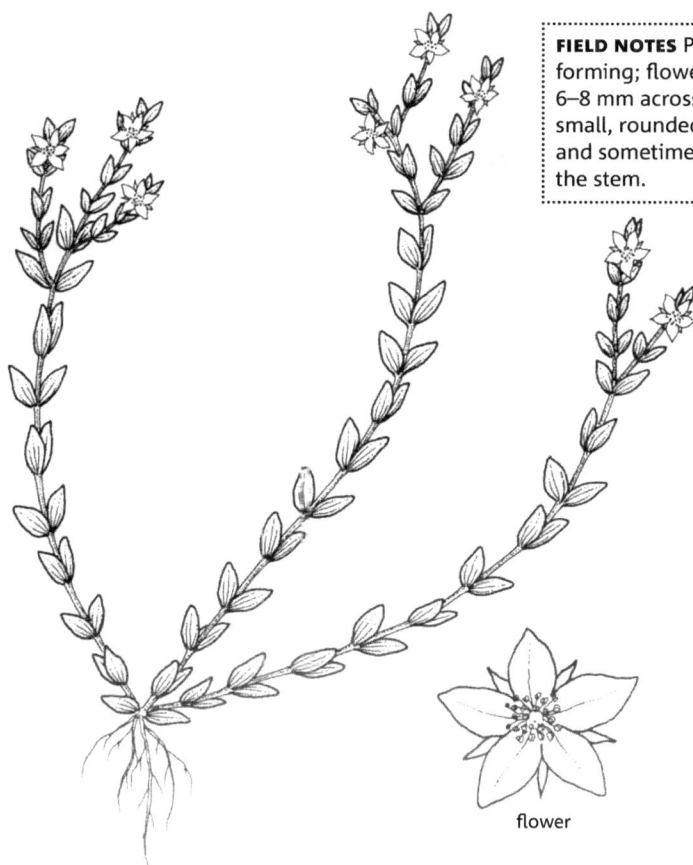

FIELD NOTES Plants mat-forming; flowers yellow, 6–8 mm across; leaves small, rounded at the tip and sometimes clasping the stem.

flower

Hypericum anagalloides Cham. & Schlecht.
BOG ST. JOHN'S-WORT

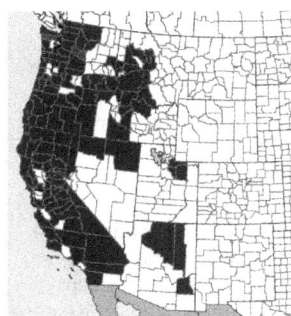

FAMILY Hypericaceae (St. John's-wort)
FLOWERING June–August
HABITAT Marshes, wet meadows, around springs.
HABIT Native annual (or sometimes perennial) herb, rooting at the nodes, often forming mats.
STEMS Sprawling or upright, slender, usually unbranched, to 40 cm tall, smooth.
LEAVES Opposite, simple, elliptic to ovate, rounded at the tip, tapering or rounded at the base, on short stalks or sometimes clasping the stem, smooth, with 5 or 7 veins, without teeth.
FLOWERS 1–few in cymes. **Sepals** 5, green, free from each other, lanceolate, to 4 mm long, unequal in size. **Petals** 5, yellow, free from each other, oval, to 4 mm long, not dotted. **Stamens** 15–20. **Pistils:** Ovary superior; **styles** 5, very short.
FRUIT Capsules ovoid, to 4 mm long, smooth, containing many seeds. i
WETLAND STATUS AW OBL | WMV OBL

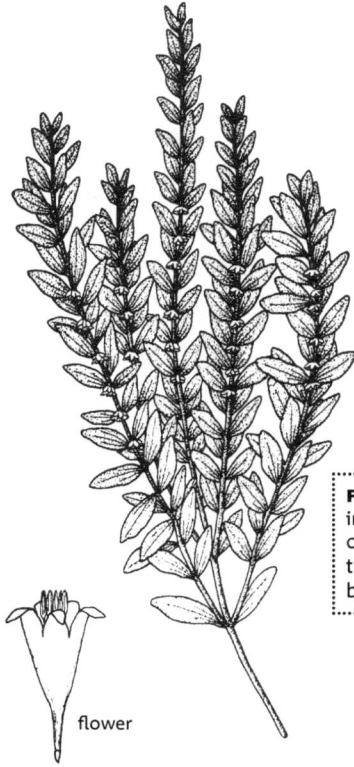

> **FIELD NOTES** Genus *Iva* with inconspicuous heads comprised only of greenish white, short, tubular flowers; bracts united at base to form a cup.

flower

Iva axillaris Pursh
SMALL-FLOWER SUMPWEED

FAMILY Asteraceae (Aster)

FLOWERING May–September

HABITAT Salt marshes, alkaline flats.

HABIT Native perennial herb with creeping rhizomes.

STEMS Upright, or lying flat before becoming upright, branched or unbranched, to 60 cm tall, smooth or hairy.

LEAVES Lower leaves opposite, upper leaves alternate, all simple, obovate to oblanceolate, to 5 cm long, to 20 mm wide, rounded or somewhat pointed at the tip, tapering to the nearly sessile or sessile base, without teeth, 3-veined, usually hairy.

FLOWERS Several crowded into nodding heads, with a single head in the axils of the leaves, each head to 6 mm across, consisting only of greenish white tubular flowers; bracts subtending each head united to form a cup. **Sepals** absent. **Petals** 5, greenish white, united to form a short, tubular flower. **Stamens** 5. **Pistils:** Ovary inferior.

FRUIT Achenes obovoid. smooth although sometimes glandular, to 3 mm long.

NOTE The outer flowers of the disk are only female, while the inner flowers have both stamens and pistils.

WETLAND STATUS AW FAC | GP FAC | WMV FAC

> **FIELD NOTES** Petals 5, blue, united to form a rotate flower, rather than a tubular one; stigmas attached to the sides of the ovary.

flower

Lomatogonium rotatum (L.) Fries
MARSH FELWORT

FAMILY Gentianaceae (Gentian)
FLOWERING August–September
HABITAT Wet meadows, bogs, often in salty habitats.
HABIT Native annual herb with fibrous roots.
STEMS Upright, usually unbranched, to 25 cm tall, smooth.
LEAVES Opposite, simple, those on the lower part of the stem oblong to oblanceolate, those on the upper part of the stem linear, to 4 cm long, smooth, without teeth.
FLOWERS Borne singly in the axils of the upper leaves, on slender, smooth stalks. **Sepals** 4–5, green, united only at the base, smooth, 8–13 mm long. **Petals** 4–5, blue or rarely white, united near the base but not forming a tube, pointed at the tip, 6–20 mm long. **Stamens** 4–5, attached to the base of the petals. **Pistils**: Ovary superior, smooth; **stigmas** attached to the sides of the ovary.
FRUIT Capsules ellipsoid to obconçoid, smooth; seeds numerous.
WETLAND STATUS AW OBL | GP OBL | WMV OBL

sepals

> **FIELD NOTES** *Lycopus* have tiny white flowers crowded in the axils of the opposite leaves. *L. uniflorus* differs by its triangular sepals, white tubers, and nearly smooth stems; leaves coarsely toothed but not lobed.

Lycopus uniflorus Michx.
NORTHERN BUGLEWEED

FAMILY Lamiaceae (Mint)

FLOWERING July–September

HABITAT Wet ground in woods, along streams, in meadows, bogs, marshes, around springs.

HABIT Native perennial herb with stolons bearing white tubers.

STEMS Upright, square, branched or unbranched, usually smooth, to 60 cm tall; stolons forming white tubers.

LEAVES Opposite, simple, lanceolate to oblong-lanceolate, to 6 cm long, pointed at the tip, tapering to the base, smooth, coarsely toothed.

FLOWERS Several crowded in dense clusters in the axils of the leaves, the clusters at maturity to 13 mm in diameter. **Sepals** 4 or 5, green, united below, the lobes triangular, smooth, rounded at the tip. **Petals** 5, unequal, united below, white, to 3 mm long. **Stamens** 2. usually exserted beyond the petals. **Pistils**: Ovary superior, 4-parted.

FRUIT Nutlets 4. brownish, about 1 mm long.

WETLAND STATUS AW OBL | GP OBL | WMV OBL

FIELD NOTES Small, succulent perennial; flowers solitary, axillary, sessile, with 5 petal-like sepals; petals absent.

flower

Lysimachia maritima (L.) Galasso, Banfi & Soldano
SEA-MILKWORT

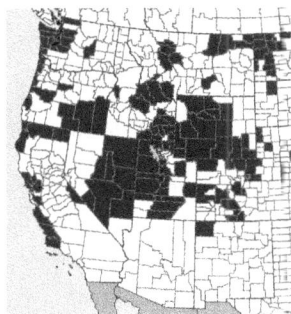

FAMILY Primrose (Primulaceae)
FLOWERING May–July
HABITAT Salt marshes.
HABIT Native perennial herb with slender rootstocks.
STEMS Upright, branched or unbranched, succulent, to 20 cm tall, smooth.
LEAVES Opposite, simple, fleshy, linear to oblong, to 13 mm long, pointed or rounded at the tip. tapering to the sessile base, smooth, without teeth.
FLOWERS Solitary in the axils of the leaves, sessile. **Sepals** 5, united below, lavender to white, to 4 mm long. **Petals** absent. **Stamens** 5. **Pistils:** Ovary superior, smooth.
FRUIT Capsules ovoid to nearly spherical, to 3 mm long, smooth, with several flattened, brown, pitted seeds.
SYNONYMS *Glaux maritima* L.
WETLAND STATUS AW FACW | GP OBL | WMV OBL

> **FIELD NOTES** Dwarf annual with rotate pink flowers; petals usually 4, shorter than the sepals.

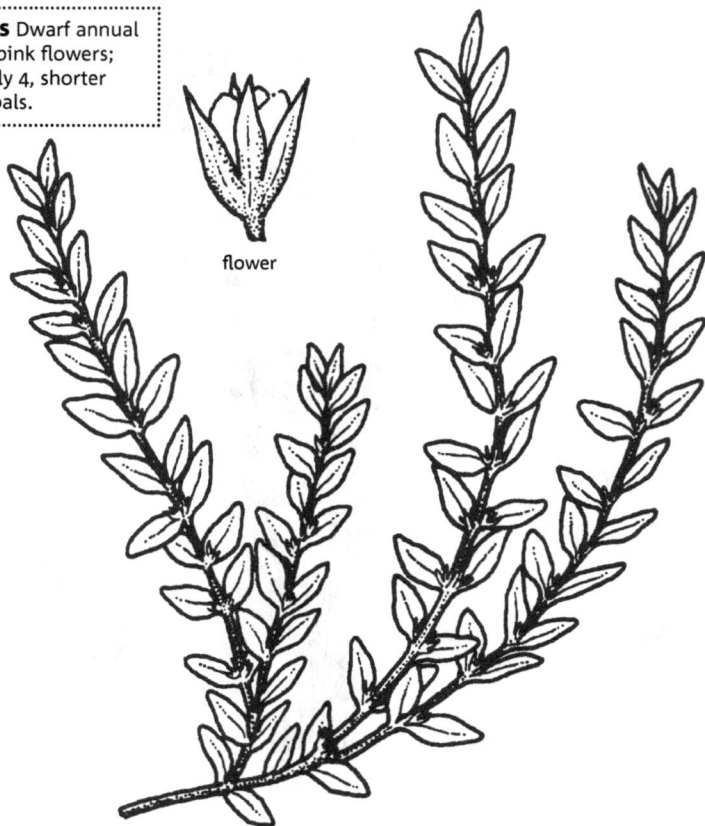

flower

Lysimachia minima (L.) U. Manns. & A. Anderb.
CHAFFWEED

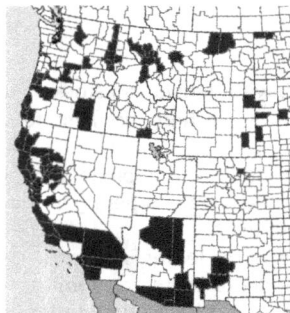

FAMILY Primrose (Primulaceae)

FLOWERING April–July

HABITAT Most moist areas, including vernal pools.

HABIT Native annual herb with fibrous roots.

STEMS Ascending, branched or unbranched, to 10 cm tall, smooth.

LEAVES Simple, alternate or opposite above, opposite near the base of the stem, spatulate to obovate. rounded or pointed at the tip, tapering to the base, without teeth, smooth.

FLOWERS Very tiny and solitary in the axils of the leaves, rotate. **Sepals** usually 4, green, united below, to 4 mm long. **Petals** usually 4, pink, united at the base, a little shorter than the sepals. **Stamens** usually 4. **Pistils**: Ovary superior, smooth.

FRUIT Capsules spherical, smooth.

SYNONYMS *Centunculus minimus* L.

WETLAND STATUS AW FACW | GP FACW | WMV FACW

> **FIELD NOTES** Tall, coarse perennial with densely flowered terminal red-purple spikes. Flowers with 6 free petals; leaves opposite; leaves and stems hairy.

Lythrum salicaria L.
PURPLE LOOSESTRIFE

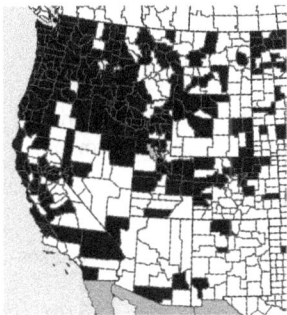

FAMILY Lythraceae (Loosestrife)

FLOWERING June–September.

HABITAT Marshes, margins of ponds, lakes, rivers, and streams.

HABIT Coarse perennial herb from a thickened rootstock.

STEMS 1-several erect, branched or unbranched, usually hairy, to 1 m or more tall.

LEAVES Opposite or sometimes in whorls of 3, simple, linear to lanceolate to oblong, pointed at the tip, rounded at the sessile base, without teeth, usually hairy, the largest to 10 cm long.

FLOWERS Crowded in terminal spikes to 50 cm long, reddish purple, the flowers subtended by green, leafy bracts. **Sepals** usually 6, green, united below to form a tube that is usually shorter than the petals, usually somewhat hairy. **Petals** usually 6, purple, free from each other, up to 13 mm long. **Stamens** usually 6. **Pistils:** Ovary superior.

FRUIT Capsules slightly longer than broad, containing numerous minute seeds.

NOTE This species, a native of Europe and formerly used as a landscape plant, is an aggressive invader of wetlands, eventually choking out much of the native vegetation.

WETLAND STATUS AW OBL | GP OBL | WMV OBL

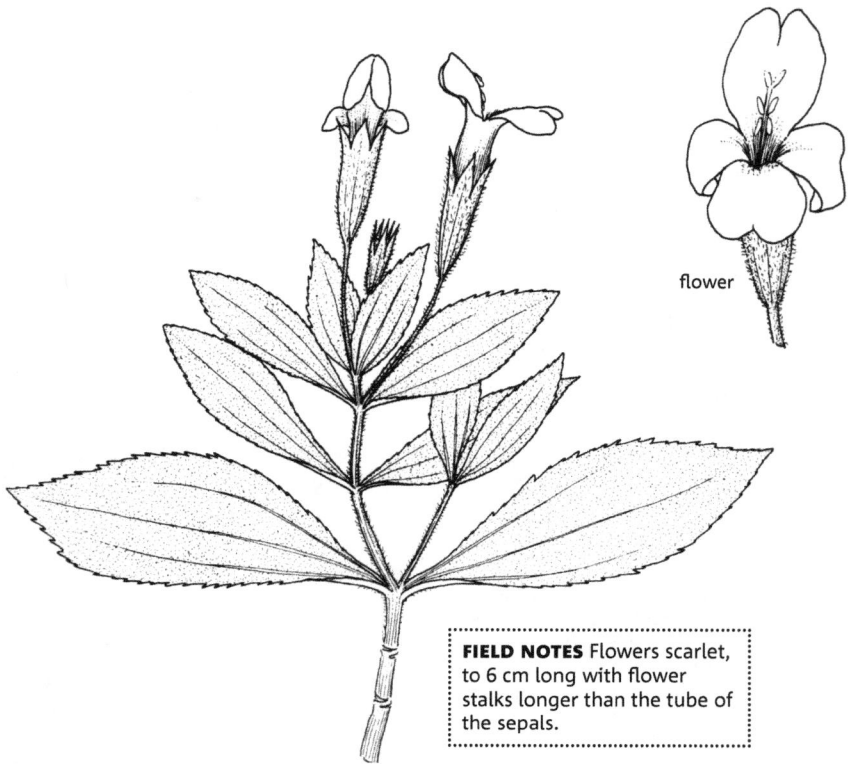

flower

> **FIELD NOTES** Flowers scarlet, to 6 cm long with flower stalks longer than the tube of the sepals.

Mimulus cardinalis Dougl. ex Benth.
SCARLET MONKEY-FLOWER

FAMILY Phrymaceae (Lopseed)

FLOWERING April–October

HABITAT Along streams, around springs, on wet cliffs, usually in the mountains.

HABIT Native perennial herb with creeping rhizomes.

STEMS Upright or ascending, branched, to 75 cm tall, sticky-hairy.

LEAVES Opposite, simple, obovate to oblong, to 8 cm long, sticky-hairy, pointed at the tip, tapering or rounded at the base, toothed, with 3–5 main veins, sessile and sometimes clasping the stem.

FLOWERS Several in racemes, showy, scarlet, to 6 cm long, on stalks 6–8 cm long. **Sepals** 5, green, united below into a tube, winged, the tube to 4 cm long, the teeth 2–6 mm long, pointed. **Petals** 5, united to form 2 lips, scarlet, with a yellow center and hairy yellow ridges. **Stamens** 4, extending beneath the upper lip of the flower. **Pistils:** Ovary superior, smooth.

FRUIT Capsules oblongoid, to 20 mm long, pointed at the tip, containing several narrow, pointed, wrinkled seeds.

NOTE This is one of our showiest wildflowers.

WETLAND STATUS AW FACW | WMV FACW

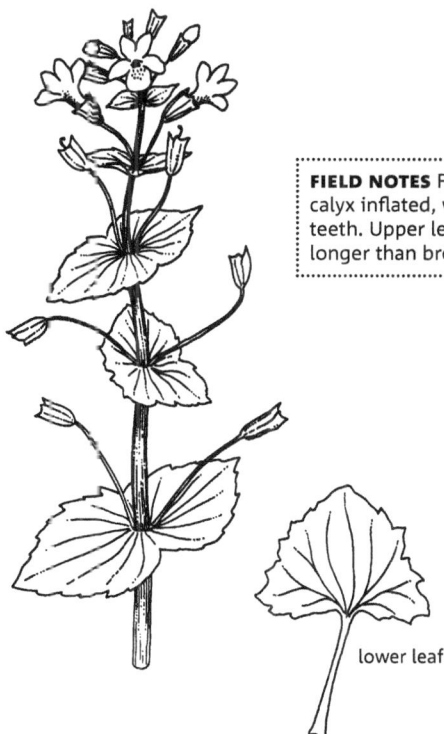

> **FIELD NOTES** Flowers yellow; calyx inflated, with pointed teeth. Upper leaves usually longer than broad.

lower leaf

Mimulus guttatus DC.
COMMON LARGE MONKEY-FLOWER

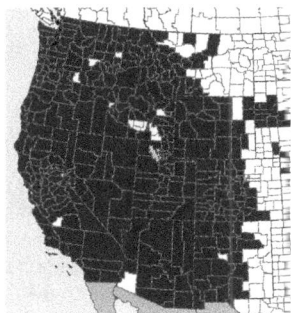

FAMILY Phrymaceae (Lopseed)

FLOWERING May–September

HABITAT Most wet, mucky places, sometimes growing in shallow standing water.

HABIT Native annual or perennial herb, sometimes rooting at the lower nodes, and sometimes bearing rhizomes and/or stolons.

STEMS Upright or ascending, branched or unbranched, to 75 cm tall, smooth or slightly hairy, sometimes hollow.

LEAVES Opposite, simple, broadly ovate to nearly round, the upper ones usually longer than broad, to 10 cm long, toothed, smooth or slightly hairy, the upper leaves sessile, the lower on stalks to 6 cm long.

FLOWERS Borne in a terminal raceme, yellow, showy. **Sepals** 5, green, united below to form a bell, to 20 mm long, each lobe triangular, but the middle one much longer than the others, and lanceolate. **Petals** 5, yellow, often spotted with red-brown, arranged to form 2 lips, one with 3 parts, the other with 2, to 30 mm long. **Stamens** 4, two longer than the others. **Pistils**: Ovary superior; **styles** 2-lobed.

FRUIT Capsules oblong to obovoid. narrowed to a short stalk at the base, to 13 mm long; seeds to 6 mm long, brown.

WETLAND STATUS AW OBL | GP OBL | WMV OBL

FIELD NOTES Plants 10–60 cm tall; sepals reddish; flowers yellow, 4–6 mm long, shed soon after opening; leaves and stems glandular-hairy.

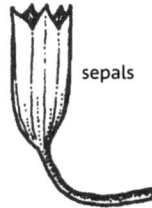

sepals

Mimulus suksdorfii Gray
SUKSDORF'S MONKEY-FLOWER

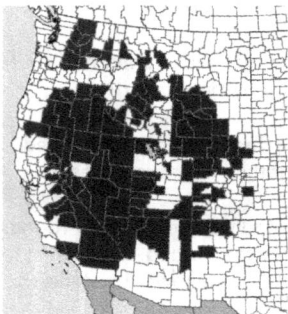

FAMILY Phrymaceae (Lopseed)

FLOWERING May–August

HABITAT Moist or dry areas in pinyon-juniper woodlands or in sagebrush.

HABIT Native annual herb with a slender taproot.

STEMS Upright or ascending, much branched, to 60 cm tall, glandular-hairy.

LEAVES Opposite, simple, linear to lanceolate, to 20 mm long, to 4 mm wide, pointed at the tip, tapering to the base, glandular-hairy, without teeth, often turning reddish, sessile or nearly so.

FLOWERS usually borne singly in the axils of the leaves, the often curved stalks 8–13 mm long. **Sepals** 5, green or reddish, united below to form a tube much longer than the lobes, glandular-hairy, the tube 3–6 mm long, the lobes less than 1 mm long, broadly ovate, not ciliate. **Petals** 5, yellow with red spots, more or less 2-lipped, united below to form a short tube, 4–8 mm long. **Stamens** 4. **Pistils:** Ovary superior, smooth; **stigmas** 2.

FRUIT Capsules ovoid, to 4 mm long, barely longer than the subtending sepals; seeds numerous, yellowish, oblongoid.

WETLAND STATUS AW FACU | GP FACW | WMV FACU

FIELD NOTES Flower heads single on each stem; sepals with very short teeth.

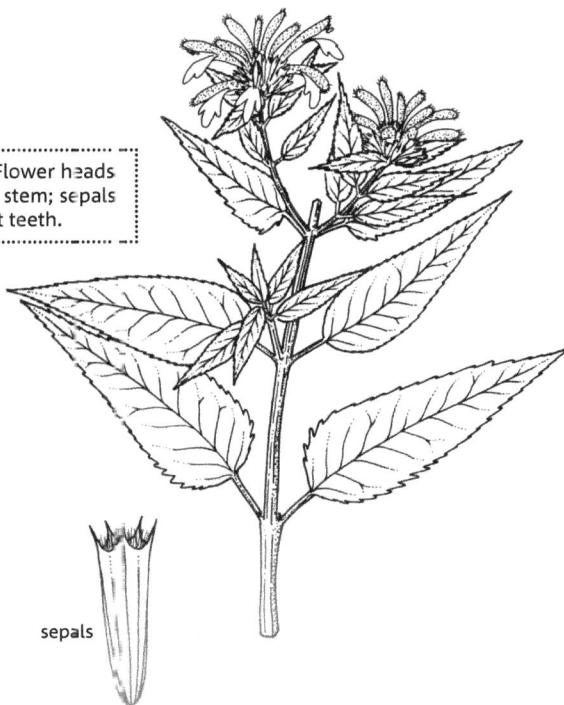

sepals

Monarda fistulosa L
WILD BERGAMOT

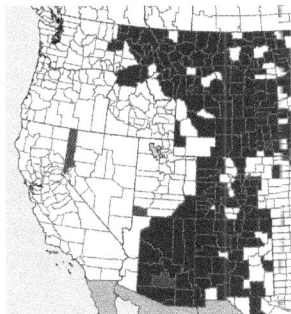

FAMILY Lamiaceae (Mint)

FLOWERING June–September

HABITAT A wide variety of habitats, including wet meadows, damp prairies, pastures, roadside ditches, along streams, around ponds, and in damp thickets.

HABIT Native, erect perennial, with slender, creeping rhizomes.

STEMS Upright, branched or unbranched, to 1 m tall, short-hairy on the upper part of the stem, usually smooth on the lower part.

LEAVES Opposite, simple, lanceolate to ovate, to 10 cm long, to 5 cm wide, pointed at the tip, truncate or tapering to the base, coarsely toothed, usually short-hairy, aromatic, dotted; leaf stalks to 2.5 cm long.

FLOWERS Numerous in single heads at the tips of the branches, each head to 4 cm across, each flower subtended by a green to pinkish bract that often curves backward; flowers to 4 cm long. **Sepals** 5, green, united below into a tube to 13 mm long, the 5 teeth spine-like, to 2 mm long. **Petals** 5, lavender to purple, united below into a tube, with 3 lobes on one side and 2 on the other, short-hairy on the outer surface. **Stamens** 2, protruding beyond the petals. **Pistils:** Ovary superior, 4-parted.

FRUIT Nutlets 4, oblong to obovoid, dark brown to black, to 2 mm long.

NOTE The nutlets are eaten by small mammals.

WETLAND STATUS AW FACU | GP UPL | WMV FACU

> **FIELD NOTES** Leaves opposite, oblong; flowers white or pink, with petals 6–8 mm long; sepals tiny unequal.

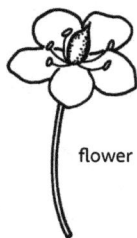

flower

Montia chamissoi (Ledeb. ex Spreng.) Greene
CHAMISSO'S MINER'S-LETTUCE

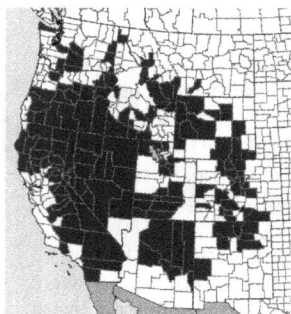

FAMILY Montiaceae (Candy-flower)
FLOWERING June–August
HABITAT Wet meadows, boggy areas, along streams.
HABIT Native perennial herb with creeping or floating stems; sometimes slender stolons bear small bulblets.
STEMS Creeping or floating, but with ascending branches to 15 cm long, smooth.
LEAVES Opposite, simple, oblong, to 5 cm long, to half as wide, smooth, without teeth or lobes.
FLOWERS Borne singly in the axils of the leaves, or 3–8 in racemes, each flower with a slender stalk eventually recurved in fruit, the stalk to 2.5 cm long. **Sepals** 2, free from each other, green, suborbicular, unequal in size, the large one about 4 mm long, persistent in fruit. **Petals** usually 5, free from each other, pink or white, 6–8 mm long. **Stamens** usually 5, attached to the base of the petals. **Pistils:** Ovary superior; **styles** 3.
FRUIT Capsules ovoid, to 2 mm long, smooth; seeds 1–3, black, less than 2 mm long, covered with very minute spines.
WETLAND STATUS AW OBL | GP OBL | WMV OBL

FIELD NOTES Sepals barely united or free from each other, papery in texture; leaves opposite.

flower with bracts

Nitrophila occidentalis (Moq.) S. Wats.
WESTERN BORAXWEED

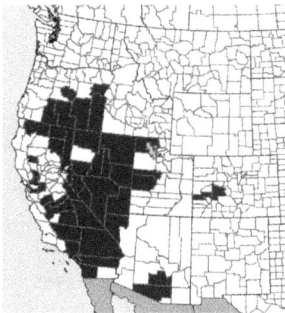

FAMILY Amaranthaceae (Amaranth)

FLOWERING May–October

HABITAT Moist alkaline habitats.

HABIT Native perennial herb with deep rootstocks.

STEMS Spreading but sometimes becoming upright, much branched, to 30 cm tall, smooth.

LEAVES Opposite, simple, linear to oblong, to 2.5 cm long, pointed at the tip. tapering to the sessile base, somewhat fleshy, smooth, without teeth.

FLOWERS Small, 1–3 in the axils of the leaves, subtended by a few short, leaf-like bracts. **Sepals** 5, green, free or nearly so from each other, oblong, to 2 mm long, pink at first, fading to straw-colored, somewhat papery in texture. **Petals** absent. **Stamens** 5, united at the base. **Pistils:** Ovary superior, smooth; **stigmas** 2.

FRUIT Ovoid, beaked, smooth, shorter than the subtending sepals, brown; seeds minute, black, shiny.

WETLAND STATUS AW FACW | WMV FACW

> **FIELD NOTES** Dwarf, matted perennial; sepals 5, no more than 2 mm long; petals usually slightly shorter.

flower with
two sepals removed

Sagina saginoides (L.) Karst.
ARCTIC PEARLWORT

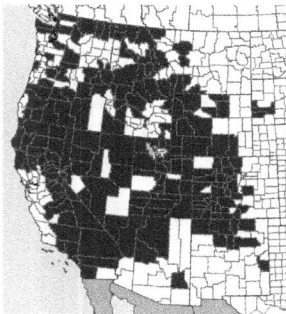

FAMILY Caryophyllaceae (Pink)

FLOWERING May–December

HABITAT Moist areas in the mountains.

HABIT Native perennial herb with a thickened root.

STEMS Lying flat on the ground to ascending, mat-forming, to 10 cm tall, smooth.

LEAVES Opposite, simple, thread-like to linear, 6–13 mm long, smooth, without teeth.

FLOWERS usually in the axils of the uppermost leaves, on slender stalks 6–13 mm long, smooth, often curved near the tip. **Sepals** 5, green, oval, to 2 mm long, rounded at the tip, green. **Petals** 5, white, oval, to 2 mm long, rounded at the tip. **Stamens** 5. **Pistils:** Ovary superior, smooth.

FRUIT Capsules ovoid, to 3 mm long, smooth.

WETLAND STATUS AW FACW | WMV FACW

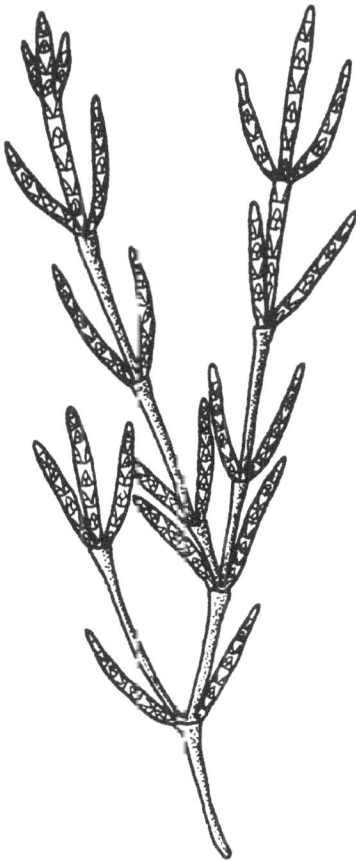

> **FIELD NOTES** Saltworts have leafless, jointed stems and flowers in fleshy cylindrical spikes. *Salicornia rubra* is an annual, with the joints of the spike longer than wide.

flower

Salicornia rubra A. Nels.
RED SALTWORT

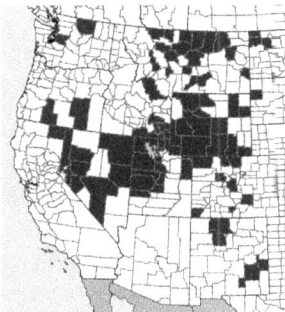

FAMILY Amaranthaceae (Amaranth)

FLOWERING July–November

HABITAT Moist alkaline or salty areas.

HABIT Native annual herb with fibrous roots.

STEMS Upright or ascending, branched from the base, to 30 cm tall, smooth, with opposite, jointed branchlets, with the joints longer than thick, usually turning reddish.

LEAVES Reduced to scales, opposite.

FLOWERS 1–7 in a group, sunken in the joints of succulent spikes, the joints longer than thick, the spikes to 6 cm long, to 3 mm thick, with the central flower higher than the others. **Sepals** united, barely with 3–4 teeth, rather fleshy. **Petals** absent. **Stamens** 1–2. **Pistils:** Ovary superior; styles 2.

FRUIT Oblongoid, more or less enclosed by the fleshy sepals; seeds yellow-brown, minutely hairy, about 2 mm long.

WETLAND STATUS AW OBL | GP OBL | WMV OBL

> **FIELD NOTES** Flowers in widely spreading cymes; styles 3; sepals smooth; leaves rough to the touch along the edges.

flower

Stellaria longifolia Muhl. ex Willd.
LONG-LEAF STARWORT

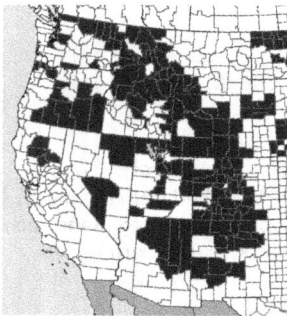

FAMILY Caryophyllaceae (Pink)

FLOWERING May–July

HABITAT Wet meadows, wet woods, fens, boggy areas, along streams.

HABIT Native sprawling perennial herb with slender roots.

STEMS Sprawling or becoming somewhat upright, square, smooth except for roughness on the angles, to 120 cm long.

LEAVES Opposite, simple, linear to narrowly lanceolate to elliptic, to 5 cm long, rarely to 8 mm wide, rough along the edges and often ciliate at the base.

FLOWERS Few in widely spreading cymes, subtended by ovate to lanceolate bracts to 6 mm long; flower stalks slender, spreading or reflexed, to 2.5 cm long. **Sepals** 5, free from each other, green, lanceolate, to 4 mm long, smooth. **Petals** 5, but appearing to be 10 because each one is deeply 2-lobed, free from each other, 4–6 mm long. **Stamens** 10. **Pistils:** Ovary superior; styles 3.

FRUIT Capsules yellow-brown to dark brown, ovoid, longer than the persistent sepals; seeds brown, oblongoid.

WETLAND STATUS AW FACW | GP FACW | WMV FACW

> **FIELD NOTES** Petals 5, white, slightly longer than the sepals and notched at tip. Flower stalks upright rather than curved. Capsules dark brown.

flower

Stellaria longipes Goldie
LONG-STALK STARWORT

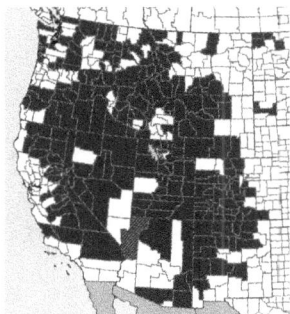

FAMILY Caryophyllaceae (Pink)

FLOWERING May–August

HABITAT Moist areas.

HABIT Native perennial, tufted herb with creeping rhizomes.

STEMS Erect or ascending, branched or unbranched, to 25 cm long, smooth or sparsely hairy.

LEAVES Opposite, simple, ascending, linear-lancolate, to 2.5 cm long, pointed at the tip, tapering to the base, smooth, without teeth.

FLOWERS 1 or few at the tip of the stem; stalks upright, not curved, slender, to 4 cm long. **Sepals** 5, green, free from each other, lanceolate to oblong, to 6 mm long, pointed at the tip, smooth. **Petals** 5, white, free from each other, a little longer than the sepals, notched at the tip. **Stamens** 10. **Pistils:** Ovary superior, smooth.

FRUIT Capsules narrowly ovoid, dark brown, longer than the sepals, smooth.

WETLAND STATUS AW FACW | GP FACW | WMV FACW

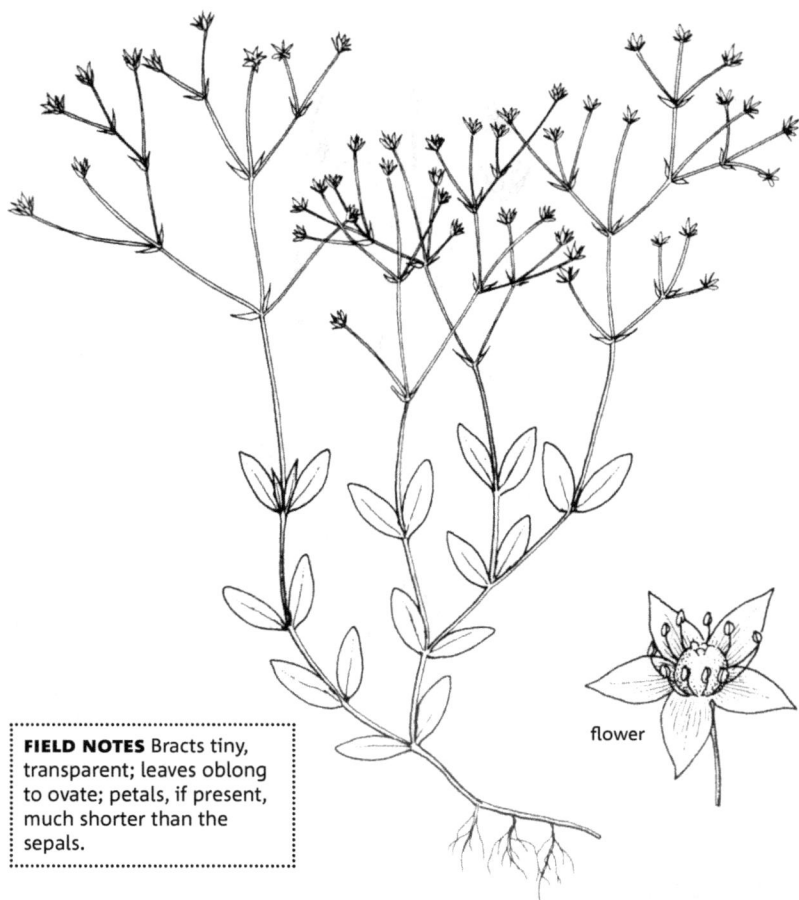

flower

> **FIELD NOTES** Bracts tiny, transparent; leaves oblong to ovate; petals, if present, much shorter than the sepals.

Stellaria umbellata Turcz. ex Kar. & Kir.
UMBELLATE STARWORT

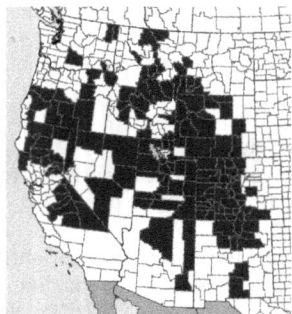

FAMILY Caryophyllaceae (Pink)
FLOWERING July–August
HABITAT Moist soil in woods and thickets.
HABIT Native perennial herb with slender rootstocks.
STEMS Sprawling to upright, weak, branched, to 30 cm long, smooth.
LEAVES Opposite, simple, oblong to ovate, to 2.5 cm long, pointed at the tip, tapering to the base, smooth except for cilia near the base.
FLOWERS Few to several in umbel-like cymes, subtended by tiny, transparent bracts; flower stalks very slender, smooth, recurved at the tip. **Sepals** 5, green with whitish borders, to 4 mm long, pointed at the tip. **Petals** 5 or absent, very tiny, much shorter than the sepals. **Stamens** 10. **Pistils:** Ovary superior; styles 3.
FRUIT Capsules oblongoid to ovoid, to 6 mm long, smooth; seeds many, light brown.
WETLAND STATUS AW FACW | GP FACW | WMV FACW

FIELD NOTES Flowers blue, in axillary racemes on stalks 6–13 mm long; leaves smooth, simple, on short stalks.

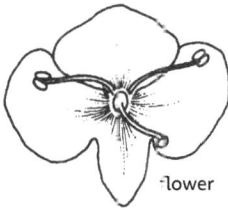

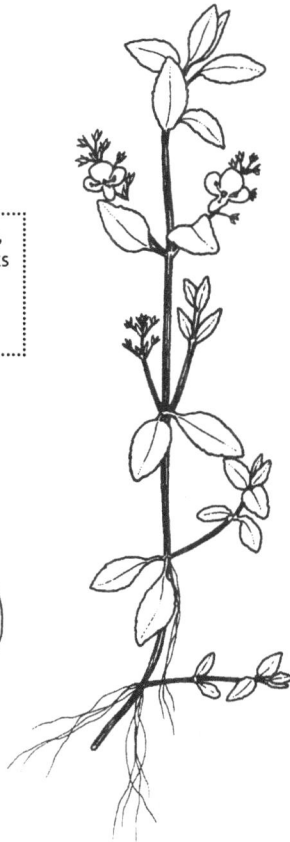

flower

Veronica americana Schwein. ex Benth.
AMERICAN SPEEDWELL

FAMILY Plantaginaceae (Plantain)

FLOWERING July–August

HABITAT Wet meadows, along streams, around lakes, around springs, often in water.

HABIT Native perennial herb with rhizomes.

STEMS Spreading to eventually upright, rooting at the nodes, to 60 cm long, smooth.

LEAVES Opposite, simple, lanceolate to ovate, to 5 cm long, to 30 mm wide, rounded or pointed at the tip, rounded at the base, toothed or without teeth, smooth.

FLOWERS 10–25 in axillary racemes, the racemes to 13 cm long; bracts small; flower stalks to 13 mm long.

Sepals 4, green, united at the base, to 6 mm long, the lobes lanceolate to ovate.

Petals 4, blue, united to form a short tube, 6–13 mm across, the lobes unequal in size.

Stamens 2, attached to the petals **Pistils:** Ovary superior, smooth.

FRUIT Capsules nearly spherical, to 4 mm in diameter, smooth, not notched at the tip; seeds numerous, very small, brownish.

WETLAND STATUS AW OBL | GP OBL | WMV OBL

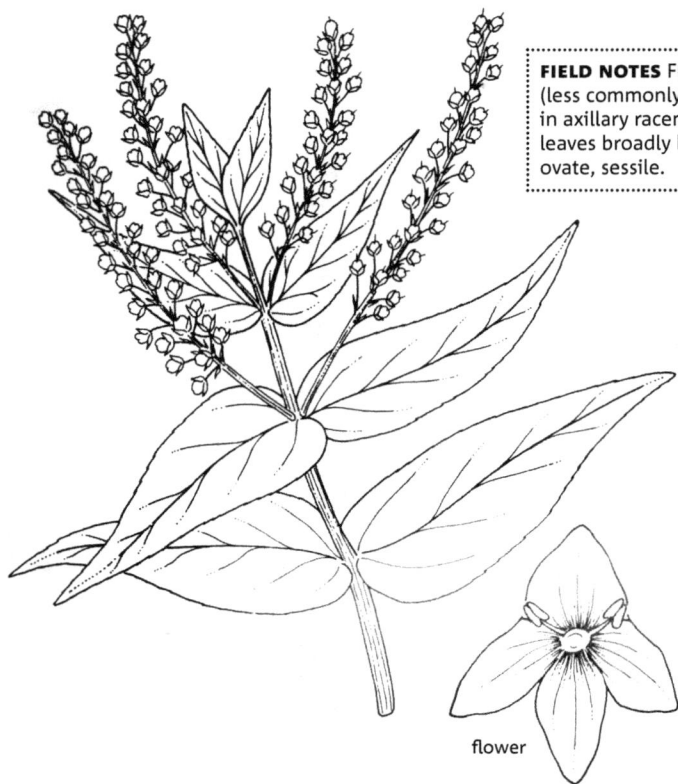

FIELD NOTES Flowers bluish (less commonly pinkish), borne in axillary racemes; upper leaves broadly lanceolate to ovate, sessile.

flower

Veronica anagallis-aquatica L.
WATER SPEEDWELL

FAMILY Plantaginaceae (Plantain)

FLOWERING June–August

HABITAT Wet areas along and in streams, wet meadows, marshes, ditches.

HABIT Introduced perennial herb from slender rhizomes.

STEMS Upright or ascending, to 60 cm long, branched or unbranched, smooth or sometimes glandular-hairy near the flowers.

LEAVES Opposite, simple, smooth, the uppermost sessile, broadly lanceolate to ovate, to 8 cm long, to 30 mm wide, pointed at the tip, rounded or heart-shaped at the base, the lowermost oblanceolate to obovate, rounded or pointed at the tip, sometimes with a short stalk.

FLOWERS Many in axillary racemes, each flower to 13 mm across, on a stalk 4–6 mm long. **Sepals** 4, green, united at the base, each segment lanceolate, to 6 mm long. **Petals** 4, bluish (less commonly pinkish), united below, the lobes unequal in size. **Stamens** 2. **Pistils:** Ovary superior.

FRUIT Capsules to 4 mm long, nearly as wide, with numerous minute seeds.

WETLAND STATUS AW OBL | GP OBL | WMV OBL

FIELD NOTES Leaves opposite, simple; petals 4, blue, united, stamens 2. Differing from other *Veronica* species in having terminal racemes, sepals equal in length, and stamens not protruding above the petals.

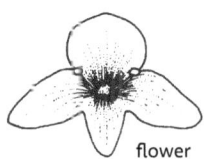

flower

Veronica wormskjoldii Roem. & J. A. Schultes
AMERICAN ALPINE SPEEDWELL

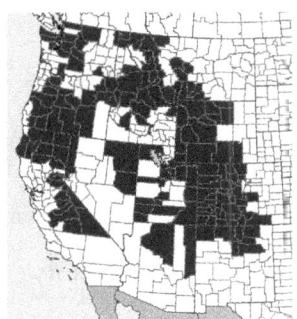

FAMILY Plantaginaceae (Plantain)
FLOWERING June–August
HABITAT Wet meadows, particularly in the high mountains.
HABIT Native perennial herb with slender rhizomes.
STEMS Upright, usually unbranched, to 30 cm tall, softly hairy or glandular-hairy.
LEAVES Opposite, simple, oval to ovate, to 30 mm long, pointed or rounded at the tip, rounded at the sessile or short-stalked base, round-toothed or without teeth, hairy.
FLOWERS Several in terminal racemes, on stalks to 6 mm long, each flower to 8 mm across, subtended by a small bract. **Sepals** 4, green, united below, rounded at the tip, to 4 mm long. **Petals** 4, blue, united below. **Stamens** 2, not protruding above the petals. **Pistils:** Ovary superior.
FRUIT Capsules ovoid, to 8 mm long, shallowly notched at the tip, bearing many minute seeds.
NOTE The seeds may be eaten by small mammals.
WETLAND STATUS AW FACW | GP FAC | WMV FACW

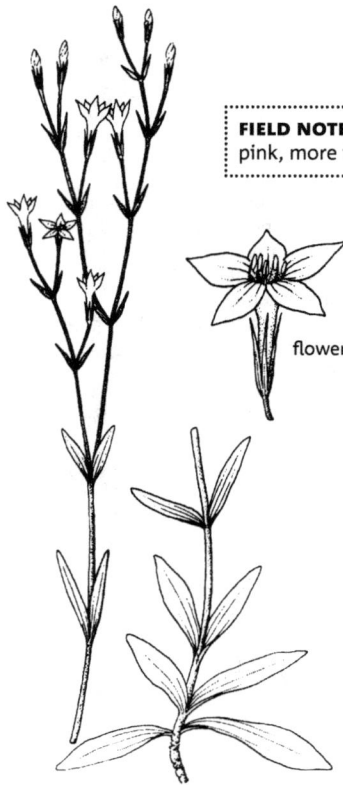

> **FIELD NOTES** Petal-lobes pink, more than 8 mm long.

flower

Zeltnera calycosa (Buckl.) G. Mans.
ARIZONA CENTAURY

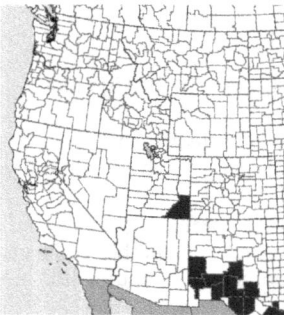

FAMILY Gentianaceae (Gentian)

FLOWERING June–September

HABITAT Along streams, in marshes.

HABIT Native annual or biennial herb with a thickened taproot.

STEMS Upright, unbranched or sparsely branched, to 60 cm tall, smooth. 4-sided, narrowly winged.

LEAVES Opposite, simple, elliptic to oblanceolate. to 8 cm long, more or less pointed at the tip, tapering to the sessile base, without teeth, smooth.

FLOWERS Few in a cyme, each flower on a slender stalk to 4 cm long. **Sepals** usually 5, united below, green, to 13 mm long, with very slender teeth. smooth. **Petals** usually 5, united below to form an elongated tube. pink, the tube to 13 mm long or longer, with a yellow center, the lobes more than 8 mm long. **Stamens** usually 5, exserted beyond the tube of the petals; anthers twisted after shedding pollen. **Pistils:** Ovary superior.

FRUIT Capsules cylindrical, 8–13 mm long, smooth, containing many small, nearly spherical, dark brown seeds.

SYNONYMS *Centaurium calycosum* (Buckl.) Fern.

WETLAND STATUS AW FACW | GP FACW | WMV FACW

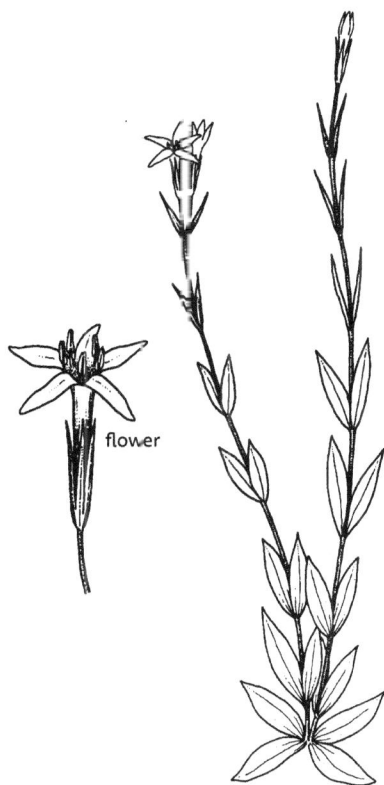

> **FIELD NOTES** *Zeltnera* flowers are pink or rose in color and form a slender tube below the 5 petal-lobes; anthers become twisted after releasing their pollen. Petal-lobes not more than 8 mm long, and all flowers borne on stalks.

flower

Zeltnera exaltata (Griseb.) G. Mans.
TALL CENTAURY

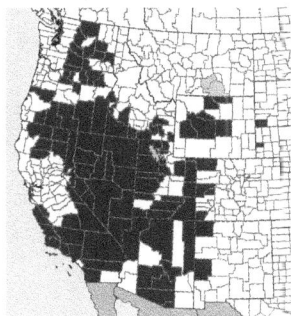

FAMILY Gentianaceae (Gentian)

FLOWERING June–August

HABITAT Along streams, in marshes, around hot springs, sometimes in alkaline soils.

HABIT Native annual herb with fibrous roots.

STEMS Upright, unbranched. slender, to 40 cm tall, smooth, often 4-sided and even narrowly winged.

LEAVES Opposite, simple, linear to lanceolate, pointed at the tip, tapering to the sessile base, to 5 cm long, to 20 mm wide, without teeth, smooth.

FLOWERS Few in a cyme, each flower on a slender stalk 3–8 cm long. Sepals 4 or 5, united below, green, to 13 mm long, with very slender teeth, smooth. Petals 4 or 5, united below to form an elongated tube, pink or rose, the tube to 13 mm long and with a greenish or yellowish center, the lobes to 8 mm long. Stamens 4 or 5, exserted beyond the tube of the petals; anthers twisted after shedding the pollen. Pistils: Ovary superior.

FRUIT Capsules cylindrical. 13–20 mm long, smooth, containing small, ovoid, dark brown seeds.

SYNONYMS *Centaurium exaltatum* (Griseb.) W. Wight ex Piper

WETLAND STATUS AW FACW | GF FACW | WMV FACW

petal

stamen

sepal

stigma
style
ovary

pistil

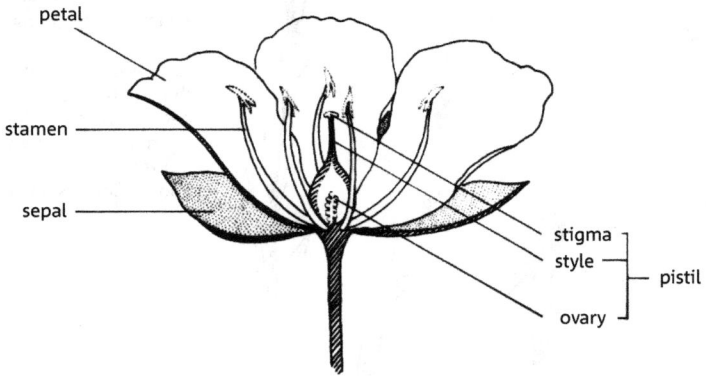

CROSS-SECTION OF A TYPICAL FLOWER

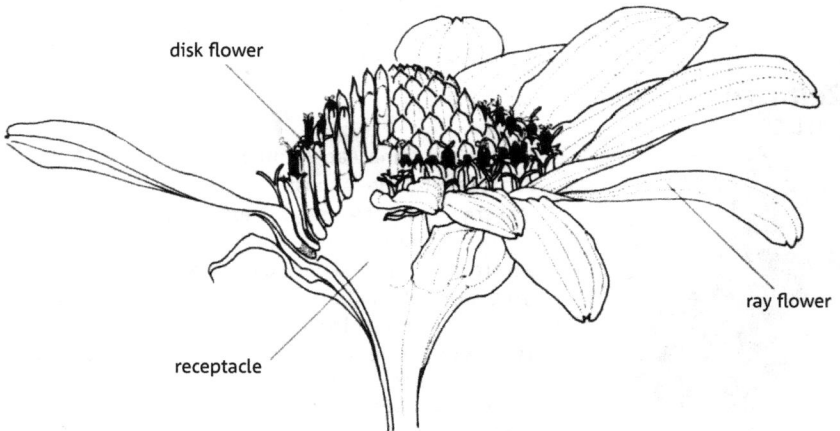

disk flower

ray flower

receptacle

CROSS-SECTION OF A TYPICAL COMPOSITE FLOWER (ASTERACEAE)

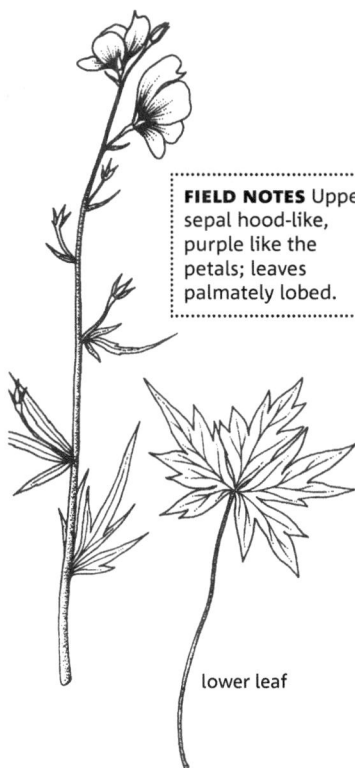

FIELD NOTES Upper sepal hood-like, purple like the petals; leaves palmately lobed.

lower leaf

Aconitum columbianum Nutt.
COLUMBIA MONKSHOOD

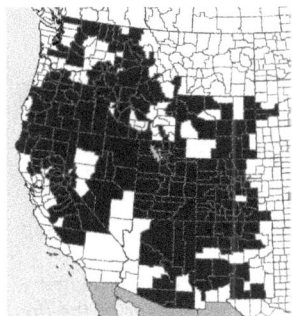

FAMILY Ranunculaceae (Buttercup)
FLOWERING June–August
HABITAT Wet meadows, moist woods.
HABIT Native perennial herb with a thickened rootstock and fibrous roots.
STEMS Upright, usually rather stout, to 2 m tall, smooth to hairy, the hairs sometimes glandular.
LEAVES Basal and alternate and cauline, palmately divided into 3–7 lobes, to 10 cm long, sometimes wider than long, smooth or hairy, the lobes toothed or more deeply cut; basal leaves with a stalk to 25 cm long; uppermost cauline leaves sessile.
FLOWERS Several in usually uncrowded racemes; racemes to 25 cm long; flower stalks to 13 mm long. **Sepals** 5, purple, usually slightly hairy, the uppermost forming a hood to 30 mm long; lateral sepals 2, oval, to 20 mm long; lower sepals 2, narrower, to 13 mm long. **Petals** 2, whitish, projecting beneath the hooded sepal and forming a coiled spur at its base. **Stamens** numerous, with broad filaments. **Pistils** 3, free from each other.
FRUIT 3 follicles, erect, smooth or hairy, to 20 mm long; seeds brown to black, delicately winged.
WETLAND STATUS AW FACW | GP FACW | WMV FACW

achene

> **FIELD NOTES** Flower heads solitary, burnt-orange, with only ray flowers; flowers fade to pink or purple as they dry; stems with milky sap.

Agoseris aurantiaca (Hook.) Greene
ORANGE-FLOWER FALSE-DANDELION

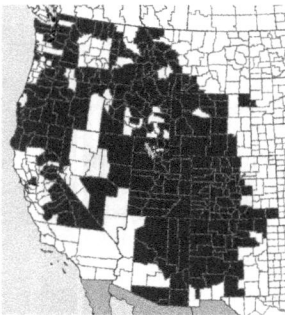

FAMILY Asteraceae (Aster)

FLOWERING June–August

HABITAT Wet meadows and fields.

HABIT Native perennial herb with a thickened rootstock.

STEMS Aerial stem bears only a flower head, to 45 cm tall, hairy, particularly just beneath the head; milky sap present.

LEAVES All basal, simple or with a few jagged segments, lanceolate to oblong, to 25 cm long, pointed at the tip, tapering to the base, more or less smooth except for the hairy midvein; milky sap present.

FLOWERS Crowded into solitary heads 2–2.5 cm across, consisting only of burnt orange ray flowers; each head subtended by several oblong to lanceolate, pointed bracts. Sepals absent. Petals 5, united to form burnt orange ray flowers 6–13 mm long. Stamens 5. Pistils: Ovary inferior.

FRUIT Achenes linear, 4–8 mm long, strongly ribbed, tapering to a slender beak, with a tuft of soft hairs at the tip of the beak.

WETLAND STATUS AW FAC | WMV FACU

> **FIELD NOTES** Plants much branched, lying flat on the ground; flowers in small, axillary clusters; each female flower has only one sepal.

flower

Amaranthus californicus (Moq.) S. Wats.
CALIFORNIA AMARANTH

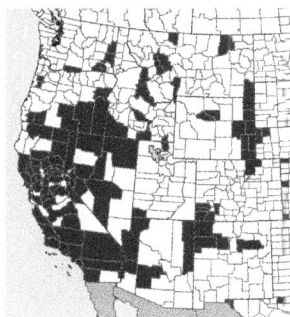

FAMILY Amaranthaceae (Amaranth)
FLOWERING July–October
HABITAT Moist mud or sand flats.
HABIT Native, mat-forming annual herb with a taproot.
STEMS Lying flat on the ground, much branched, forming mats to 45 cm across, often tinged with red.
LEAVES Alternate, simple, spatulate to obovate, to 20 mm long, pale green, often with a white border, smooth; leaf stalks slender, to 20 mm long.
FLOWERS Male and female flowers borne separately but on the same plant in small axillary clusters; bracts lance-olate, slender-tipped, about 1 mm long. **Sepals** 2 or 3 in the male flowers, free from each other, lanceolate, greenish; 1 in the female flower, about 1 mm long. **Petals** absent. **Stamens** 1–2. **Pistils**: Ovary superior.
FRUIT Nearly spherical, red or purple, to 2 mm in diameter; seeds round, red-brown.
WETLAND STATUS AW FACW | GP FACW | WMV FACW

FIELD NOTES Leaves broad, heart-shaped at base; inflorescence to 8 cm wide, with 4–8 conspicuous white bracts.

Anemopsis californica (Nutt.) Hook. & Arn.
YERBA MANSA

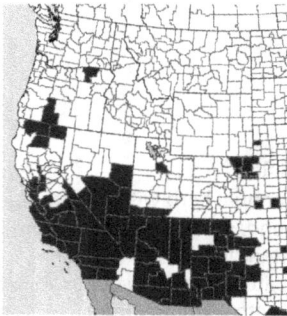

FAMILY Saururaceae (Lizard's-tail)

FLOWERING March–September

HABITAT Wet meadows, marshes, swamps, along streams, sometimes in alkaline areas.

HABIT Native perennial herb with thick, creeping, aromatic rhizomes.

STEMS Upright, to 45 cm tall, woolly, bearing a terminal group of flowers, one large leaf on the stem, and a few basal leaves.

LEAVES Basal leaves elliptic to oblong, to 15 cm long, rounded at the tip, heart-shaped at the base, somewhat hairy, on stalks to 15 cm long; one stem leaf ovate, pointed at the tip, clasping at the base, with 1–3 smaller leaves in the axil.

FLOWERS Few, terminal in spikes, surrounded by 4–8 white, petal-like bracts to 4 cm long, the entire structure to 8 cm across. **Sepals** absent. **Petals** absent. **Stamens** 6 or 8. **Pistils** 3 or 4. free or united at the base, smooth, with the ovary superior.

FRUIT Capsules cone-like, rusty-colored, with numerous seeds.

NOTE Despite the absence of sepals and petals, this species has showy, anemone-like flowers because of the conspicuous white bracts.

WETLAND STATUS AW OBL | GP FACW | WMV OBL

> **FIELD NOTES** Flower head bracts with a conspicuous black spot near their middle.

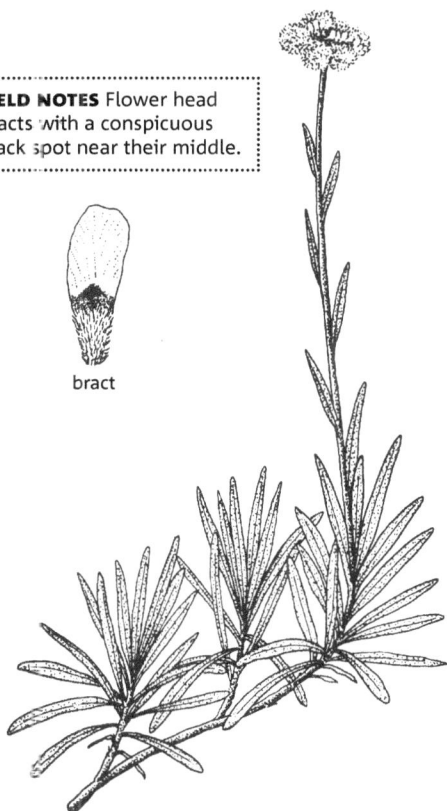

bract

Antennaria corymbosa E. Nels.
FLAT-TOP PUSSY-TOES

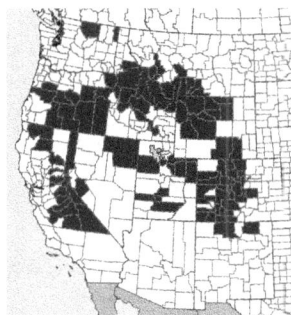

FAMILY Asteraceae (Aster)

FLOWERING June–August

HABITAT Wet meadows in the mountains.

HABIT Native perennial, mat-forming herb from a slender rootstock.

STEMS Both spreading stolons and upright stems present, to 25 cm long, hairy.

LEAVES Basal and alternate on the stem, simple, oblanceolate to spatulate, to 5 cm long, to 6 mm wide, pointed at the tip, tapering to the base, hairy but greenish.

FLOWERS Several crowded together into heads, the male and female flowers borne on separate plants; heads several in a terminal cluster, subtended by bracts; bracts woolly, with a conspicuous black spot near the middle. **Sepals** absent. **Petals** white, those of the male flowers united into thread-like rays, those of the female flowers united into a tube with 5 teeth. **Stamens** 5. **Pistils**: Ovary inferior; **styles** in the female flowers deeply 2-cleft.

FRUIT Achenes minutely hairy, topped by slender bristles.

WETLAND STATUS AW FAC | WMV FAC

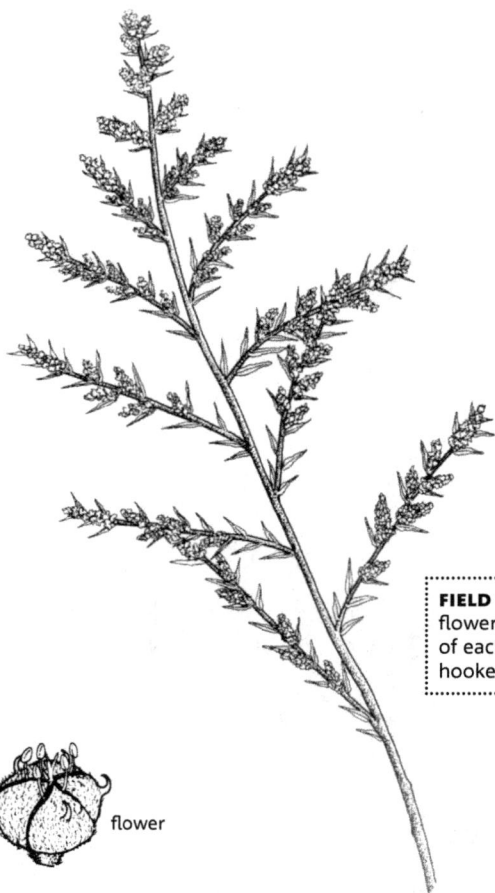

FIELD NOTES Annual herb; flowers inconspicuous; tip of each sepal with a short, hooked spine.

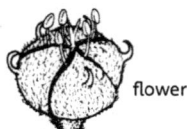

flower

Bassia hyssopifolia (Pallas) Kuntz
FIVE-HORN SMOTHERWEED

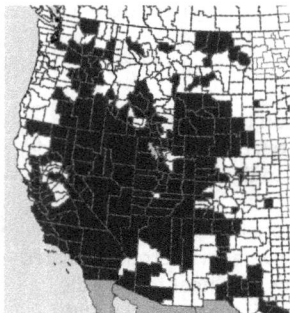

FAMILY Amaranthaceae (Amaranth)
FLOWERING July–September
HABITAT Alkaline flats.
HABIT Introduced annual herb with fibrous roots.
STEMS usually upright, much branched, to 20 cm tall, smooth or hairy.
LEAVES Alternate, simple, linear to linear-lanceolate, to 5 cm long, pointed at the tip, tapering to the base, hairy, without teeth.

FLOWERS Several crowded into terminal or axillary spikes. **Sepals** 5, green, united below, hairy, each lobe about 2 mm long, with a short, hooked spine at or near the tip. **Petals** absent. **Stamens** 5. **Pistils:** Ovary superior; **stigmas** 2.

FRUIT Flattened, brown to black, to 2 mm across, enclosed by the enlarged, hairy sepals.

WETLAND STATUS AW FACU | GP FACU | WMV FACU

> **FIELD NOTES** Annual herb; flowers greenish, petals absent, borne in the axils of the leaves; leaves without teeth.

flower

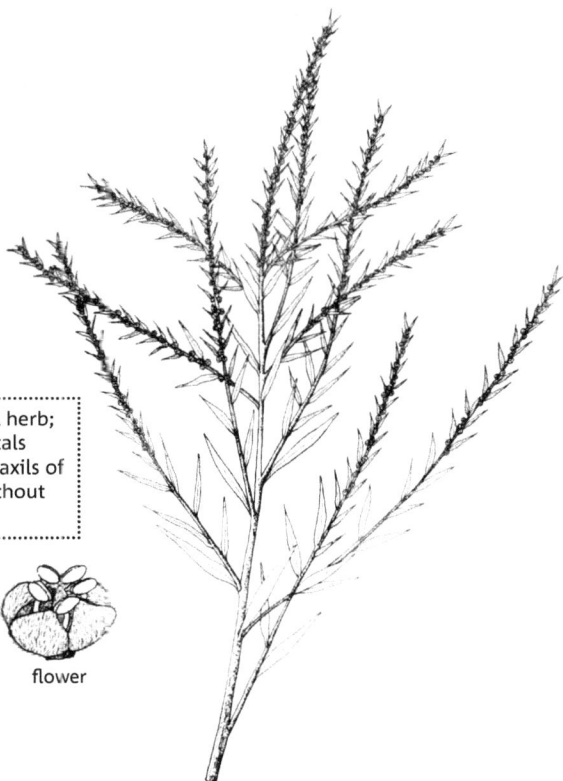

Bassia scoparia (L.) A. J. Scott
MEXICAN SUMMER-CYPRESS

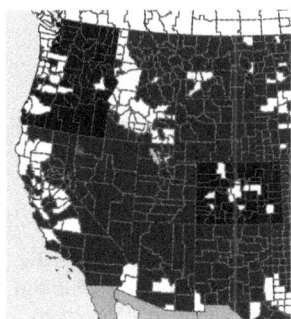

FAMILY Amaranthaceae (Amaranth)

FLOWERING July–October

HABITAT Disturbed areas.

HABIT Introduced annual herb with fibrous roots.

STEMS Upright, branched, to 1 m tall, usually hairy but occasionally nearly smooth.

LEAVES Alternate, simple, linear to lanceolate to obovate, to 10 cm long, to 13 mm wide, flat, hairy or smooth, pointed or rounded at the tip, tapering to the base.

FLOWERS Borne in clusters in the axils of the leaves. Sepals 5, green, united to form a minute cup, ciliate, curving over the fruit at maturity. Petals absent. Stamens 5. Pistils: Ovary superior; styles usually 2, rarely 3.

FRUIT Spherical but somewhat flattened, containing an obovoid seed 3–4 mm long, the seed brown or black, not shiny, smooth or granular.

NOTE This introduced species is also known as fireweed as plants turn reddish in autumn. This species varies in degree of hairiness and in its size and shape.

SYNONYMS *Kochia scoparia* (L.) Schrad.

WETLAND STATUS AW FAC | GP FACU | WMV FAC

> **FIELD NOTES** Basal leaves long-stalked; stems unbranched; raceme solitary, spike-like.

flower and bract

Bistorta bistortoides (Pursh) Small
AMERICAN BISTORT

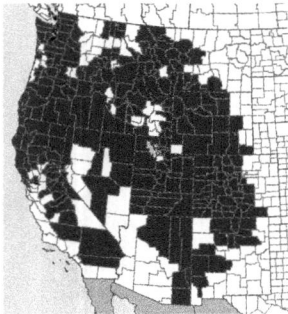

FAMILY Polygonaceae (Buckwheat)

FLOWERING May–August

HABITAT Along streams, in wet meadows, particularly in the mountains.

HABIT Native perennial herb from a thick, fleshy root-stock.

STEMS Upright, unbranched. to 45 cm tall, smooth.

LEAVES Basal and alternate, smooth, without teeth, the lowermost oblong to oblanceolate, pointed or rounded at the tip, to 25 cm long, to 5 cm wide, on stalks to 20 cm long; uppermost leaves similar but sessile.

FLOWERS Many crowded into a terminal, solitary, cylindrical, spike-like raceme; raceme to 8 cm long, to 20 mm wide; flower stalk 3–8 mm long. **Sepals** 5, united at the base, pink or white, petal-like, to 6 mm long. **Petals** absent. **Stamens** usually 5–9, exserted beyond the sepals. **Pistils:** Ovary superior.

FRUIT Achenes triangular, pale brown, smooth, shiny, to 4 mm long.

SYNONYMS *Polygonum bistortoides* Pursh

WETLAND STATUS AW FACW | GP OBL | WMV FACW

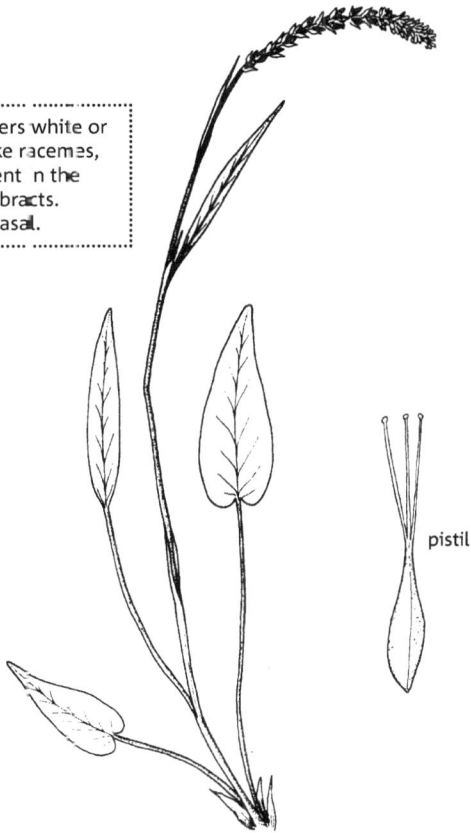

> **FIELD NOTES** Flowers white or pinkish, in spike-like racemes, with bulblets present in the axils of the lowest bracts. Leaves almost all basal.

pistil

Bistorta vivipara (L.) Delarbre
VIVIPAROUS KNOTWEED

FAMILY Polygonaceae (Buckwheat)
FLOWERING June–August
HABITAT Moist habitats, sometimes in alkaline soil.
HABIT Native perennial herb with an erect rhizome.
STEMS Upright, unbranched, to 30 cm tall, smooth.
LEAVES Almost all basal, linear to oblong-lanceolate, to 10 cm long, to 20 mm wide, rounded or pointed at the tip, more or less heart-shaped at the base, smooth, on stalks as long as the blades; leaves on the stem few, alternate, linear-lanceolate, the uppermost sessile.
FLOWERS Several in a single spike-like raceme to 6 cm long, the lower bracts bearing reddish bulblets in their axils. **Sepals** 5, white or pink, united at the base. **Petals** absent. **Stamens** usually 5, exserted beyond the sepals. **Pistils:** Ovary superior; **styles** 3.
FRUIT Achenes triangular, dark brown, not shiny.
NOTE Most of the flowers fail to produce viable seeds.
SYNONYMS *Polygonum viviparum* L.
WETLAND STATUS AW FAC | GP FACW | WMV FAC

> **FIELD NOTES** Pods strictly
> erect, slender, 6–13 cm
> long; basal leaves toothed
> but not lobed; stems
> smooth or only slightly
> hairy .

Boechera stricta (Graham) Al-Shehbaz
CANADIAN ROCKCRESS

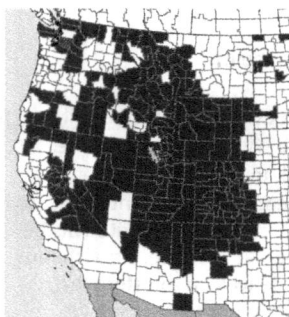

FAMILY Brassicaceae (Mustard)

FLOWERING June–August

HABITAT Rocky woods, moist slopes.

HABIT Native biennial herb with a thickened rootstock.

STEMS Upright, branched or unbranched, to 60 cm tall, smooth or only slightly hairy, sometimes glaucous.

LEAVES Both cauline and basal, linear-lanceolate to oblanceolate, pointed at the tip, tapering to the base, to 8 cm long, smooth or sparsely hairy, often toothed, the basal leaves with stalks, the cauline leaves sessile and sometimes clasping.

FLOWERS Several in a terminal cluster, each on a smooth stalk to 2.5 cm long. **Sepals** 4, green, free from each other, to 6 mm long. **Petals** 4, white, rarely pinkish, free from each other, 6-12 mm long. **Stamens** 6. **Pistils:** Ovary superior.

FRUIT Pods numerous, crowded, strictly erect, 5–13 cm long; seeds oblongoid, winged at each end.

NOTE The seeds are eaten by small mammals.

SYNONYMS *Arabis drummondii* Gray

WETLAND STATUS AW FACU | GP FACU | WMV FACU

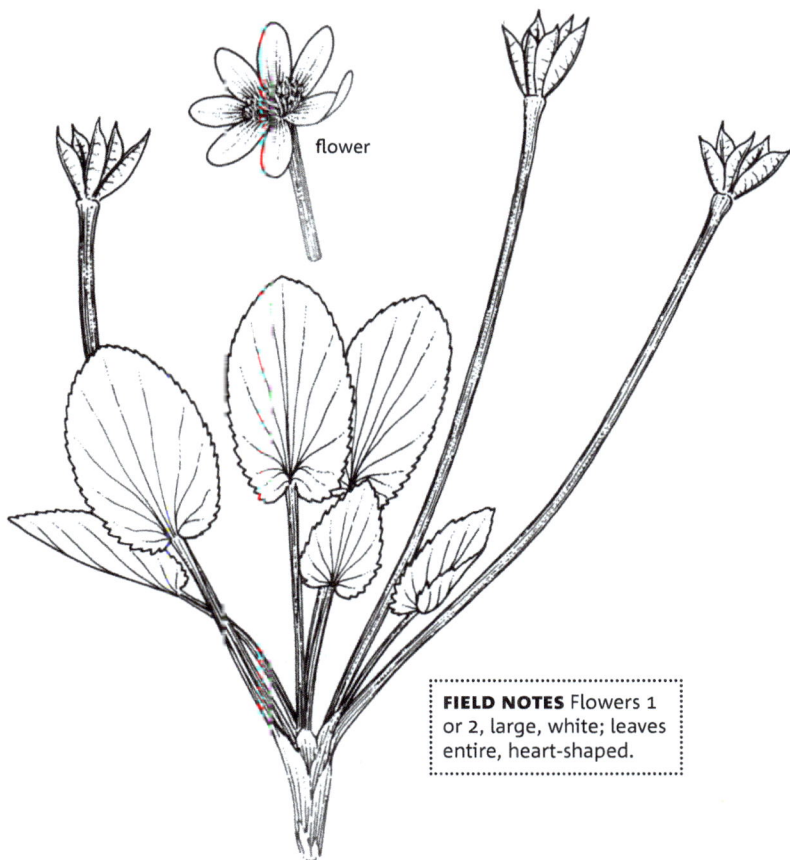

flower

> **FIELD NOTES** Flowers 1 or 2, large, white; leaves entire, heart-shaped.

Caltha leptosepala DC.
WHITE MARSH-MARIGOLD

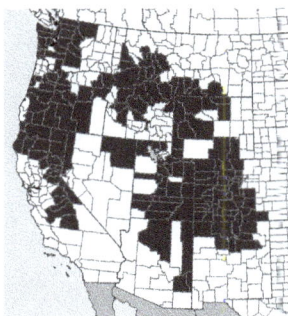

FAMILY Ranunculaceae (Buttercup)

FLOWERING June–August

HABITAT Wet meadows and bogs in the mountains.

HABIT Native perennial herb with thick fibrous roots.

STEMS Upright, either bearing no leaves and one flower or 1 leaf and 2 flowers, to 25 cm tall, smooth, somewhat succulent.

LEAVES Basal or with only one on the stem, ovate to orbicular, to 8 cm long, rounded at the tip, deeply heart-shaped at the base, regularly toothed, smooth, on stalks.

FLOWERS Solitary or in pairs, to 6 cm across, white. Sepals 6–12, white, petal-like, free from each other, to 2.5 cm long. **Petals** absent. **Stamens** numerous. **Pistils** many, each with a superior ovary.

FRUIT Follicles several, to 13 mm long, containing very many seeds.

NOTE This plant begins to flower after the first thaw in the spring.

WETLAND STATUS AW OBL | GP OBL | WMV OBL

flower

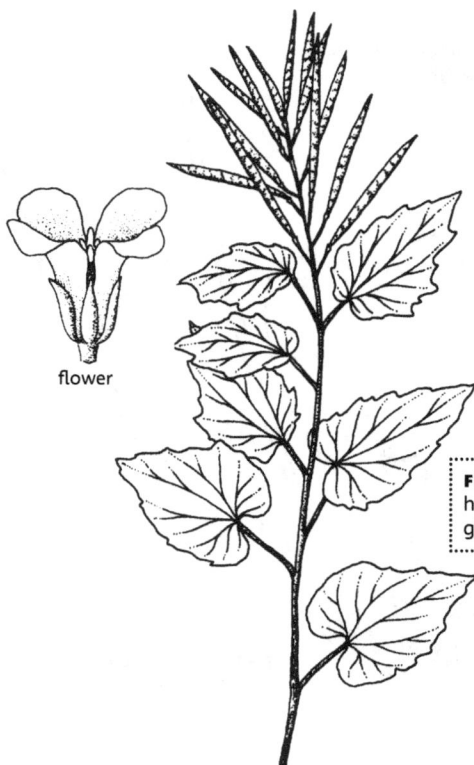

> **FIELD NOTES** Leaves simple, heart-shaped at base, margins wavy.

Cardamine cordifolia Gray
HEART-LEAF BITTERCRESS

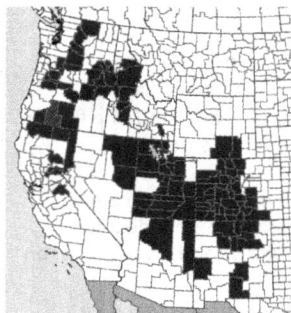

FAMILY Brassicaceae (Mustard)

FLOWERING June–August

HABITAT Along streams, in fens.

HABIT Native perennial herb from a slender rootstock.

STEMS Upright, to 45 cm tall, smooth or with some soft hairs.

LEAVES Alternate, simple, ovate to nearly round, to 8 cm long, often nearly as broad, pointed or rounded at the tip, heart-shaped at the base, smooth or with soft hairs, wavy or toothed along the edges, on slender stalks.

FLOWERS Several in dense racemes, each flower on a slender, spreading stalk to 20 mm long. **Sepals** 4, green, free from each other, about 3 mm long. **Petals** 4, white, free from each other, about 8 mm long. **Stamens** 6. **Pistils:** Ovary superior.

FRUIT Pods elongated, very narrow, to 4 cm long, about 1 mm wide, containing 8–12 seeds.

WETLAND STATUS AW FACW | GP OBL | WMV FACW

bract flower

> **FIELD NOTES** Bracts, showy,
> bright scarlet, subtending
> the 25–35 mm long petals;
> lower lip of petals short,
> only about 2 mm long.

Castilleja miniata Dougl. ex Hook.
SCARLET INDIAN-PAINTBRUSH

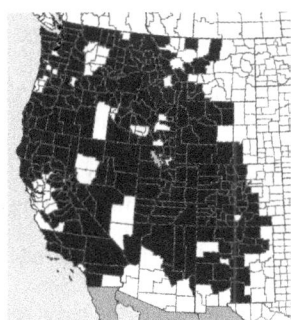

FAMILY Orobanchaceae (Broom-rape)
FLOWERING May–September
HABITAT Along streams, wet meadows, thickets, forest openings.
HABIT Native perennial herb with thickened rootstocks.
STEMS Upright, branched or unbranched, to 75 cm tall, smooth or somewhat hairy.
LEAVES Alternate, simple, lanceolate, to 6 cm long, to 20 mm wide, without teeth, or the uppermost leaves with a pair of lobes, smooth or somewhat hairy.
FLOWERS Crowded together in a spike, each flower subtended by a scarlet bract; uppermost bracts with a pair of lobes. **Sepals** 4, united into a cleft tube, usually scarlet, to 30 mm long. **Petals** 5, greenish with red edges, united below into a tube, to 4 cm long, 2-lipped, the upper lip as long as the tube, the lower lip about 2 mm long. **Stamens** 4. **Pistils:** Ovary superior.
FRUIT Capsules ovoid, about 13 mm long.
NOTE The bracts are sometimes orange or red-purple.
WETLAND STATUS AW FACW | GP FAC | WMV FACW

> **FIELD NOTES** Plant usually in alkaline habitats; leaves red or red-tinged, smooth; sepals 3.

Chenopodium rubrum L.
COAST-BLITE GOOSEFOOT

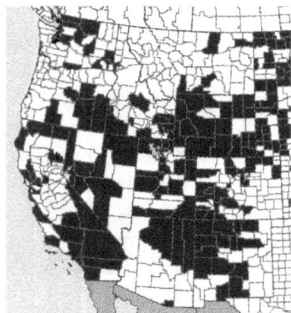

FAMILY Amaranthaceae (Amaranth)

FLOWERING June–September

HABITAT readily distinguish this species from all others.

HABITAT Salt marshes, alkaline areas.

HABIT Native annual herb with a taproot.

STEMS Upright or lying flat, branched or unbranched, to 1 m long, smooth.

LEAVES Alternate, simple, ovate to oblong, to 13 cm long, usually nearly as wide, pointed at the tip, tapering to the base, the lowest coarsely jagged toothed and stalked, the upper without teeth and sessile, red or red-tinged, smooth.

FLOWERS Crowded into spikes forming a panicle. **Sepals** 3, green, united at the base, to 2 mm long. **Petals** absent. **Stamens** usually 5. **Pistils:** Ovary superior.

FRUIT Minute, spherical, about 1 mm in diameter, smooth, the seeds dark brown and shiny.

NOTE The seeds of this plant are eaten by rodents and other small mammals.

SYNONYMS *Oxybasis rubra* (L.) S. Fuentes, Uotila & Borsch

WETLAND STATUS AW FACW | GP OBL | WMV FACW

WESTERN WETLAND FLORA

> **FIELD NOTES** Leaves simple;
> bracts toothless or have
> only shallow lobes; petals
> shorter than the sepals.

flower

Chloropyron maritimum (Benth.) Heller
SALTMARSH BIRD'S-BEAK

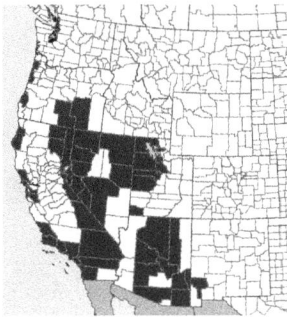

FAMILY Orobanchaceae (Broom-rape)

FLOWERING June–August

HABITAT Salt marshes, alkaline meadows, hot springs.

HABIT Native annual herb with yellow roots.

STEMS Upright or spreading, much branched, to 50 cm tall, sticky-hairy to nearly smooth.

LEAVES Alternate, simple, oblong, narrowly lanceolate to lanceolate, to 2.5 cm long, pointed at the tip, tapering to the base, bluish green, hairy or occasionally nearly smooth, usually without teeth.

FLOWERS Crowded in terminal spikes, the flowers subtended by bracts that appear sepal-like; bracts without teeth or only shallowly lobed, narrowly oblong, bluish green, to 2.5 cm long. **Sepals** United to form a single structure split nearly to the base, 13–25 mm long, green, notched at the tip. **Petals** 5, united to form 2 lips, 2–2.5 cm across, white or yellow, striped with purple or red, and tipped with yellow, pink, or purple, minutely hairy on the back. **Stamens** 4, attached to the petals, the anthers with a tuft of hairs at the base. **Pistils**: Ovary superior.

FRUIT Capsules ellipsoid, 8–13 mm long; seeds to 2 mm long, curved, brown.

NOTE The increasing rarity of subsp. *maritimus* along the coast has resulted in its listing as a Federally endangered plant. Plants of the Great Basin (subsp. *canescens*), which have usually toothless bracts and capsules with more seeds, are more common and are not Federally listed.

SYNONYMS *Cordylanthus maritimus* Nutt. ex Benth.

WETLAND STATUS AW OBL | WMV OBL

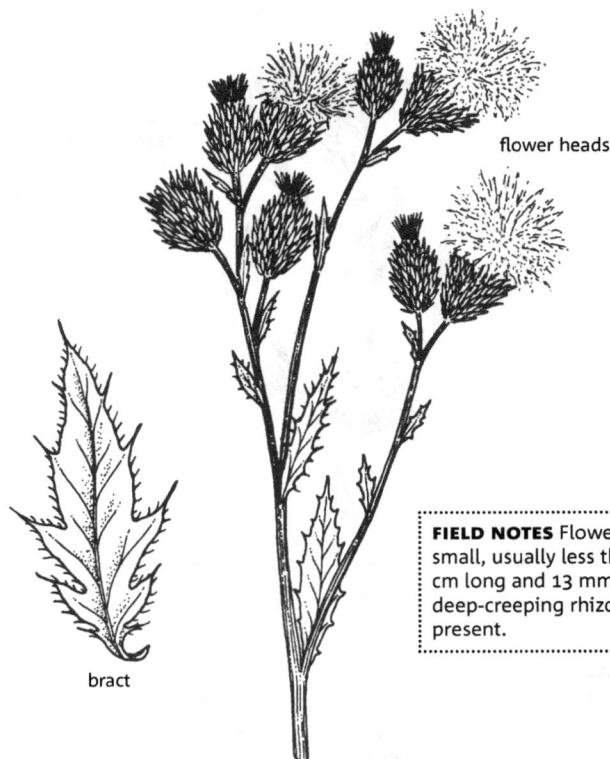

flower heads

> **FIELD NOTES** Flower heads small, usually less than 2.5 cm long and 13 mm wide; deep-creeping rhizomes present.

bract

Cirsium arvense (L.) Scop.
CANADIAN THISTLE

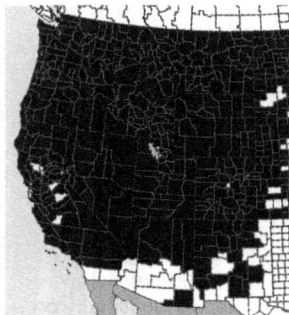

FAMILY Asteraceae (Aster)

FLOWERING June–August

HABITAT Bottomland areas, ditches, old fields, disturbed areas.

HABIT Introduced perennial herb from a thickened, deep, creeping rootstock.

STEMS Upright, branched, to 1 m tall, usually covered with white, cobwebby hairs.

LEAVES Alternate, simple, usually pinnately lobed, to 15 cm long, to 8 cm wide, smooth or white-hairy, the lobes bearing sharp spines; leaf stalk to 13 mm long or absent.

FLOWERS Many crowded together into a head, with several heads per plant; each head less than 2.5 cm long, less than 13 mm across, subtended by 5–6 rows of small spine-tipped, green bracts; flowers all tubular. **Sepals** absent. **Petals** 5, pink or purple, forming a tube to 2.5 cm long. **Stamens** 5. **Pistils:** Ovary inferior.

FRUIT Achenes pale brown, topped by a cluster of white or gray plumose bristles to 2.5 cm long.

NOTE The female and male flowers are usually borne in separate heads on separate plants.

WETLAND STATUS AW FACU | GP FACU | WMV FAC

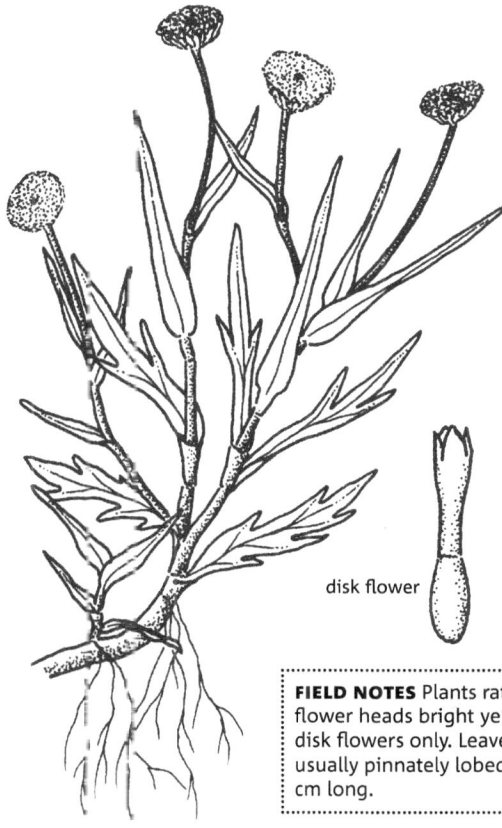

disk flower

> **FIELD NOTES** Plants rather fleshy; flower heads bright yellow, of disk flowers only. Leaves smooth, usually pinnately lobed and to 8 cm long.

Cotula coronopifolia L.
BRASSBUTTONS

FAMILY Asteraceae (Aster)

FLOWERING March–December

HABITAT Salt marshes, along streams, often in mud.

HABIT Introduced perennial herb, rooting at the lower nodes.

STEMS Ascending to spreading, rather fleshy, branched, to 30 cm long, smooth.

LEAVES Alternate, simple but usually pinnately lobed, linear to oblong, to 8 cm long, smooth, sessile or even slightly clasping at the base.

FLOWERS Borne in heads, with usually 1 head from the axils of the uppermost leaves, the head to 13 mm across, bright yellow, consisting only of disk flowers; bracts oblong, with 3–5 veins. **Sepals** absent. **Petals** 4, bright yellow, united to form tubular disk flowers. **Stamens** 5. **Pistils**: Ovary inferior, smooth.

FRUIT Achenes winged or unwinged, to 2 mm long, without a tuft of soft bristles at the tip.

WETLAND STATUS AW OBL | WMTV OBL

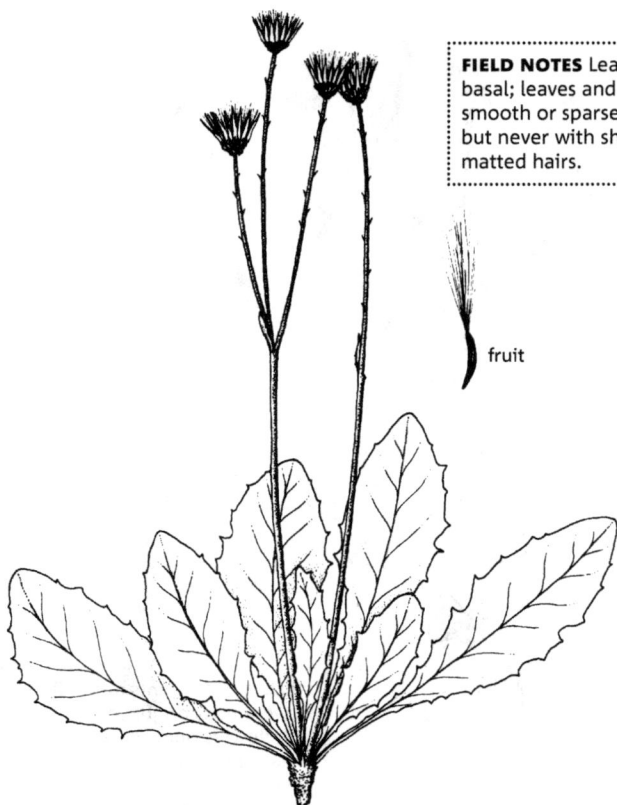

> **FIELD NOTES** Leaves all basal; leaves and stems smooth or sparsely hairy, but never with short, curly, matted hairs.

fruit

Crepis runcinata (James) Torr. & Gray
DANDELION HAWKSBEARD

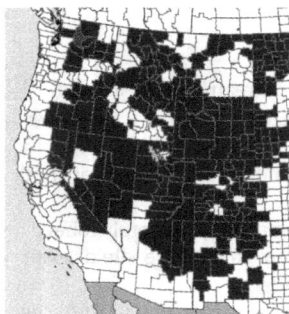

FAMILY Asteraceae (Aster)

FLOWERING June–July

HABITAT Alkaline meadows.

HABIT Native perennial herb with a thickened, fleshy rootstock.

Stem Upright, smooth or sometimes hairy but not with curly, matted hairs, to 45 cm tall, bearing only the flower heads and a few bracts; milky sap present.

LEAVES All basal, oblong to obovate, to 25 cm long, to 8 cm wide, rounded at the tip, tapering or abruptly contracted to the leaf stalk, smooth or somewhat hairy, but not with curly, matted hairs, toothed along the edges.

FLOWERS Many borne in heads on a leafless stem, the heads to 2.5 cm high, consisting entirely of ray flowers; bracts surrounding each head narrowly lanceolate, pointed at the tip, usually glandular-hairy. **Sepals** absent. **Petals** 5, orange, united to form rays 13–20 mm long. **Stamens** 5. **Pistils**: Ovary inferior.

FRUIT Achenes 6–8 mm long, tapering to a short point at the tip, pale brown to dark brown, with at least 10 ribs, topped by a cluster of fine bristles.

WETLAND STATUS AW FACU | GP FAC | WMV FACU

> **FIELD NOTES** Plants gray-hairy, usually sprawling; flowers solitary, white, borne in axils of upper leaves; petals and sepals about the same length.

flower

Cressa truxillensis Kunth
SPREADING ALKALI-WEED

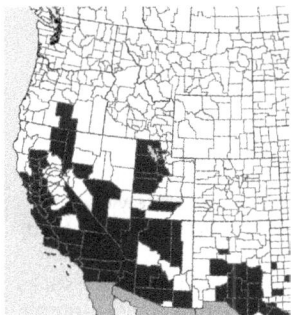

FAMILY Convolvulaceae (Morning-glory)

FLOWERING May–October

HABITAT Moist alkaline and salty habitats.

HABIT Native perennial herb with a slender taproot.

STEMS Upright or lying flat, much branched, gray-hairy, to 20 cm long.

LEAVES Alternate, simple, oblong to ovate, to 13 mm long, to 8 mm wide, rounded or somewhat pointed at the tip, tapering to the usually sessile base, gray-hairy, without teeth.

FLOWERS Solitary in the axils of the upper leaves, subtended by bracteoles; stalks very short to 8 mm long. **Sepals** 5, green, united below, hairy, 4–6 mm long, the lobes ovate. **Petals** 5, white, united below into a short tube 4–6 mm long, the lobes shorter, ovate, spreading or turned downward. **Stamens** 5, exserted beyond the petals. **Pistils**: Ovary superior, hairy; **styles** 2.

FRUIT Capsules ovoid, 6–8 mm long, hairy, usually 1-seeded; seed ovoid, dull brown.

NOTE Highly variable, with some plants upright and others lying on the ground.

WETLAND STATUS AW FACW   GP FACW | WMV FACW

FIELD NOTES Sepals violet-purple, to 13 mm long. Stems smooth, glaucous, hollow, and tend to be woody at the base. Plants may be 1–2 m tall.

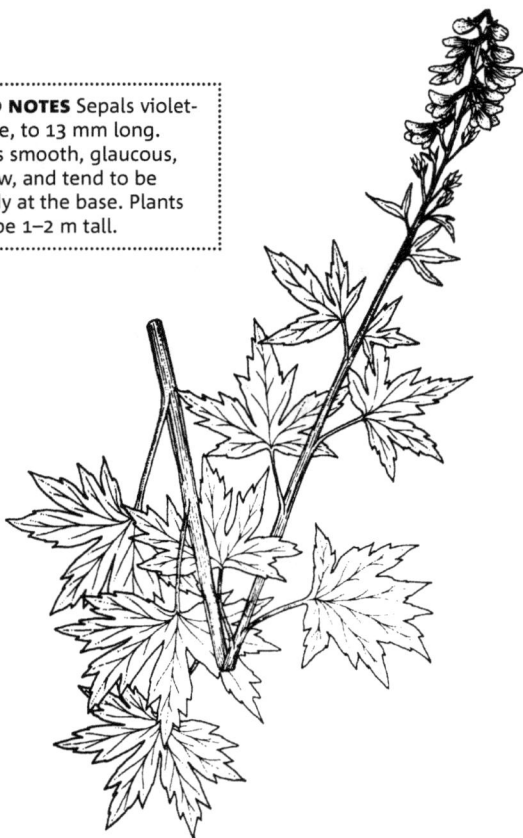

Delphinium glaucum S. Wats.
TOWER LARKSPUR

FAMILY Ranunculaceae (Buttercup)

FLOWERING July–September

HABITAT Wet meadows, along streams.

HABIT Perennial herb with a stout, woody rootstock.

STEMS Coarse, upright, hollow, branched, to 2 m tall, woody at the base, smooth, glaucous.

LEAVES Alternate, to 20 cm across, palmately divided into several segments, the segments lobed or toothed, smooth or hairy.

FLOWERS Many in racemes to 45 cm long; flower stalks to 5 cm long; upper bracts leafy. **Sepals** 5, violet-purple, to 13 mm long, minutely hairy, one of them prolonged backward into a spur, the spur 8–13 mm long. **Petals** 4, 2 of them larger than the other 2, narrow, oblong, notched and purple at the tip. **Stamens** numerous. **Pistils** 1–5, each with a superior ovary.

FRUIT Follicles 1–5, smooth or minutely hairy, upright, 13–20 mm long; seeds ovoid, straw-colored, to 3 mm long.

WETLAND STATUS AW FACW | GP FACW | WMV FACW

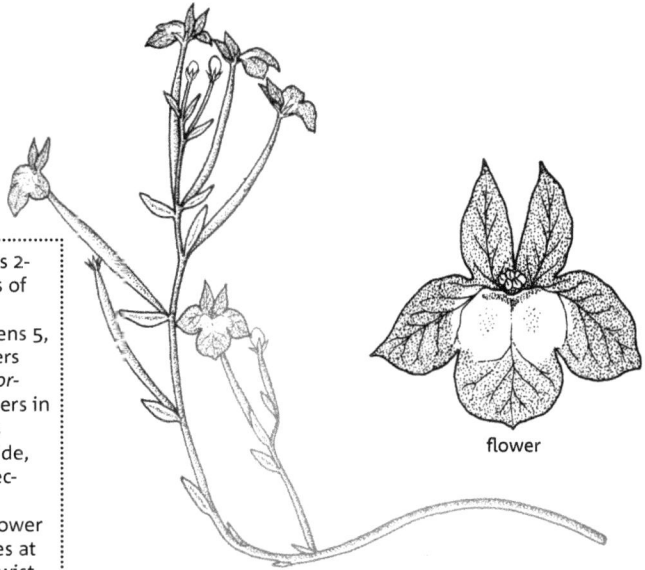

flower

> **FIELD NOTES** Flowers 2-lipped, borne in axils of reduced uppermost leaves (bracts); stamens 5, united by their anthers and filaments. *D. bicornuta* differs from others in the genus by flowers more than 13 mm wide, by the 2 purple projections at base of the middle petal of the lower lip, and by the bristles at tips of anthers that twist around each other.

Downingia bicornuta Gray
DOUBLE-HORN CALICO-FLOWER

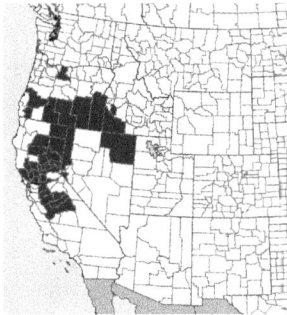

FAMILY Campanulaceae (Bellflower)

FLOWERING April–July

HABITAT Edge of ponds, along streams, in roadside ditches, wet depressions.

HABIT Native annual herb with fibrous roots.

STEMS Upright to ascending or even sprawling, branched or unbranched, hollow, to 25 cm tall, smooth.

LEAVES Alternate, simple, narrowly lanceolate, to 2.5 cm long, pointed at the tip, tapering to the base, smooth, without teeth.

FLOWERS Borne singly in the axils of the uppermost reduced leaves (bracts), the bracts green, narrowly lanceolate, to 20 mm long; flowers sessile, but appearing to be stalked because of the very slender floral tube. **Sepals** 5, green, united below to form a slender floral tube, the lobes 4–6 mm long, narrowly lanceolate, smooth. **Petals** 5, united to form 2 lips, 13–20 mm long, the upper lip darker blue than the lower, the lower lip paler blue with two large yellow or green blotches outlined in white or yellow, with 2 purple projections at the base of the middle petal. **Stamens** 5, not attached to the petals, with the filaments and anthers united, the anthers bearing 2 bristles at the tip that twist around each other. **Pistils:** Ovary inferior, borne at the base of the slender floral tube.

FRUIT Capsules linear, terete, to 5 cm long, smooth or with some stiff short hairs, twisted below, opening by means of vertical slits; seeds to 1 mm long, pale brown with dark brown tips.

WETLAND STATUS AW OBL | WMV OBL

> **FIELD NOTES** Flowers to 8 mm wide; petals white or pale blue or pink; the lower lip petal with yellow blotches, dotted with purple.

flower

Downingia laeta (Greene) Greene
GREAT BASIN CALICO-FLOWER

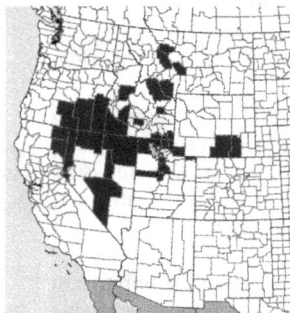

FAMILY Campanulaceae (Bellflower)

FLOWERING May–August

HABITAT Vernal pools, edge of ponds and lakes, in roadside ditches.

HABIT Native annual herb with fibrous roots.

STEMS Upright to ascending or even spreading, unbranched, hollow, to 20 cm tall, smooth.

LEAVES Alternate, simple, narrowly lanceolate, to 2.5 cm long, to 4 mm wide, pointed or sometimes more or less rounded at the tip, tapering to the base, smooth, without teeth, the lowest leaves withering early, the uppermost reduced to bracts.

FLOWERS Borne singly in the axils of the uppermost reduced leaves (bracts), the bracts green, narrowly lanceolate, to 20 mm long; flowers sessile, but appearing to be stalked because of the very slender floral tube. **Sepals** 5, green, united below to form a slender floral tube, the lobes to 8 mm long, narrowly lanceolate, smooth. **Petals** 5, united to form 2 lips, to 8 mm long, white to pale blue or pink, the lower lip with 2 yellow blotches dotted with purple. **Stamens** 5, not attached to the petals, with the filaments and anthers united, the anthers bearing untwisted bristles. **Pistils:** Ovary inferior, borne at the base of the slender floral tube.

FRUIT Capsules linear, terete, to 5 cm long, usually smooth; seeds pale brown.

NOTE As the vernal pools dry up. this species may spread rapidly and form dense colorful patches.

WETLAND STATUS AW OBL | WMV OBL

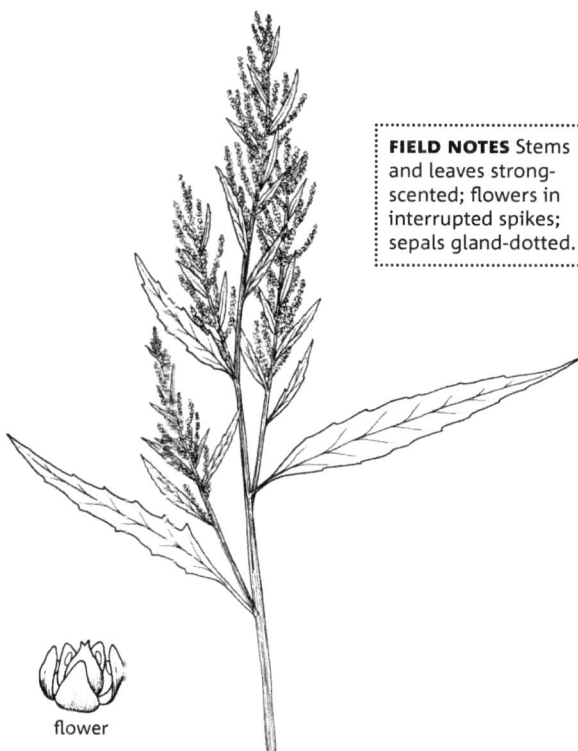

> **FIELD NOTES** Stems and leaves strong-scented; flowers in interrupted spikes; sepals gland-dotted.

flower

Dysphania ambrosioides (L.) Mosyakin & Clemants
AMERICAN WORMSEED

FAMILY Amaranthaceae (Amaranth)

FLOWERING June–December

HABITAT Old fields, disturbed areas, particularly in low moist areas.

HABIT introduced annual or perennial herb with a tap-root.

STEMS Upright, much branched, to 1 m tall, glandular-hairy or nearly smooth, strongly scented.

LEAVES Alternate, simple, lanceolate to oblong, to 10 cm long, pointed at the tip, shallowly or coarsely toothed or not toothed at all, glandular-hairy or nearly smooth, strongly scented.

FLOWERS Many crowded into elongated or spherical spikes; spikes often interrupted, usually without bracts. **Sepals** 5, green, united below, about 1 mm long, gland-dotted. **Petals** absent. **Stamens** 5. **Pistils** Ovary superior; **stigmas** 2.

FRUIT Flattened, enclosed by the sepals. 1-seeded; seed spherical, less than 1 mm in diameter, dark brown.

NOTE This species is native to tropical America. The plant has been used in the past medicinally.

SYNONYMS *Chenopodium ambrosioides* L.

WETLAND STATUS AW FAC | GP FAC | WMV FAC

FIELD NOTES Flowers large, white or pinkish; petals hairy on the inner surface for about half their length; flower stalks usually with purple glands.

flower

Geranium richardsonii Fisch. & Trautv.
RICHARDSON'S CRANE'S-BILL

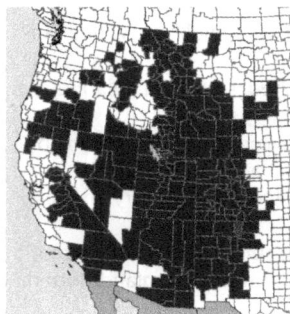

FAMILY Geraniaceae (Geranium)
FLOWERING July–August
HABITAT Wet meadows, along streams, mostly in the mountains.
HABIT Native perennial herb with fibrous roots.
STEMS Upright, branched or unbranched, to 1 m tall, smooth or somewhat hairy.
LEAVES Basal and alternate, simple but deeply palmately 5- to 7-lobed, to 15 cm wide, smooth or sparsely hairy, the lowest leaves on long stalks.
FLOWERS Several, 2–4 cm across, borne on stalks usually bearing purple glands. **Sepals** 5, green, free from each other, 6–13 mm long, tipped with a short awn. **Petals** 5, white or pinkish, free from each other, 13–20 mm long, each petal with purple veins and hairy on the inner surface for about half its length. **Stamens** 10. **Pistils**: Ovary deeply 5-lobed, sparsely hairy.
FRUIT Elongated, to 6 mm long, sparsely hairy, with the style persistent as a beak.
WETLAND STATUS AW FACU | GP FAC | WMV FAC

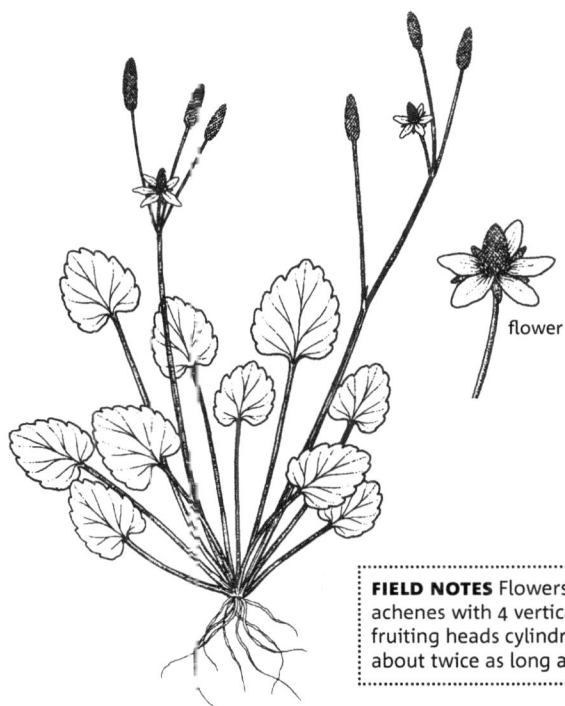

flower

> **FIELD NOTES** Flowers yellow; achenes with 4 vertical stripes; fruiting heads cylindrical, about twice as long as wide.

Halerpestes cymbalaria (Pursh) Greene
ALKALI BUTTERCUP

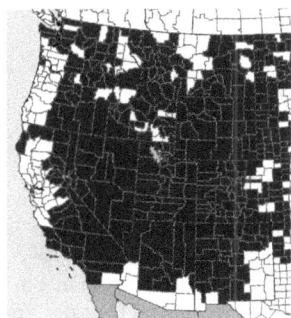

FAMILY Ranunculaceae (Buttercup)

FLOWERING May–September

HABITAT Along streams, in marshes, around springs, sometimes in alkaline areas.

HABIT Native perennial herb with fibrous roots and slender stolons.

STEMS Upright, branched or unbranched, to 30 cm tall, smooth.

LEAVES Basal and alternate, simple, ovate to somewhat kidney-shaped, rounded at the tip, rounded or heart-shaped at the base, smooth, shallowly toothed or shallowly 3-lobed, to 5 cm long, to 2.5 cm wide; stalks to 8 cm long.

FLOWERS 1–few at the tip of the stem, on stalks to 4 cm long. **Sepals** 5, greenish yellow, free from each other, elliptic, to 6 mm long, smooth, falling away early. **Petals** usually 5, yellow, free from each other, obovate, to 8 mm long, to 3 mm wide. **Stamens** 10–30. **Pistils** very numerous in a cylindrical head, the head 13–20 mm long, 3–8 mm thick; ovaries superior.

FRUIT Achenes very numerous in a cylindrical head, each achene oblong but tapering to the base, to 3 mm long, with 4 vertical stripes, with a very short, straight beak.

NOTE The achenes are eaten by small birds.

SYNONYMS *Cyrtorhyncha cymbalaria* (Pursh) Britt., *Ranunculus cymbalaria* Pursh

WETLAND STATUS AW OBL | GP OBL | WMV OBL

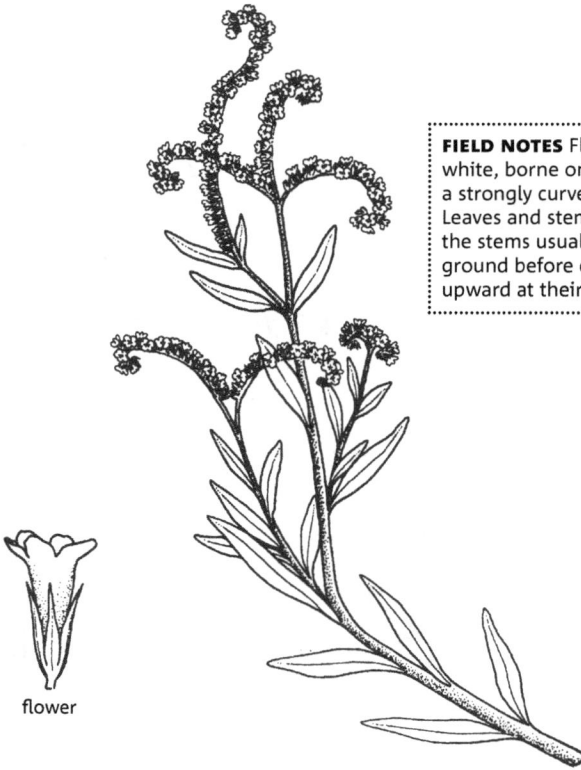

FIELD NOTES Flowers white, borne on one side of a strongly curved axis. Leaves and stems smooth, the stems usually lie on the ground before curving upward at their tips.

flower

Heliotropium curassavicum L.
SEASIDE HELIOTROPE

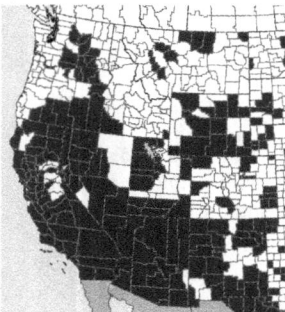

FAMILY Boraginaceae (Borage)

FLOWERING May–October

HABITAT Usually in moist, salty areas.

HABIT Native perennial herb with creeping rootstocks.

STEMS Sprawling along the ground before curving upward at their tips, to 75 cm long, usually somewhat fleshy, smooth and often glaucous.

LEAVES Alternate, simple, succulent, linear to oblanceolate to obovate, or the lowest ones reduced to scales, to 6 cm long, to 20 mm wide, rounded or somewhat pointed at the tip, tapering to the base, sessile or on short stalks, smooth, glaucous, without teeth.

FLOWERS Several borne on one side of a strongly curved axis, without bracts. **Sepals** 5, green, united below, to 3 mm long, smooth. **Petals** 5, united below to form a short tube, white, sometimes with a purple center, to 3 mm long. **Stamens** 5, attached to the tube of the petals. **Pistils:** Ovary superior, 4-parted, smooth.

FRUIT Nutlets 4, ovoid, to 3 mm long.

NOTE There is much variation in this species, including some of the flowers occasionally having a purple center. There is also variation in leaf shape and leaf size.

WETLAND STATUS AW FACU | GP OBL | WMV OBL

> **FIELD NOTES** Plants hollyhock-like, 1–2 m tall; petals 5, white or rose, notched at tip, about 2.5 cm long. Leaves 5- to 7-lobed, coarsely round-toothed, heart-shaped at the base.

flower

Iliamna rivularis (Dougl. ex Hook.) Greene
STREAMBANK WILD HOLLYHOCK

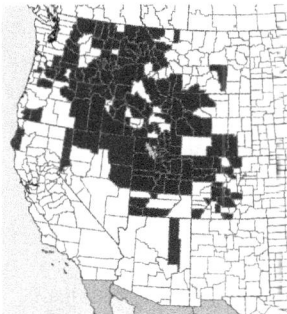

FAMILY Malvaceae (Mallow)
FLOWERING June–August
HABITAT Along streams, in wet meadows, on moist slopes.
HABIT Native shrubby perennial from thickened root-stocks.
STEMS Upright, to 2 m tall, hairy, with some of the hairs star-shaped (stellate).
LEAVES Alternate, simple, 5- to 7-lobed, to 20 cm long, to 15 cm wide, heart-shaped at the base, the lobes triangular, coarsely round-toothed, hairy.
FLOWERS Few to several in axillary clusters and terminal racemes; bracts 3, narrow to thread-like, 4–6 mm long; flower stalks to 20 mm long. **Sepals** 5, green, united below, 6–8 mm long, with star-shaped hairs. **Petals** 5, white or rose, free from each other, about 2.5 cm long, notched at the tip. **Stamens** numerous, borne on a column, the column with star-shaped hairs. **Pistils** usually 5 or more, the ovaries superior.
FRUIT Usually 5 or more, ovoid to ellipsoid, pointed at the tip. to 20 mm long, densely covered with star-shaped hairs, each with 2–4 seeds; seeds minutely hairy.
NOTE There is variation in the size of the leaves.
WETLAND STATUS AW FACW | WMV FAC

FIELD NOTES Plants with basal leaves only, these elliptic or oblong on long stalks; flowers solitary, white, on leafless stalks.

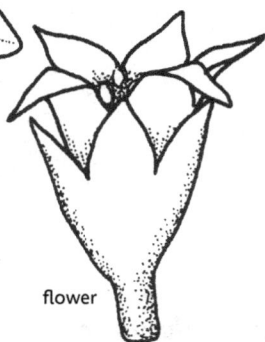

flower

Limosella aquatica L.
NORTHERN MUDWORT

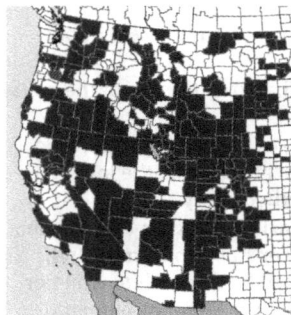

FAMILY Plantaginaceae (Plantain)

FLOWERING May–November

HABITAT Along streams, in ponds, sometimes in water as much as 15 cm deep.

HABIT Native perennial herb with stolons.

STEMS Only the flower-bearing stem present, 13–25 mm long, smooth.

LEAVES All in a basal tuft, the blades often floating, oblong to elliptic, to 4 cm long, to 13 mm wide, on long stalks 2–4 times longer than the blade, smooth, without teeth.

FLOWERS Solitary at the tip of leafless stalks. **Sepals** 5, green with purple spots, united below to form a short tube, smooth, to 3 mm long. **Petals** 5, white, united below to form a short tube, the lobes nearly equal in size and spreading, about 2 mm long. **Stamens** 4. **Pistils:** ovary superior, smooth.

FRUIT Capsules ovoid to spherical, to 3 mm long, smooth, with dark brown seeds.

NOTE The seeds are sometimes eaten by waterfowl.

WETLAND STATUS AW OBL | GP OBL | WMV OBL

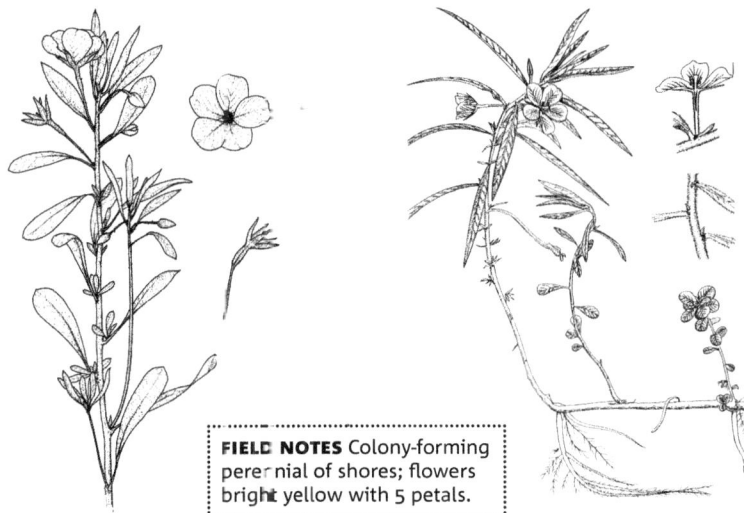

FIELD NOTES Colony-forming perennial of shores; flowers bright yellow with 5 petals.

Ludwigia grandiflora (Michx.) Greuter & Burdet
LARGE-FLOWER PRIMROSE-WILLOW

FAMILY Onagraceae (Evening-prmrose)

FLOWERING May–October

HABITAT Streambanks, shores, slow-moving rivers, ponds; plants typically rooted at or near the shore and extend into open water to form large, dense floating mats.

HABIT Introduced perennial herb; plants may be somewhat woody at base.

STEMS prostrate to erect, to about 1 m long; flowering shoots erect, green to reddish, often hairy; occasionally with thick white spongy roots at floating nodes.

LEAVES alternate, 3–12 cm long, leaf stalk 1–20 mm long; blade narrowly elliptic or oblanceolate to widely obovate, entire, smooth to hairy. In young growth, leaves usually rounded, growing rosette-like around the hairy stem. Once flowering begins, the leaves lengthen, becoming much lanceolate to elliptic in shape.

FLOWERS yellow, large, 2–5 cm wide, arising from the leaf axils on stalks to about 3 cm long. **Sepals** 5. **Petals** usually 5 (rarely 6), to 20 mm long. **Stamens** 10 (rarely 12) in 2 unequal sets. **Pistils:** Ovary inferior.

FRUIT Capsules cylindrical, 13–25 mm long, divided into 5 chambers. Seeds numerous.

NOTE Native of Central and South America; invasive due to its rapid growth covering large areas and forming dense mats. Reproduction by seed and from floating plant fragments.

SYNONYMS *Jussiaea grandiflora* Michx., *Ludwigia hexapetala* (Hook. & Arn.) Zardini, Gu & Raven

WETLAND STATUS AW OBL | GP OBL | WMV OBL

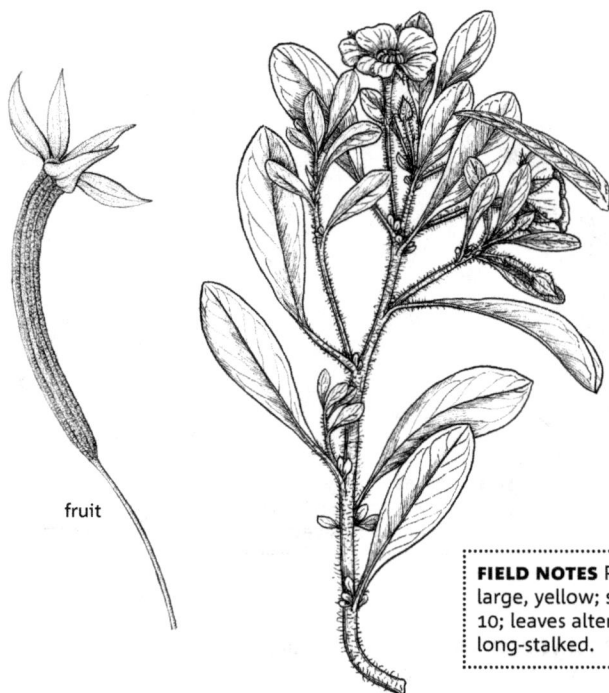

fruit

> **FIELD NOTES** Petals 5, large, yellow; stamens 10; leaves alternate, long-stalked.

Ludwigia peploides (Kunth) Raven Floating PRIMROSE-WILLOW

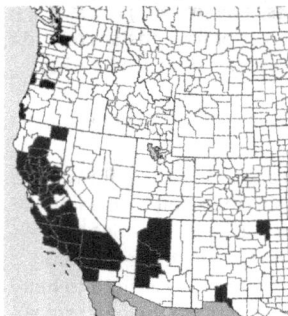

FAMILY Onagraceae (Evening-prmrose)

FLOWERING May-September

HABITAT Shallow ponds, ditches, lakes, pools, streams, canals, usually in shallow water.

HABIT Native perennial herb with floating or creeping stems that root at the nodes, often forming mats in shallow water.

STEMS Creeping or floating, smooth or slightly hairy, to 1 m long.

LEAVES Alternate, simple, oval to obovate to elliptic, rounded or pointed at the tip, narrowed to the base, to 10 cm long, to 5 cm wide, without teeth, smooth or less commonly sparsely hairy on the lower surface; leaf stalks often as long as the blade.

FLOWERS Solitary in the axils of the upper leaves, to 2.5 cm across, borne on long, smooth stalks up to 7 cm long. **Sepals** 5, green, united below and also to the ovary, the lobes to 13 mm long. **Petals** 5, yellow, free from each other, to 2 cm long. **Stamens** 10. **Pistils**: Ovary inferior.

FRUIT Capsules elongated, cylindric, smooth, to 5 cm long, to 4 mm wide, with the sepals persistent; seeds smooth, embedded in the capsule.

NOTE This species tends to be aggressive and can form dense mats in shallow water (listed as a noxious weed in Oregon and Washington).

SYNONYMS *Jussiaea diffusa* auct. non Forssk., *Jussiaea repens* L.

WETLAND STATUS AW OBL | GP OBL | WMV OBL

FIELD NOTES *Mertensia* have usually bluish, tubular flowers, and toothless, alternate leaves. *M. ciliata* differs by its larger size (often at least 60 cm tall), with leaves smooth on both surfaces, and sepals less than 4 mm long.

flower

Mertensia ciliata (James ex Torr.) G. Don
STREAMSIDE BLUEBELLS

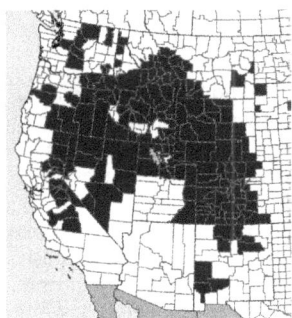

FAMILY Boraginaceae (Borage)

FLOWERING June–August

HABITAT Wet meadows, along streams, particularly in the mountains.

HABIT Native perennial herb from a thickened rootstock.

STEMS Upright, to 1.5 m tall, smooth, sometimes bluish.

LEAVES Alternate, simple, elliptic to ovate to lanceolate, to 15 cm long, to 5 cm wide, more or less pointed at the tip, tapering or less commonly rounded at the base, smooth, the basal leaves larger and on long stalks, the upper leaves progressively smaller and without stalks.

FLOWERS Several clustered in small cymes, each cyme subtended by a small bract. Sepals 5, green, united only at the base, to 4 mm long, smooth except for short cilia along the edges. **Petals** 5, blue, united to form a tube, to 20 mm long. **Stamens** 5, attached to the tube of the petals. **Pistils:** Ovary superior. 4-parted.

FRUIT Nutlets 4. somewhat veiny

WETLAND STATUS AW FACW | GP FACW | WMV FACW

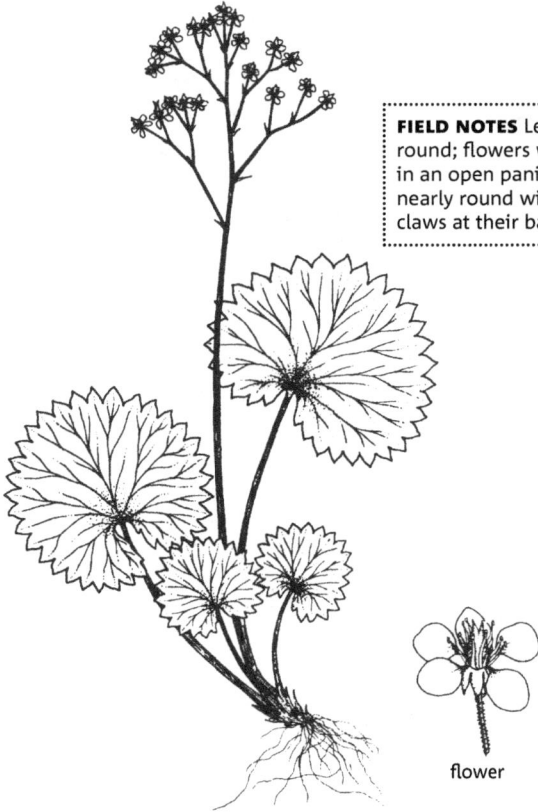

FIELD NOTES Leaves nearly round; flowers white, borne in an open panicle; petals nearly round with slender claws at their base.

flower

Micranthes odontoloma (Piper) Heller
BROOK SAXIFRAGE

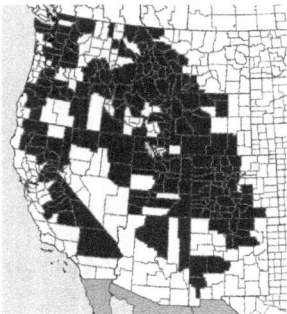

FAMILY Saxifragaceae (Saxifrage)

FLOWERING July–August

HABITAT Along streams in the mountains.

HABIT Native perennial herb with rhizomes.

STEMS Bearing only flowers and no leaves, to 15 cm long, smooth or glandular-hairy near the tip.

LEAVES All basal, nearly round, to 10 cm across, heart-shaped at the base, coarsely toothed, smooth; leaf stalks slender, to 20 cm long.

FLOWERS Few to several in an open panicle on a leafless stem, the branchlets of the panicle sometimes purplish and minutely glandular-hairy. **Sepals** 5, green or purple, united at the base, lanceolate to oblong, to 2 mm long, usually smooth. **Petals** 5, white with 2–3 yellow dots near base, free from each other, nearly round but with a slender claw at the base, each petal to 4 mm long. **Stamens** 10, the filaments broadened and petal-like. **Pistils:** 2, united at base, the ovary superior.

FRUIT Capsules usually 2, beaked, more or less purplish, 4–8 mm long.

SYNONYMS *Saxifraga arguta* auct. non D. Don

WETLAND STATUS AW FACW | GP FACW | WMV FACW

WESTERN WETLAND FLORA

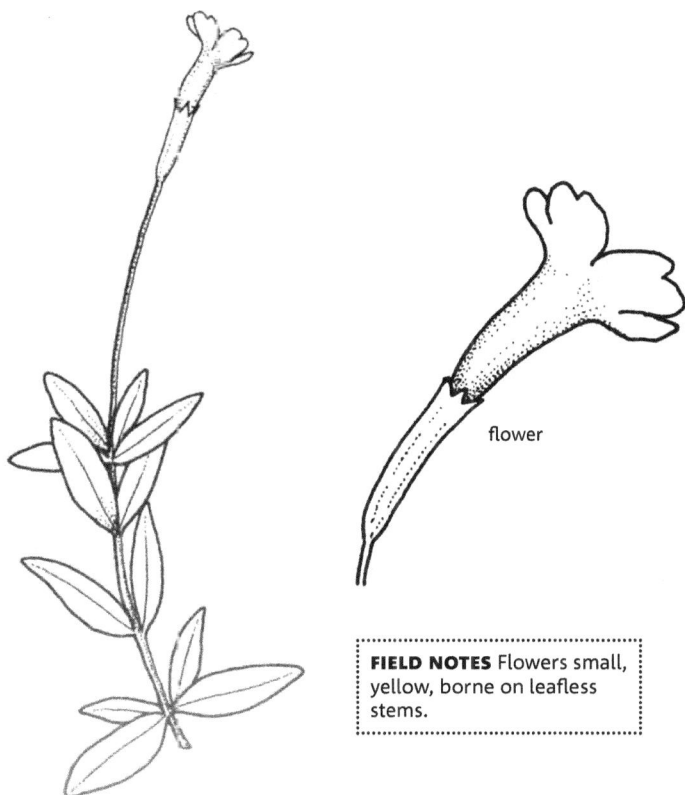

flower

> **FIELD NOTES** Flowers small, yellow, borne on leafless stems.

Mimulus primuloides Benth.
PRIMROSE MONKEY-FLOWER

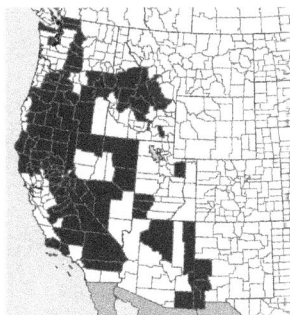

FAMILY Phrymaceae (Lopseed)

FLOWERING June–August

HABITAT Wet meadows, along streams.

HABIT Native perennial herb with short rhizomes and slender stolons.

STEMS Upright, bearing only a solitary flower on a stalk to 8 cm long, usually smooth.

LEAVES All clustered at the base of the plant, oblanceolate, to 4 cm long, to 13 mm wide, pointed at the tip, tapering to the base, shallowly toothed, hairy.

FLOWERS Solitary at the tip of a slender stalk. **Sepals** 5, green, united below to form a tube 6–8 mm long and with reddish ribs, smooth, the lobes 2 mm long. **Petals** 5, yellow with reddish dots, united below to form a tube 8–13 mm long, the lobes notched. **Stamens** 4. **Pistils:** Ovary superior.

FRUIT Capsules ovoid, pointed at the tip. 6–8 mm long, smooth.

NOTE The petals fall away shortly after the flowers open.

SYNONYMS *Erythranthe primuloides* (Benth.) G. L. Nesom & N. S. Fraga

WETLAND STATUS AW FACW | GP FACW | WMV FACW

FIELD NOTES Petals 5, deeply
and intricately divided, giving
the appearance of a snowflake.

flower

Mitella pentandra Hook.
FIVE-POINT BISHOP'S-CAP

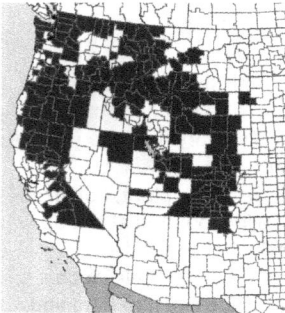

FAMILY Saxifragaceae (Saxifrage)

FLOWERING May–July

HABITAT Moist woods, bogs.

HABIT Native perennial herb with short rootstocks.

STEMS Upright, slender, to 30 cm tall, glandular-hairy,
bearing only a raceme of flowers.

LEAVES All basal, shallowly and palmately 5-lobed,
heart-shaped at the base, to 8 cm long and about as
broad, hairy, with some of the hairs glandular, toothed;
stalks to 10 cm long, hairy or smooth.

FLOWERS 8–20 in an uncrowded raceme, on short
stalks. **Sepals** 5, green, united below, the lobes more or
less triangular, rounded at the tip, usually curved downward. **Petals** 5, white, free from
each other, finely divided into thread-like segments. **Stamens** 5. **Pistils**: Ovary more or
less inferior; styles 2.

FRUIT Capsules ovoid, 2-beaked, to 4 mm across, with numerous smooth and shiny
seeds.

SYNONYMS *Pectiantia pentandra* (Hook.) Rydb.

WETLAND STATUS AW FACW | GP FAC | WMV FAC

> **FIELD NOTES** Plants annual, branched; leaves alternate, hastate; flowers small, greenish, with only 1 sepal and 1 stamen.

flower

Monolepis nuttalliana (J. A. Schultes) Greene
NUTTALL'S POVERTY-WEED

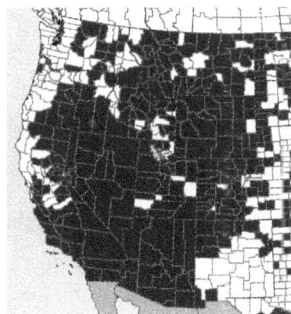

FAMILY Amaranthaceae (Amaranth)

FLOWERING April–September

HABITAT Alkaline soils.

HABIT Native annual herb with fibrous roots.

STEMS Sprawling and ascending, much branched, to 25 cm tall, succulent, mealy when young, but becoming smooth.

LEAVES Alternate, simple, hastate, to 5 cm long, mealy at first, becoming smooth.

FLOWERS Many in small, dense, sessile, reddish clusters, the male and female often borne separately on separate plants. **Sepals** 1, spatulate to obovate. about 1 mm long. **Petals** absent. **Stamens** 1. **Pistils:** Ovary superior; **styles** 2.

FRUIT Ovoid, pitted, about 1 mm across, containing 1 dark seed.

NOTE The seeds are sometimes eaten by birds.

SYNONYMS *Blitum nuttallianum* ..A. Schultes

WETLAND STATUS AW FAC | GP FAC | WMV FAC

FIELD NOTES Plants small; achenes slender, the beak spreading, giving the spike of fruits a jagged appearance.

flower

Myosurus apetalus C. Gay
BRISTLY MOUSETAIL

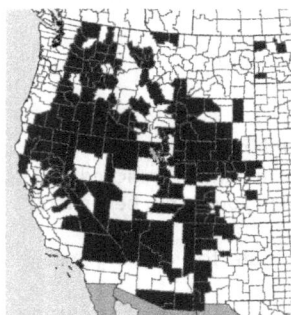

FAMILY Ranunculaceae (Buttercup)

FLOWERING April–July

HABITAT Moist to wet places.

HABIT Native annual herb with fibrous roots.

STEMS Upright, often curved, slender, to 10 cm long, smooth, without leaves.

LEAVES All basal, linear to narrowly spatulate, to 8 cm long, smooth.

FLOWERS Very tiny, densely crowded into a terminal cylindrical spike to 13 mm long. **Sepals** usually 5, green, oblong, to 2 mm long, spurred at the base. **Petals** 5 (sometimes absent), greenish yellow, very narrow, about as long as the sepals, with a nectar-bearing pit on its surface. **Stamens** 5–25. **Pistils** many, borne on a cylindrical receptacle, each with a superior ovary.

FRUIT Achenes ellipsoid, smooth, with a slender, spreading beak that projects outward from the spike.

NOTE The achenes are eaten by small birds and mammals.

SYNONYMS *Myosurus aristatus* auct. non Benth.

WETLAND STATUS AW OBL | GP FACW | WMV OBL

FIELD NOTES Petals yellow, to 5 cm long; stigmas 4-parted; leaves narrow, less than 1/4 as wide as long.

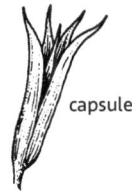

capsule

Oenothera elata Kunth
HOOKER'S EVENING-PRIMROSE

FAMILY Onagraceae (Evening-primrose)
FLOWERING June–September
HABITAT Along streams, in marshes, on wet cliffs.
HABIT Native biennial herb with a taproot.
STEMS Upright, branched, to 2 m tall, hairy, with some of the hairs gland-tipped.
LEAVES Alternate as well as clustered at the base of the plant, simple, elliptic to lanceolate, usually pointed at the tip. hairy, the basal leaves often coarsely toothed, to 18 cm long, the leaves on the stem finely toothed, to 10 cm long, the basal leaves with long stalks, the stem leaves sessile or with short stalks.

FLOWERS Many crowded in a dense showy cluster to 40 cm long; bracts lanceolate, hairy. **Sepals** 4, green to reddish, united below to form a floral tube, the tube to 5 cm long, the lobes to 5 cm long, hairy. **Petals** 4, yellow but fading to reddish, free from each other, obovate, to 5 cm long. **Stamens** 8, to 4 cm long. **Pistils:** Ovary inferior, hairy. **FRUIT** Capsules cylindrical, to 5 cm long, to 6 mm thick, hairy, containing several reddish brown seeds.
SYNONYMS *Oenothera hookeri* Torr. & Gray
WETLAND STATUS AW FACW | GP FACW | WMV FACW

FIELD NOTES Flowering stems with no leaves; capsule woody, winged in upper half.

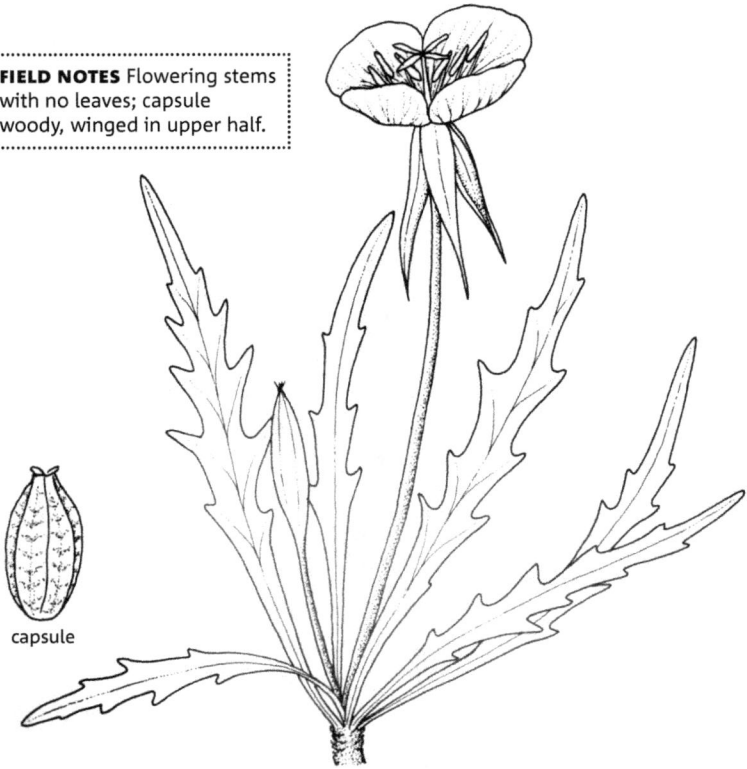

capsule

Oenothera flava (A. Nels.) Garrett
YELLOW EVENING-PRIMROSE

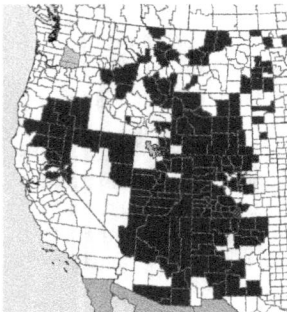

FAMILY Onagraceae (Evening-primrose)
FLOWERING May–July
HABITAT Low, dry depressions.
HABIT Native perennial herb with a thick taproot.
STEMS Only the flower-bearing stem present, upright, to 30 cm tall, smooth or hairy, the hairs sometimes glandular.
LEAVES All basal, narrowly oblong to oblanceolate, irregularly toothed or lobed, to 20 cm long, to 5 cm wide, smooth or glandular-hairy; leaf stalk slightly winged.
FLOWERS Solitary at the tip of a leafless stem, opening in the evening, to nearly 5 cm across. **Sepals** 4, green becoming purplish, more or less free from each other at the tip and pointing downward, becoming tubular below, the tube to nearly 8 cm long. **Petals** 4, pale yellow, free from each other, 13-25 mm long. **Stamens** 6. **Pistils:** Ovary inferior; stigmas 4.
FRUIT Capsules more or less woody, winged in the upper half, ovoid, 15–40 mm long, with minutely granular, obovoid seeds with a wing-like margin.
WETLAND STATUS AW FAC | GP FACW | WMV FAC

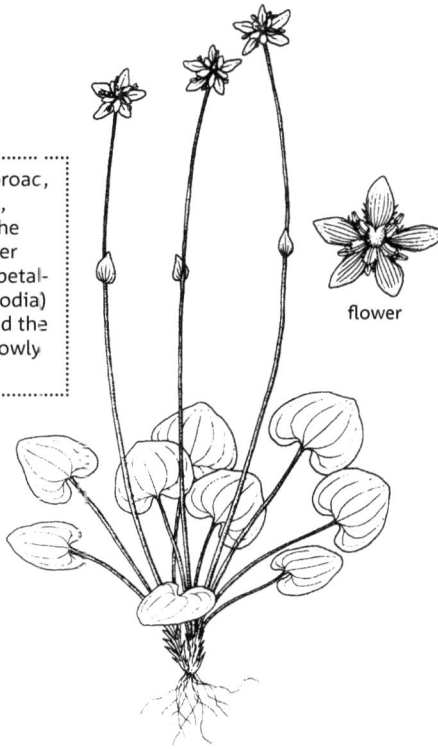

> **FIELD NOTES** Leaves broad, heart-shaped; petals 5, white, fringed along the sides. Differs from other *Parnassia* by its small petal-like structures (staminodia) between the petals and the stamens that are shallowly lobed at the tip.

flower

Parnassia fimbriata Koenig
FRINGED GRASS-OF-PARNASSUS

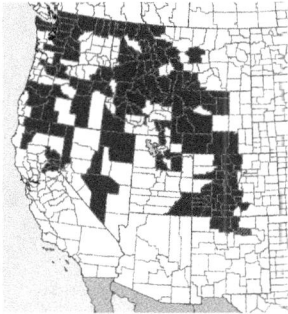

FAMILY Parnassiaceae (Grass-of-Parnassus)

FLOWERING July–September

HABITAT Bogs, around springs.

HABIT Native perennial herb with a short, thickened rootstock.

STEMS All stems bearing only a single flower and no leaves (1 bract present), to 40 cm tall, smooth.

LEAVES All basal, orbicular, to 5 cm across, deeply heart-shaped at the base, smooth, without teeth; leaf stalks to 15 cm long, smooth.

FLOWERS Solitary on a leafless stem, although a single leaf-like bract usually present about halfway up stem, the bract ovate, sessile and clasping the stem, smooth, to 13 mm long. **Sepals** 5, united below, green, the lobes 4–6 mm long, elliptic to ovate, smooth. **Petals** 5, free from each other, white but with conspicuous veins, obovate, white, about 13 mm long, fringed along the edges; small petal-like structures (staminodia) between the petals and the stamens lobed at the tip, to 6 mm long. **Stamens** 5. **Pistils**: Ovary superior or somewhat inferior, 3- to 4-parted.

FRUIT Capsules 3- or 4-parted nearly to base, 8–13 mm long, smooth, containing many tiny winged seeds.

WETLAND STATUS AW OBL | GP OBL | WMV OBL

> **FIELD NOTES** Petals unfringed, with 3–13 veins.

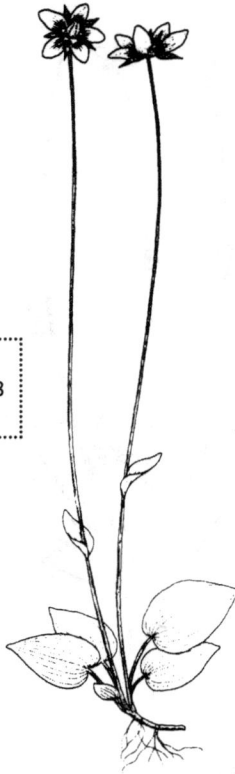

Parnassia palustris L.
NORTHERN GRASS-OF-PARNASSUS

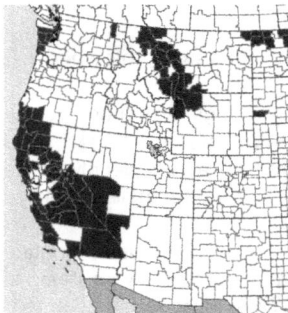

FAMILY Parnassiaceae (Grass-of-Parnassus)
FLOWERING July–October
HABITAT Wet meadows and other moist places.
HABIT Native perennial herb with short rootstocks.
STEMS Upright, slender, unbranched, to 45 cm tall, smooth, bearing a single leaf (often called a bract) about half the way up the stem.
LEAVES All basal except for the single, ovate, sessile leaf (or bract) about halfway up the stem; basal leaves ovate, heart-shaped to tapering to the base, to 4 cm long, smooth, without teeth, on stalks to 10 cm long.
FLOWERS Solitary on the stem, to 2.5 cm across. **Sepals** 5, green, united below to form a short floral tube, lanceolate to ovate, pointed at the tip, 6–13 mm long, with 5–7 veins. **Petals** 5, white, free from each other, ovate to obovate, to 13 mm long, not fringed, with 3–13 veins. **Stamens** 5 fertile, with many sterile stamens present consisting of slender, gland-tipped filaments to 8 mm long. **Pistils:** Ovary more or less superior or slightly inferior.
FRUIT Capsules ovoid, to 13 mm long, subtended by the persistent sepals, with numerous tiny, angular seeds.
WETLAND STATUS AW OBL | GP OBL | WMV OBL

FIELD NOTES Flowers purple, the lip prolonged into a curved beak resembling trunk of an elephant; leaves deeply pinnately divided.

flower

Pedicularis groenlandica Retz.
ELEPHANT'S-HEAD LOUSEWORT

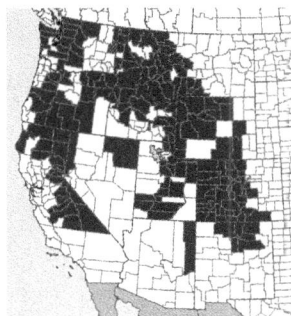

FAMILY Orobanchaceae (Broom-rape)

FLOWERING June–September

HABITAT Wet meadows, damp woods, along streams, particularly in the mountains.

HABIT Native perennial herb with a thickened rootstock.

STEMS Upright, to 60 cm tall, usually unbranched but often several together from the base, smooth.

LEAVES Nearly all basal but a few alternate, simple but deeply pinnately divided nearly to the middle, to 30 cm long, to 8 cm wide, smooth, each lobe of the leaf toothed, the lowest leaves stalked, the upper smaller and without a stalk.

FLOWERS Several crowded together into a spike, the spike to 20 cm long, each flower subtended by narrow, often deeply lobed bracts. **Sepals** 5, united below, green, the tips about 1 mm long, triangular, the united part with white veins and hairy on the inner surface. **Petals** 5, forming 2 lips, violet to purple, to 8 mm long, the upper petal prolonged into a curved beak to 20 mm long.

FRUIT Capsules 8–10 cm long, asymmetrical, smooth; seeds 3–4 mm long, winged, smooth, with a prominent venation pattern.

WETLAND STATUS AW OBL | GP OBL | WMV OBL

FIELD NOTES Spikes ovoid, less than 5 cm long; flowers red or pinkish red; leaves usually rounded or heart-shaped at base.

water surface

flower

Persicaria amphibia (L.) S. F. Gray p.p.
WATER SMARTWEED

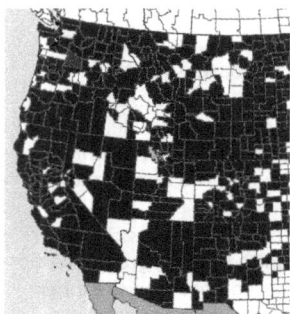

FAMILY Polygonaceae (Buckwheat)

FLOWERING June–August.

HABITAT Marshes, around ponds and lakes; in shallow water or or water's edge.

HABIT Native erennial herb with rhizomes.

STEMS Erect, branched or unbranched, smooth or hairy, to 1 m long.

LEAVES Alternate, simple, broadly lanceolate to oblong, pointed or rounded at the tip, usually rounded or heart-shaped at the base, without teeth, smooth or hairy, to 20 cm long, 4–10 cm broad, with a sheath sometimes bearing bristles. **Sepals** usually 5, red or pinkish red, united at the base. **Petals** absent. **Stamens** usually 8. **Pistils:** Ovary superior.

FRUIT Achenes pale brown, not shiny, not triangular, 1–3 mm long.

NOTE Waterfowl eat the achenes of this plant.

WETLAND STATUS AW OBL | GP OBL | WMV OBL

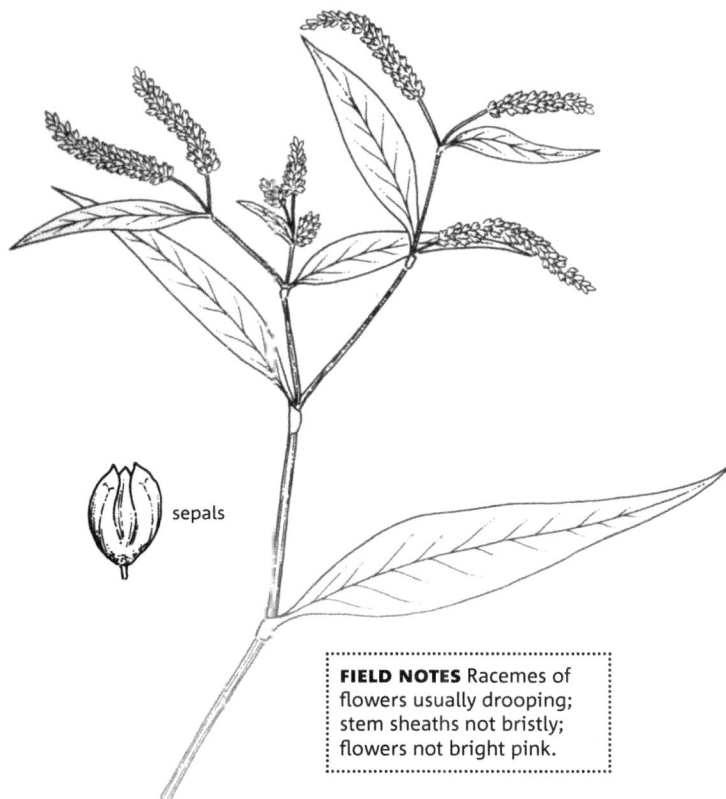

sepals

> **FIELD NOTES** Racemes of
> flowers usually drooping;
> stem sheaths not bristly;
> flowers not bright pink.

Persicaria lapathifolia (L.) S. F. Gray
WILLOW-WEED

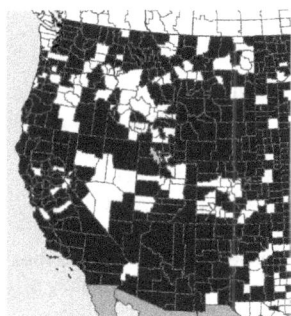

FAMILY Polygonaceae (Buckwheat)

FLOWERING July–October

HABITAT Moist soil, wet meadows, roadside ditches, often in disturbed areas.

HABIT Native, erect annual herb from a taproot.

STEMS Upright, sometimes rather stout, to 1 m tall, smooth, the sheaths not bearing bristles at the tip.

LEAVES Alternate, simple, narrowly to broadly lanceolate, to 20 cm long, to 4 cm wide, pointed at the tip, tapering to the base, usually smooth, often with glandular dots on the lower surface.

FLOWERS Many in few to several drooping racemes, the racemes to 8 cm long, to 13 mm broad, their stalks sometimes with stalked glands. **Sepals** 6, partly united, white, greenish, or pale-pink, petal-like, 3-veined. **Petals** absent. **Stamens** usually 9. **Pistils**: Ovary superior; **styles** 2, free to the base.

FRUIT Achenes lenticular, shiny, 3–4 mm long.

NOTE The achenes are eaten by waterfowl.

SYNONYMS *Polygonum lapathifolium* L.

WETLAND STATUS AW FACW | GP OBL | WMV FACW

> **FIELD NOTES** Plants annual; sheaths bristly; flowers pinkish, in dense, erect spikes.

achene

Persicaria maculosa S. F. Gray
LADY'S THUMB

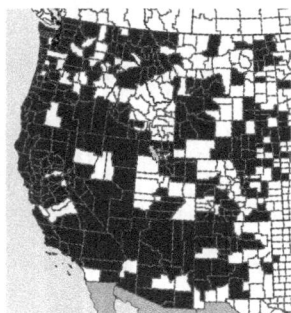

FAMILY Polygonaceae (Buckwheat)

FLOWERING June–November

HABITAT Moist, disturbed areas.

HABIT Introduced annual herb with a taproot.

STEMS Upright or ascending, branched, to 1 m tall, usually smooth; sheaths bristly at the tip.

LEAVES Alternate, simple, lanceolate to linear-lanceolate, to 13 cm long, to 20 mm wide, pointed at the tip, tapering to the nearly sessile base, usually smooth.

FLOWERS Many crowded in dense racemes, the racemes erect, to 4 cm long, on smooth stalks. **Sepals** usually 5, united below, usually pinkish, to 4 mm long. **Petals** absent. **Stamens** 6–9. **Pistils:** Ovary superior; **styles** usually 2.

FRUIT Achenes triangular or sometimes flattened, 3–4 mm long, smooth, shiny.

NOTE The distinctive blotch present on many of the leaves of the lady's thumb may also be present in other species of *Persicaria*. The achenes are eaten by waterfowl and small mammals.

SYNONYMS *Polygonum persicaria* L.

WETLAND STATUS AW FACW | GP FACW | WMV FACW

> **FIELD NOTES** Sepals white with black dots; stem sheaths bristly.

sepals

Persicaria punctata (Ell.) Small
DOTTED SMARTWEED

FAMILY Polygonaceae (Buckwheat)

FLOWERING July–October

HABITAT Wet soil, wet meadows, marshes, roadside ditches, around lakes and ponds.

HABIT Native, erect perennial, but with the lower branches sometimes rooting at the nodes, usually bearing rhizomes and stolons.

STEMS Upright, usually rather slender, to 1 m tall, but usually shorter, usually smooth, the sheath bearing bristles at the top.

LEAVES Alternate, simple, elliptic to lanceolate, to 10 cm long, to 20 mm wide, pointed at the tip, tapering to the base, usually smooth, or occasionally strigose on the lower surface.

FLOWERS Many in arching or erect, interrupted racemes, the racemes to 10 cm long, about 6 mm broad. **Sepals** 6, partly united, white to greenish white, petal-like, the surface covered with black dots **Petals** absent. **Stamens** usually 9. **Pistils:** Ovary superior; styles 2 or 3.

FRUIT Achenes black, shiny, lenticular to 3-angled.

NOTE The achenes are eaten by waterfowl.

SYNONYMS *Polygonum punctatum* Ell.

WETLAND STATUS AW OBL | GP OBL | WMV OBL

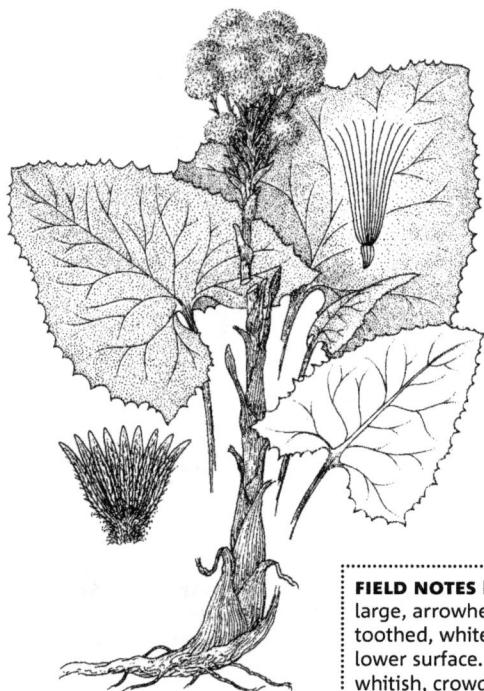

> **FIELD NOTES** Basal leaves large, arrowhead-shaped, toothed, white-woolly on lower surface. Flowers whitish, crowded together into heads.

Petasites frigidus var. *sagittatus* (Banks ex Pursh) Cherniawsky
ARROW-LEAF SWEET COLTSFOOT

FAMILY Asteraceae (Aster)

FLOWERING April–June

HABITAT Wet meadows, bogs, particularly in high mountains.

HABIT Native perennial herb with creeping rhizomes.

STEMS Upright, unbranched, to 45 cm tall, smooth or hairy, bearing a few toothless, parallel-veined bracts with a dilated tip.

LEAVES All basal, arrowhead-shaped, to 30 cm long, to 25 cm wide, toothed, smooth on the upper surface, white-woolly on the lower surface, on long stalks. (What appears to be narrow, toothless leaves on the stem are actually bracts.)

FLOWERS Many borne in heads, with several heads crowded at the tip of the stem; some flowers ray-like and white, others tubular and white. **Sepals** absent. **Petals** 5, some united to form white rays, others united to form a whitish tube. **Stamens** 5. **Pistils:** Ovary inferior, smooth.

FRUIT Achenes very narrow, with a tuft of bristles at the tip.

SYNONYMS *Petasites sagittatus* (Banks ex Pursh) Gray

WETLAND STATUS AW FACW | GP FAC | WMV FACW

petals

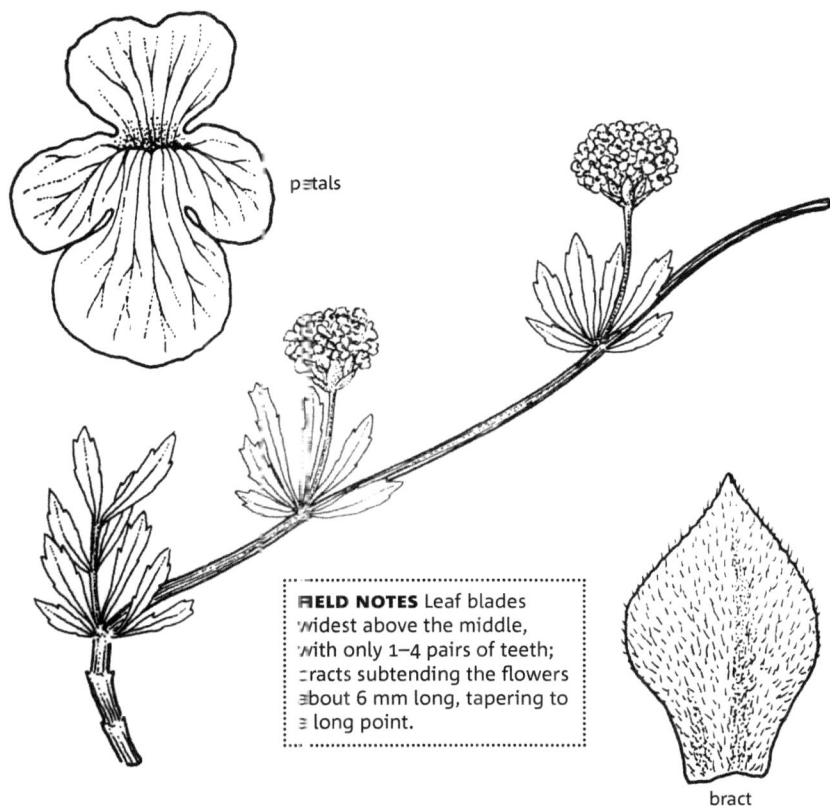

FIELD NOTES Leaf blades widest above the middle, with only 1–4 pairs of teeth; bracts subtending the flowers about 6 mm long, tapering to a long point.

bract

Phyla cuneifolia (Torr.) Greene
WEDGE-LEAF FROGFRUIT

FAMILY Verbenaceae (Verbena)

FLOWERING May–September

HABITAT Moist prairies.

HABIT Native, creeping perennial herb, rooting at the nodes.

STEMS Creeping to ascending, to 1 m long, with appressed hairs.

LEAVES Opposite, simple, narrowly oblanceolate, widest above the middle, to 2.5 cm long, to 8 mm wide, pointed at the tip, smooth or sparsely hairy, with 1–4 pairs of teeth.

FLOWERS Several borne in rounded to cylindrical heads, the heads to 2.5 cm long, to 13 mm wide, on a stalk to 6 cm long; bracts about 6 mm long, tapering to a long point. **Sepals** 4, united below, green, very small. **Petals** 4, united below to form a tube, white to purplish, the tube to 6 mm long. **Stamens** 4, attached to the tube of the petals. **Pistils:** Ovary superior, shallowly 4-parted.

FRUIT Borne in pairs, enclosed by the sepals.

NOTE The seeds may be eaten by small birds and other animals.

WETLAND STATUS AW FAC | GP FAC | WMV FAC

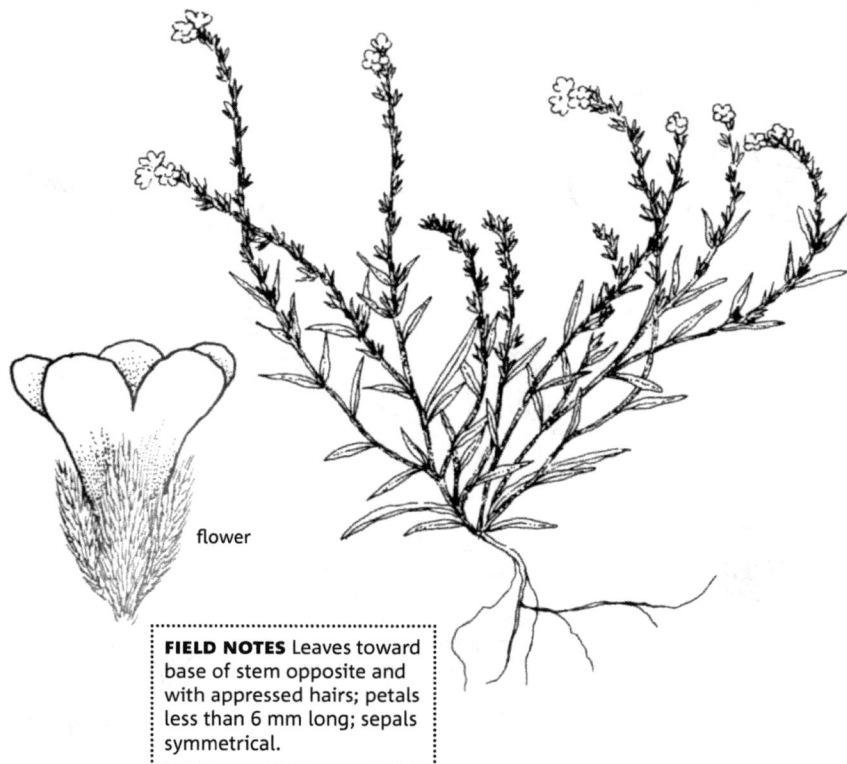

flower

> **FIELD NOTES** Leaves toward
> base of stem opposite and
> with appressed hairs; petals
> less than 6 mm long; sepals
> symmetrical.

Plagiobothrys scouleri (Hook. & Arn.) I. M. Johnst.
SCOULER POPCORN-FLOWER

FAMILY Boraginaceae (Borage)

FLOWERING May–August

HABITAT Moist, poorly drained soil, including alkaline habitats.

HABIT Native, prostrate annual with a taproot or fibrous roots.

STEMS Usually prostrate but turning upward at the tip, to 20 cm long, usually with appressed hairs.

LEAVES Lower leaves opposite, upper leaves alternate, linear, to 8 cm long, to 6 mm wide, with appressed hairs, without teeth.

FLOWERS Borne on one side of the upper end of the stem, the stem usually upturned and even slightly coiled at the tip. **Sepals** 5, symmetrical, green, united below, hairy, to 4 mm long. **Petals** 5, white, united below, usually a little longer than the sepals. **Stamens** 5. **Pistils:** Ovary superior, 4-parted, hairy.

FRUIT Nutlets usually 4, ovoid to lanceoloid, to 3 mm long, wrinkled, usually somewhat bristly.

WETLAND STATUS AW FACW | WMV FACW

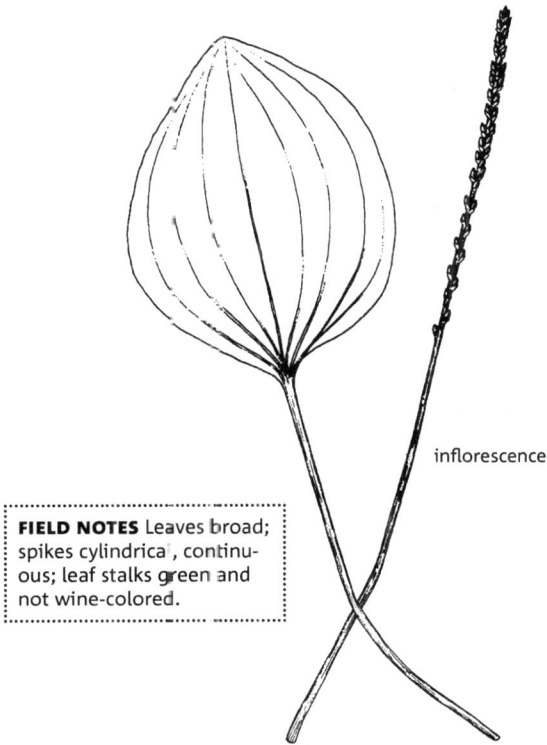

inflorescence

FIELD NOTES Leaves broad; spikes cylindrical, continuous; leaf stalks green and not wine-colored.

Plantago major L.
COMMON PLANTAIN

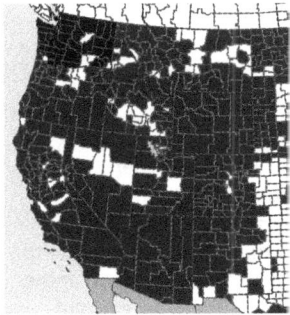

FAMILY Plantaginaceae (Plantain)

FLOWERING May–October

HABITAT Disturbed areas, often becoming an aggressive weed in lawns.

HABIT Introduced perennial herb with a thickened rootstock and with fibrous roots.

STEMS None except for the leafless stalk that bears the cylindrical spike of flowers.

LEAVES All basal, broadly elliptic to ovate, to 15 cm long, to 10 cm wide, pointed or occasionally more or less rounded at the tip, rounded or heart-shaped at the base, usually with some hairs, conspicuously veiny; leaf stalks green, not wine-colored.

FLOWERS Crowded into cylindrical spikes at the tip of leafless stems; spikes to 30 cm long, less than 13 mm thick; each flower subtended by a bract to 4 mm long. **Sepals** 4, united below, green. **Petals** 4, united below, nearly transparent, the lobes about 1 mm long and turned downward. **Stamens** 4. **Pistils:** Ovary superior; **stigmas** 2-lobed.

FRUIT Capsules more or less spherical, to 4 mm in diameter, smooth, containing 6–30 seeds; seeds ovoid, about 1 mm long, brown or black, conspicuously net-veined.

NOTE This species can be an unsightly weed in lawns.

WETLAND STATUS AW FAC | GP FAC | WMV FAC

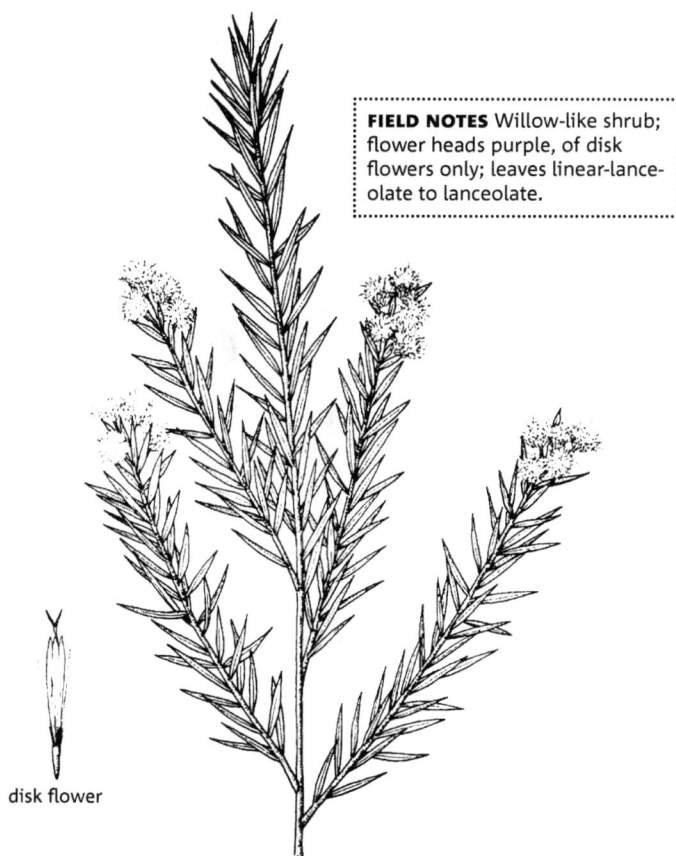

> **FIELD NOTES** Willow-like shrub; flower heads purple, of disk flowers only; leaves linear-lanceolate to lanceolate.

disk flower

Pluchea sericea (Nutt.) Coville
ARROW-WEED

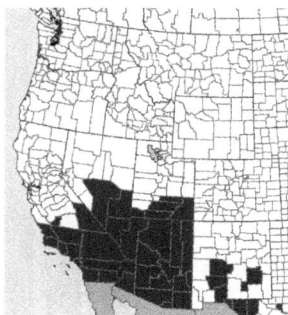

FAMILY Asteraceae (Aster)
FLOWERING March–December
HABITAT Along waterways, often in the desert.
HABIT Native willow-like shrub to 5 m tall.
STEMS Slender, smooth, grayish.
LEAVES Alternate, simple, leathery, linear-lanceolate to lanceolate, to 5 cm long, to 8 mm wide, pointed at the tip, tapering to the base, silvery-silky hairy, without teeth, with 1 vein.
FLOWERS Crowded into heads, each head consisting only of purple disk flowers; bracts subtending each head ovate to lanceolate, leathery, to 8 mm long. **Sepals** absent. **Petals** 5, united to form purple disk flowers. **Stamens** 5. **Pistils:** Ovary inferior, smooth.
FRUIT Achenes smooth, with a tuft of soft bristles at the tip, each bristle somewhat swollen at the tip.
WETLAND STATUS AW FACW | GP FAC | WMV FACW

flower

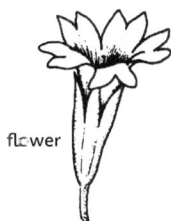

> **FIELD NOTES** Flowers white to lilac; petals deeply lobed; leaves white-mealy on underside.

Primula incana M.E. Jones
AMERICAN PRIMROSE

FAMILY Primulaceae (Primrose)
FLOWERING June–August
HABITAT Wet meadows, along streams, in swamps.
HABIT Native perennial herb with fibrous roots.
STEMS Upright, without leaves, unbranched, to 40 cm tall, bearing only flowers.
LEAVES All basal, oblanceolate, to 10 cm long, to 2.5 cm wide, more or less rounded at the tip, tapering to the base, white-mealy on the lower surface, with or without a few teeth.
FLOWERS 3–12 at the tip of a leafless stem, subtended by bracts 6–13 mm long. **Sepals** 5, green, united below, 8–13 mm long, the tube about twice as long as the lobes. **Petals** 5, white to lilac, united below, deeply lobed, 8–13 mm long. **Stamens** 5, attached to the tube of the petals. **Pistils**: Ovary superior, smooth.
FRUIT Capsules ovoid, smooth, to 13 mm long; seeds numerous.
NOTE The leaves may be browsed by deer.
WETLAND STATUS AW FACW | GP FACW | WMV OBL

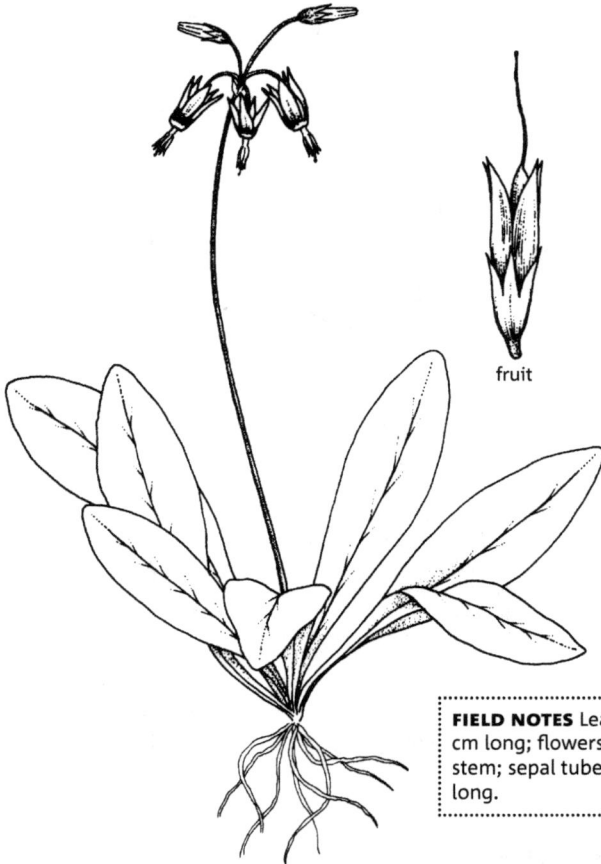

fruit

> **FIELD NOTES** Leaves to 20 cm long; flowers 5–15 per stem; sepal tube to 4 mm long.

Primula pauciflora (Greene) A.R. Mast & Reveal
DARK-THROAT SHOOTING-STAR

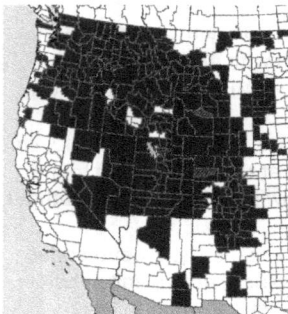

FAMILY Primulaceae (Primrose)
FLOWERING May–July
HABITAT Wet meadows.
HABIT Native perennial herb with a short crown and fleshy roots.
STEMS Underground as a short crown.
LEAVES All basal, oblanceolate to elliptic, to 20 cm long, rounded or pointed at the tip, tapering to a short stalk, without teeth, or sometimes with a wavy edge, smooth.
FLOWERS 5–15 per stem, nodding, the cluster subtended by very small bracts. **Sepals** 5, green, united below, the tube to 4 mm long, the lobes longer than the tube. **Petals** 5, united at the base, turned backward, purple, to 20 mm long. **Stamens** 5. **Pistils:** Ovary superior.
FRUIT Capsule oblong-ovoid, smooth, to 20 mm long.
SYNONYMS *Dodecatheon pauciflorum* Greene, *Dodecatheon pulchellum* (Raf.) Merr.
WETLAND STATUS AW FACW | GP FACW | WMV FACW

disk flower

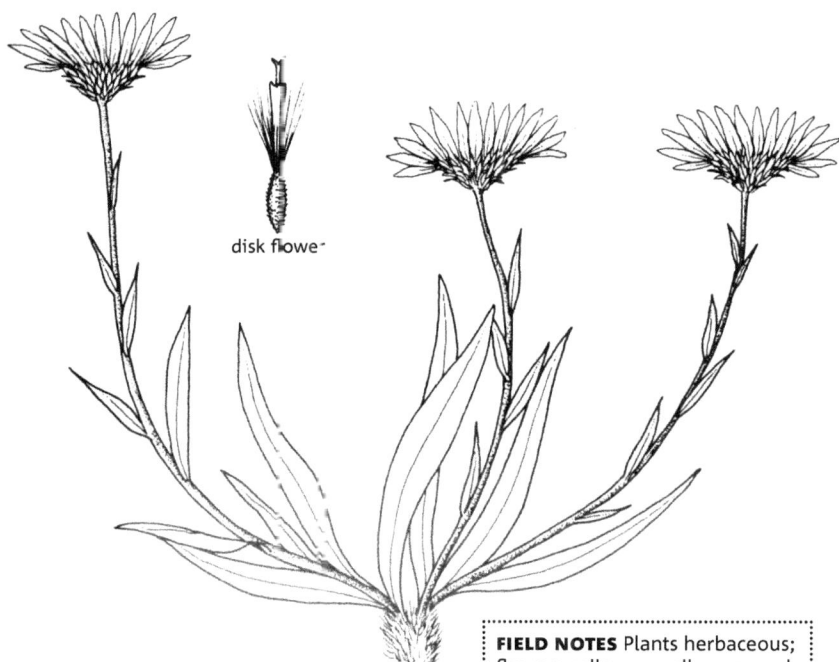

> **FIELD NOTES** Plants herbaceous; flowers yellow, usually arranged in corymbs; bracts green-tipped; achenes covered with silky hairs.

Pyrrocoma lanceolata (Hook.) Greene
LANCE-LEAF GOLDENWEED

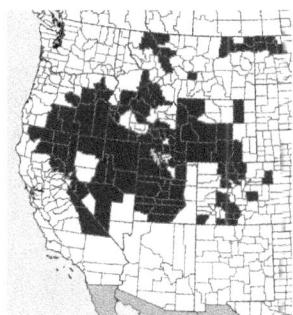

FAMILY Asteraceae (Aster)

FLOWERING June–August

HABITAT Moist meadows, alkaline flats.

HABIT Native perennial herb from a thickened rootstock and taproot.

STEMS Ascending to upright, usually unbranched. to 45 cm tall, smooth or hairy.

LEAVES Mostly basal, oblanceolate. to 15 cm long, to 20 mm wide, pointed at the tip, tapering to the base, with spiny teeth along the edge or without teeth, smooth to woolly-hairy.

FLOWERS Many crowded into heads to 2.5 cm across, each head bearing 15–35 yellow ray flowers and a disk of yellow tubular flowers; bracts surrounding each head narrow, smooth or hairy, green-tipped. **Sepals** absent. **Petals** 5, yellow, some of them united to form rays, others united to form short tubes that comprise the disk. **Stamens** 5. **Pistils:** Ovary inferior.

FRUIT Achenes narrowly ellipsoid, silky-hairy, bearing a tuft of brownish hairs at the tip.

SYNONYMS *Haplopappus lanceolatus* (Hook.) Torr. & Gray

WETLAND STATUS AW FAC | GP FACU | WMV FAC

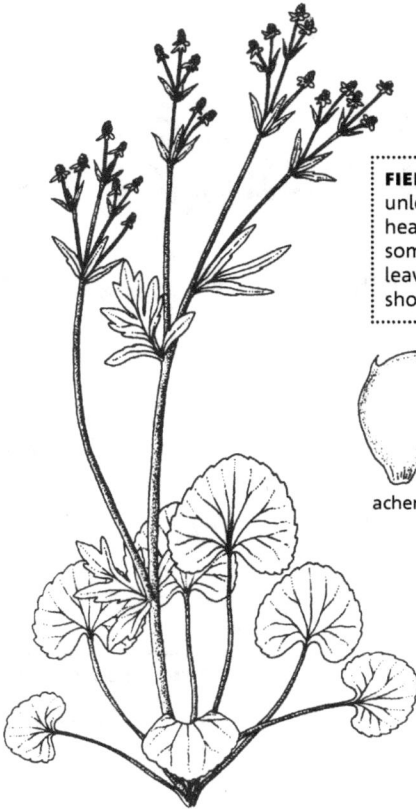

> **FIELD NOTES** Basal leaves unlobed and undivided, heart-shaped at their base; some or all of the cauline leaves 3- or 5-parted; petals shorter than the sepals.

achene

Ranunculus abortivus L.
KIDNEY-LEAF BUTTERCUP

FAMILY Ranunculaceae (Buttercup)

FLOWERING March–July

HABITAT Moist woods, common in floodplains.

HABIT Native biennial or perennial herb with thread-like rhizomes.

STEMS Upright, branched, hollow, to 50 cm tall, usually smooth but occasionally somewhat hairy.

LEAVES Basal leaves all simple, unlobed and undivided although sometimes with rounded teeth, to 6 cm long, heart-shaped at the base, smooth, on stalks to 15 cm long; cauline leaves 3- or 5-parted, sessile or nearly so, smooth.

FLOWERS Several, to 8 mm across. **Sepals** 5, free from each other, greenish yellow, to 4 mm long, smooth or with short, stiff hairs. **Petals** 5, free from each other, yellow, to 3 mm long, shorter than the sepals. **Stamens** 15–30. **Pistils** many in each flower, each with a superior ovary.

FRUIT Achenes many in a head, the head to 8 mm long, each achene obovoid, to 2 mm long, with a minute beak.

WETLAND STATUS AW FACW | GP FAC | WMV FACW

FIELD NOTES Petals bright yellow, about 3 times as long as the sepals; leaves smooth, lanceolate.

achene

Ranunculus alismifolius Geyer ex Benth.
WATER-PLANTAIN BUTTERCUP

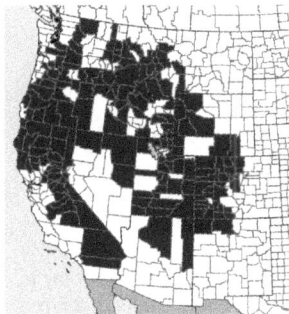

FAMILY Ranunculaceae (Buttercup)

FLOWERING April–June

HABITAT Around lakes, along streams, wet meadows, roadside ditches.

HABIT Native perennial herb with several thickened roots.

STEMS Upright, often branched, usually rather stout, to 75 cm tall, smooth, hollow.

LEAVES Basal leaves and alternate cauline leaves both present, lanceolate, rounded or pointed at the tip, tapering to a long leaf stalk, to 10 cm long, to 4 cm wide, usually without teeth or lobes, smooth.

FLOWERS 1–few at the tips of the branches, to nearly 2.5 cm across on smooth stalks to 10 cm long. **Sepals** 5, yellow-green, free from each other, about 3 mm long, to 1/3 as long as the petals. **Petals** 5, bright yellow, free from each other, obovate, 3 times longer than the sepals. **Stamens** numerous. **Pistils** 30–50 in a rather compact cluster, each with a superior ovary, smooth.

FRUIT Achenes 30–50 in a small spherical head, each achene obovoid, to 3 mm long, with a short, curved or straight beak.

NOTE The leaves are reminiscent of those of *Alisma graminea*. The achenes are eaten by small birds and small mammals.

WETLAND STATUS AW FACW | GF OBL | WMV FACW

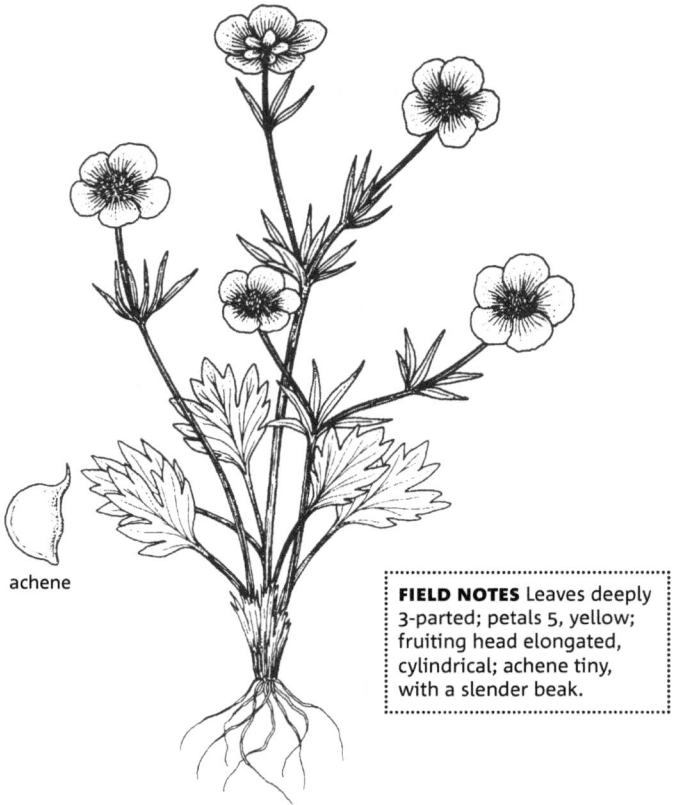

achene

> **FIELD NOTES** Leaves deeply 3-parted; petals 5, yellow; fruiting head elongated, cylindrical; achene tiny, with a slender beak.

Ranunculus eschscholtzii Schlecht.
ESCHSCHOLTZ BUTTERCUP

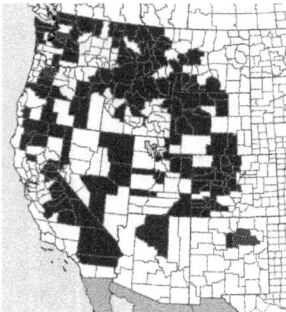

FAMILY Ranunculaceae (Buttercup)

FLOWERING July–August

HABITAT Moist meadows, moist rocky areas.

HABIT Native perennial herb with a thickened crown and fibrous roots.

STEMS Upright, usually unbranched, to 30 cm tall, smooth.

LEAVES Mostly basal with a few on the stem, circular in outline, deeply 3-parted, to 4 cm wide; stalks to 8 cm long.

FLOWERS 1–few at the tip of the stems, on stalks to 10 cm long. **Sepals** 5, yellow tinged with lavender, free from each other, 4–8 mm long, smooth or sparsely hairy. **Petals** 5, yellow, free from each other, obovate, 8–13 mm long. **Stamens** 20–40. **Pistils** many in an ovoid head, the ovaries superior.

FRUIT Achenes many in an ovoid head 13–20 mm long, 6–8 mm thick, each achene oblongoid, about 1 mm long, with a slender beak.

NOTE The achenes are eaten by small animals.

WETLAND STATUS AW FAC | GP FACW | WMV FACW

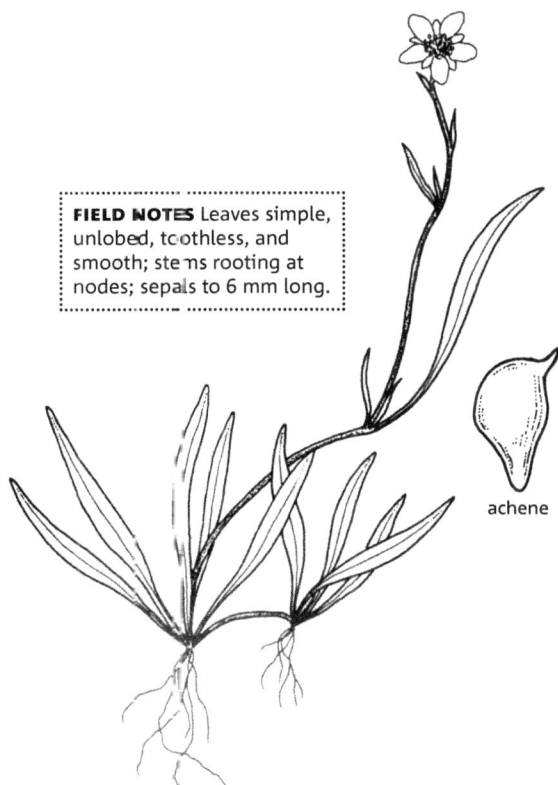

> **FIELD NOTES** Leaves simple, unlobed, toothless, and smooth; stems rooting at nodes; sepals to 6 mm long.

achene

Ranunculus flammula L.
CREEPING SPEARWORT

FAMILY Ranunculaceae (Buttercup)

FLOWERING July–August

HABITAT Marshes, wet meadows.

HABIT Native, sprawling perennial herb rooting at the nodes.

STEMS Sprawling, rooting at the nodes, usually smooth, slender, to 45 cm long.

LEAVES Simple, often clustered at the rooting nodes, narrowly spatulate to oblanceolate, to 6 cm long, to 8 mm wide, pointed at the tip, tapering to the base, smooth or nearly so, without teeth or lobes.

FLOWERS 1–few clustered at the rooting nodes, on stalks to 8 cm long. **Sepals** 5, greenish yellow, free from each other, to 6 mm long, sometimes turned downward. **Petals** 5 or 10, bright yellow, free from each other, a little longer than the sepals. **Stamens** 20–30. **Pistils** many in a rounded head, each with a superior ovary.

FRUIT Achenes 10–25. clustered in a rounded head, each achene to 2 mm long, with a short, curved beak.

NOTE The achenes are eaten by waterfowl.

WETLAND STATUS AW OBL | GP FACW | WMV FACW

> **FIELD NOTES** Stems and leaves completely smooth; basal leaves usually lobed and coarsely toothed; achenes in round heads 10-25 mm in diameter.

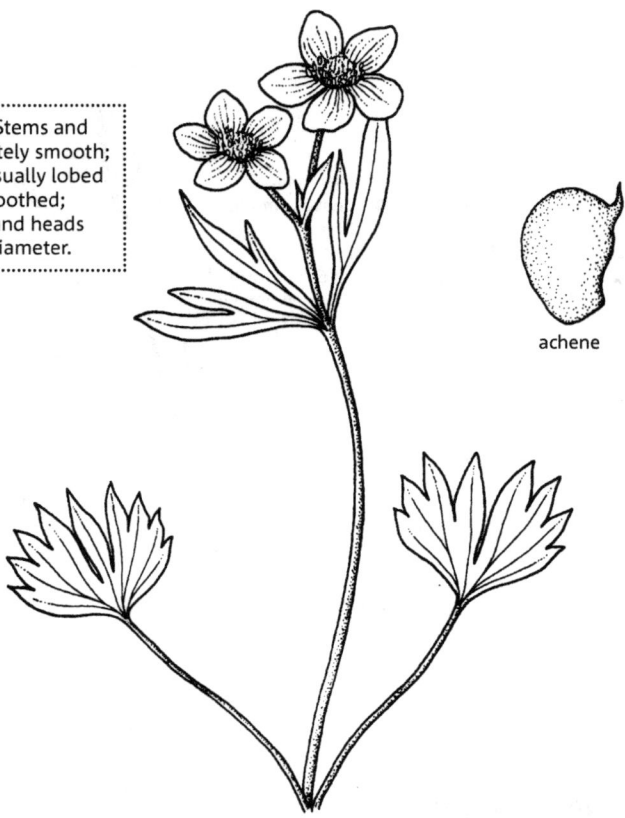

achene

Ranunculus glaberrimus Hook.
SAGEBRUSH BUTTERCUP

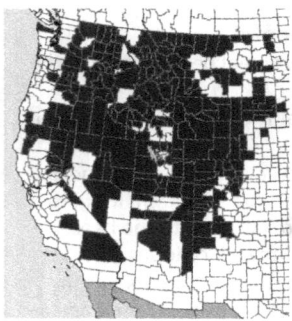

FAMILY Ranunculaceae (Buttercup)

FLOWERING April–June

HABITAT Moist woods, prairies, meadows; also drier sagebrush habitats.

HABIT Native perennial herb with fleshy roots.

STEMS Lying flat to upright, usually unbranched. to 15 cm long, smooth or rarely hairy.

LEAVES Mostly all basal, simple, spherical to obovate, often lobed and coarsely toothed, to 5 cm long, to 30 mm wide, usually smooth, on stalks to 8 cm long; leaves on stem fewer, alternate, narrower, usually 3-parted.

FLOWERS 1–6 at the tips of usually leafless stalks. **Sepals** 5, green or purplish, free from each other, elliptic, 4–8 mm long, smooth or hairy. **Petals** 5, yellow, free from each other, obovate, to 20 mm long. **Stamens** 40–80. **Pistils** many, free from each other on a conical receptacle, each with a superior ovary.

FRUIT Many achenes crowded into a spherical head, the head 10–25 mm in diameter; each achene obovoid. to 2 mm long, usually hairy, with a short, straight beak.

NOTE The achenes are eaten by birds.

WETLAND STATUS AW FAC | GP FAC | WMV FACU

FIELD NOTES Petals yellow, slightly shorter than the sepals; achenes flattened, borne in a cylindrical head; terminal lobe of each leaf stalked.

achene

Ranunculus pensylvanicus L. f.
PENNSYLVANIA BUTTERCUP

FAMILY Ranunculaceae (Buttercup)

FLOWERING July–August

HABITAT Wet meadows, marshes, roadside ditches.

HABIT Native perennial herb with thickened roots.

STEMS Upright, branched, to 60 cm tall, with spreading hairs.

LEAVES Basal and alternate, deeply 3-lobed, the terminal lobe stalked, with all the lobes coarsely toothed, hairy.

FLOWERS Few near the tips of the stems, borne on short, hairy stalks. **Sepals** 5, green, free from each other, to 4 mm long. **Petals** 5, yellow, free from each other, mostly about 3 mm long, never as long as the sepals. **Stamens** numerous. **Pistils** several crowded together, each with a superior ovary.

FRUIT Several achenes crowded into a cylindrical head to 20 mm long; each achene flattened, nearly spherical, to 3 mm long, with a flat, pointed, straight or curved beak.

NOTE The achenes are eaten by birds.

WETLAND STATUS AW FACW | CP FACW | WMV FACW

FIELD NOTES Basal leaves hastate (the leaves having a pair of basal lobes that project at right angles to the main axis of the blade); valves on the fruit the same length as the achene.

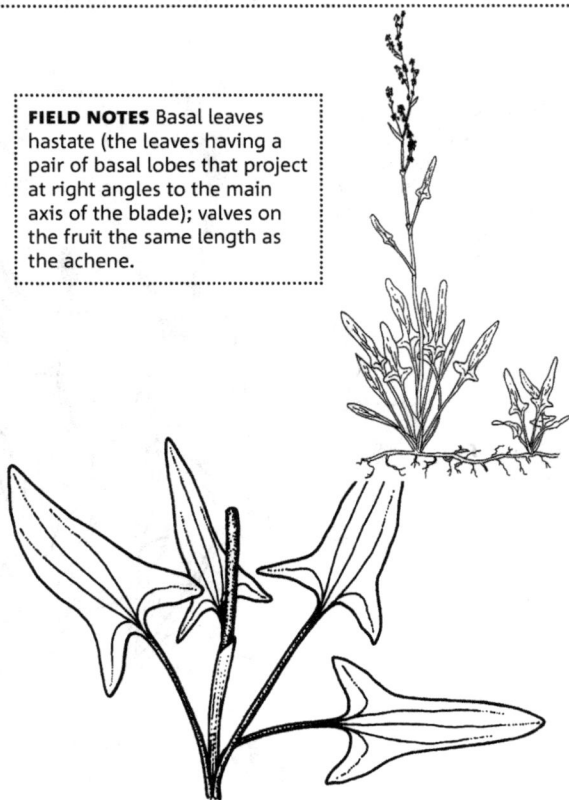

inflorescence

Rumex acetosella L.
SHEEP SORREL

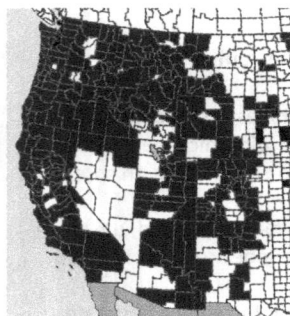

FAMILY Polygonaceae (Buckwheat)

FLOWERING April–September

HABITAT Disturbed areas, often in acid soil.

HABIT Introduced perennial herb from slender rhizomes.

STEMS Spreading to ascending, usually unbranched, to 40 cm long, smooth.

LEAVES Mostly basal, hastate, to 5 cm long, pointed at the tip, smooth, on long stalks.

FLOWERS Very small, in rings (whorls) around the upper part of the stem, the inflorescence slender, to 15 cm long, yellow or red, with small or no bracts; male and female flowers usually on separate plants. **Sepals** 6, green to yellow to reddish, about 1 mm long, united at their base. **Petals** absent. **Stamens** 6. **Pistils**: Ovary superior; styles 3.

FRUIT Achenes ellipsoid, about 1 mm long, surrounded by 3 valves of about the same length.

NOTE This species is native to Europe. The 3 valves that surround the achene are the persistent inner 3 sepals.

WETLAND STATUS AW FACU | GP FAC | WMV FACU

> **FIELD NOTES** Leaves heart-shaped at base; valves not more than 2.5 cm long, wart-like tubercles absent.

fruit

Rumex occidentalis S. Wats.
WESTERN DOCK

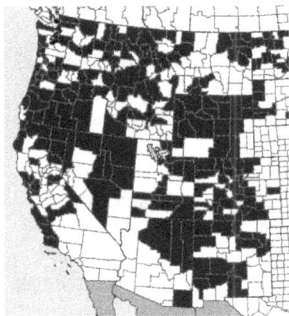

FAMILY Polygonaceae (Buckwheat)
FLOWERING August–September
HABITAT Marshes.
HABIT Native perennial herb with a taproot.
STEMS Upright, stout, branched or unbranched, to 1.4 m tall, smooth.
LEAVES Alternate and basal, lanceolate to lance-ovate, to 45 cm long, rounded or pointed at the tip, heart-shaped at the base, wavy along the margin, smooth, on long stalks; upper leaves smaller.
FLOWERS Borne in whorls usually arranged in a dense panicle, the panicle to 60 cm long; flower stalks 6–20 mm long. **Sepals** 6, the outer 3 united at the base, green but becoming rose during fruiting. **Petals** absent. **Stamens** 6 **Pistils**: Ovary superior; styles 3.
FRUIT Nutlets triangular, closely enclosed by the inner 3 sepals (valves), the valves to 6 mm long, veiny, rose-colored, without bristles or wart-like tubercles, more or less heart-shaped at the base.
NOTE The fruits are eaten by waterfowl.
WETLAND STATUS AW FACW | GF OBL | WMV FACW

> **FIELD NOTES** Leaves mostly basal, oblong to elliptic, without teeth; stems unbranched; leaves and stems smooth and glaucous.

disk flower

Senecio hydrophilus Nutt.
WATER GROUNDSEL

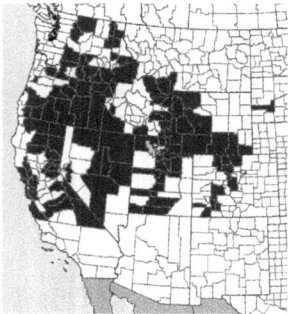

FAMILY Asteraceae (Aster)

FLOWERING May–August

HABITAT Along streams, in marshes, swamps, sometimes in alkaline habitats.

HABIT Native perennial herb with a thickened rootstock and fibrous roots.

STEMS Upright, unbranched, hollow, to 20 cm long, purplish, smooth, glaucous.

LEAVES Mostly basal, thick, oblong to elliptic, to 20 cm long, to 5 cm wide usually without teeth, smooth, glaucous; stem leaves few, smaller, sessile; leaf stalks of basal leaves elongated, winged.

FLOWERS Many crowded into heads, with several heads crowded into an inflorescence; each head to 2.5 cm across, consisting of 4–8 ray flowers (sometimes more), and a small central disk of disk flowers; bracts subtending each head 8–13, usually black-tipped. **Sepals** absent. **Petals:** some of them united to form yellow rays, others united to form yellow tubular flowers in the center of the head. **Stamens** 5. **Pistils:** Ovary inferior, smooth.

FRUIT Achenes smooth, with fine bristles at the tip.

NOTE The achenes have some importance as food for waterfowl.

WETLAND STATUS AW OBL | GP OBL | WMV OBL

> **FIELD NOTES** Leaves triangular, conspicuously toothed.

disk flower

Senecio triangularis Hook.
ARROW-LEAF GROUNDSEL

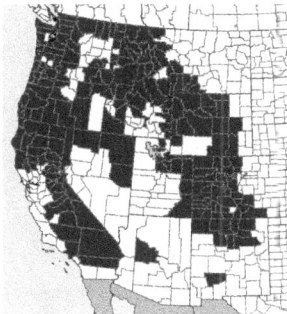

FAMILY Asteraceae (Aster)

FLOWERING July–September

HABITAT Along streams, in wet meadows, particularly in the higher mountains.

HABIT Native perennial herb from a thickened rootstock and with fibrous roots.

STEMS Upright, usually unbranched, to 1 m tall, smooth or less commonly with soft hairs, usually several growing from the base of the plant.

LEAVES Alternate, simple, distinctly triangular, pointed at the tip, heart-shaped or truncate to the base, conspicuously toothed, smooth or rarely with soft hairs, to 20 cm long, usually much smaller, to 13 cm wide, the lower on long stalks, the uppermost much smaller and without stalks

FLOWERS Many crowded together into heads, with several heads arranged to form a flat-topped cluster, each head 2.5–3 cm across, subtended by 9–13 black-tipped bracts, all flowers with rays. Rays yellow, to 12 per head, to 13 mm long or a little longer. **Sepals** absent. **Petals** 5, united to form a strap-shaped ray. **Stamens** 5. **Pistils**: Ovary inferior.

FRUIT Achenes crowded into small heads, each achene smooth, several-veined, to 5 mm long, bearing several soft, white bristles.

WETLAND STATUS AW FACW | CP OBL | WMV FACW

flowering branch and leaf

> **FIELD NOTES** Flowers purple; leaves hairy; usually stems and leaves usually glaucous.

Sidalcea neomexicana Gray
NEW MEXICO CHECKER-MALLOW

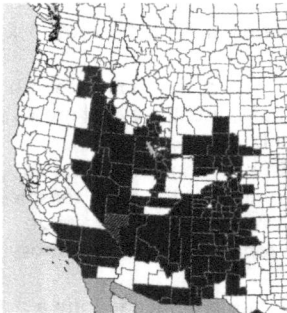

FAMILY Malvaceae (Mallow)

FLOWERING May–August

HABITAT Wet meadows in the mountains.

HABIT Native perennial herb with a thickened rootstock.

STEMS Upright, to 75 cm tall, smooth or hairy, usually glaucous.

LEAVES Basal and alternate, hairy, orbicular, sometimes glaucous, to 8 cm across, the basal leaves 5- to 9-lobed, with rounded teeth on the lobes, the upper leaves deeply 3- to 5-parted.

FLOWERS Several in a raceme, subtended by bracts 8–13 mm long; flower stalks smooth to hairy. **Sepals** 5, green, united below, 4–6 mm long, hairy, the teeth pointed. **Petals** 5, purple, free from each other, 13–20 mm long. **Stamens** numerous on a central column. **Pistils**: Ovary superior.

FRUIT 5- to 9-parted, to 3 mm long, tipped with a hairy beak.

NOTE This species is sometimes grown as an ornamental because of its large, handsome flowers, with the petals various shades of purple.

WETLAND STATUS AW FACW | GP FACW | WMV FACW

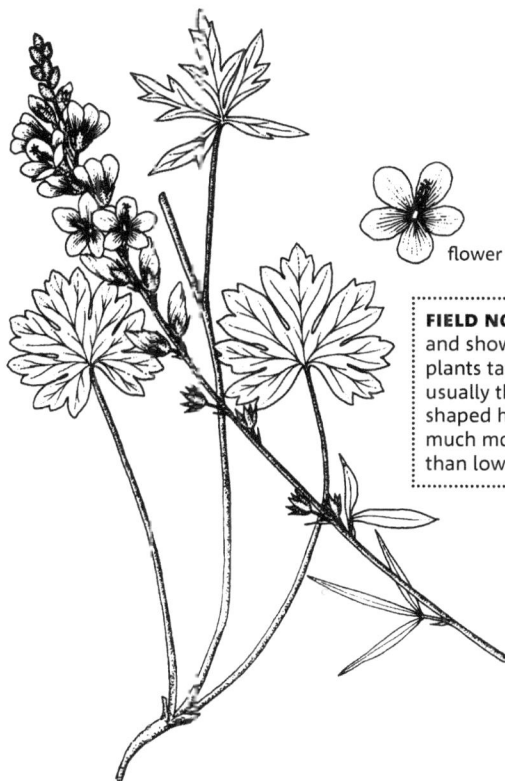

flower

> **FIELD NOTES** Flowers large and showy pink to rose-purple; plants taprooted; stems (and usually the leaves) with star-shaped hairs; upper leaves much more deeply divided than lower leaves.

Sidalcea oregana (Nutt. ex Torr. & Gray) Gray
OREGON CHECKER-MALLOW

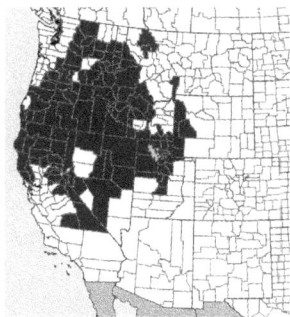

FAMILY Malvaceae (Mallow)

FLOWERING June–September

HABITAT Wet meadows, along streams, marshes, sometimes in sagebrush.

HABIT Native perennial herb with a taproot.

STEMS Upright, branched, to 1.5 m tall, usually with star-shaped hairs, sometimes glaucous.

LEAVES Alternate, palmately divided, the lower leaves deeply lobed, with each lobe again divided, the upper leaves more deeply lobed with narrow divisions, to 10 cm across, with star-shaped hairs; lower leaves on long stalks.

FLOWERS Many in a spike-like raceme, to 5 cm across, pink to rose-purple; bracts usually absent; flower stalks to 13 mm long. **Sepals** 5, green, united below, to 8 mm long, the lobes more or less triangular with star-shaped hairs. **Petals** 5, pink to rose-purple, free from each other, to 2.5 cm long. **Stamens** many, united with the styles of the pistils to form a column, the column hairy. **Pistils** 5, the filaments united with the stamens to form a column, the ovaries superior.

FRUIT Up to 3 mm long, veiny, with a very short beak.

WETLAND STATUS AW FACW | WMV FACW

> **FIELD NOTES** Plants annual; spines absent; flowers usually perfect, subtended by very small bracts; sepals unequal.

fruit

Suaeda calceoliformis (Hook.) Moq.
PURSH SEEPWEED

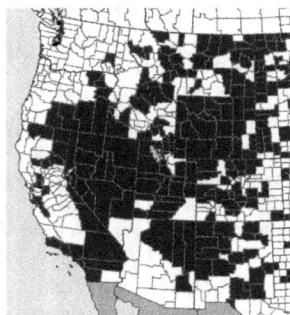

FAMILY Amaranthaceae (Amaranth)

FLOWERING July–October

HABITAT Damp saline or alkaline soils.

HABIT Native annual herb with fibrous roots.

STEMS Upright to spreading, branched or unbranched, to 60 cm long, smooth, usually bluish.

LEAVES Alternate, simple, linear, to nearly 5 cm long, to 4 mm wide, much smaller near the inflorescence, smooth, to 3 mm long.

FLOWERS Small, green, crowded into slender spikes, each flower in the axil of a bract; bracts broadly lanceolate, smooth, to 3 mm long. **Sepals** 5, green, deeply divided, unequal in size and shape, to 2 mm long. **Petals** absent. **Stamens** 5. **Pistils:** Ovary superior; **stigmas** 2–5.

FRUIT Fruits flattened, surrounded by the persistent sepals; seeds black, somewhat veiny, about 1 mm long.

SYNONYMS *Suaeda depressa* auct. non (Pursh) S. Wats.

WETLAND STATUS AW FACW | GP FACW | WMV FACW

> **FIELD NOTES** Genus *Suaeda* is
> has 5 greenish sepals sub-
> tended by bracts shorter than
> the sepals; flowers usually
> with both stamens and pistils.

fruit

Suaeda occidentalis (S. Wats.) S. Wats.
WESTERN SEEPWEED

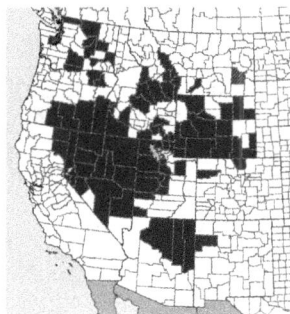

FAMILY Amaranthaceae (Amaranth)
FLOWERING July–September
HABITAT Alkaline soils.
HABIT Native annual herb with fibrous roots.
STEMS Spreading to upright, branched or unbranched,
to 45 cm long, sometimes glaucous.
LEAVES Alternate, simple, linear, to 20 mm long, about
1 mm wide, pointed at the tip, tapering to the base.
FLOWERS Several in short clusters in the axils of the
leaves, subtended by bracts shorter than the sepals.
Sepals 5, green, united below, the lobes rounded at the
tip **Petals** absent. **Stamens** 5. **Pistils:** Ovary superior;
styles 2.
FRUIT Nearly spherical, enclosed by the sepals; seeds black, shiny.
WETLAND STATUS AW FACW | GP FACW | WMV FACW

> **FIELD NOTES** Flower heads many in leafy panicles, violet, blue, or white; bracts subtending each flower head are ciliate with a green tip.

disk flower

Symphyotrichum chilense (Nees) Nesom
COMMON CALIFORNIA ASTER

FAMILY Asteraceae (Aster)

FLOWERING June–October

HABITAT Moist or dry fields, along streams.

HABIT Native perennial herb with a thickened rootstock.

STEMS Upright, branched, to 1 m tall, usually hairy.

LEAVES Lowest leaves obovate to oblanceolate, to 13 cm long, to 4 cm wide, pointed at the tip, tapering to a sessile or clasping base or to a winged stalk, rough on the upper surface and edges, toothed or untoothed; middle leaves lanceolate to linear-lanceolate, to 10 cm long, to 2.5 cm wide, sessile, rough to the touch, with or without teeth.

FLOWERS Many borne in heads 1–2.5 cm across, the heads borne in leafy panicles; each head subtended by bracts 6–8 mm long, narrowly oblong, whitish with a green tip, ciliate. **Sepals** absent. **Petals** 5, some of them united to form 20–35 violet, blue, or white rays to 13 mm long, others united to form yellow tubular flowers that form a central disk. **Stamens** 5. **Pistils**: Ovary inferior, hairy.

FRUIT Achenes oblanceoloid, hairy, with soft white bristles at the tip.

NOTE The achenes are eaten by birds.

SYNONYMS *Aster chilensis* Nees

WETLAND STATUS AW FAC | WMV FAC

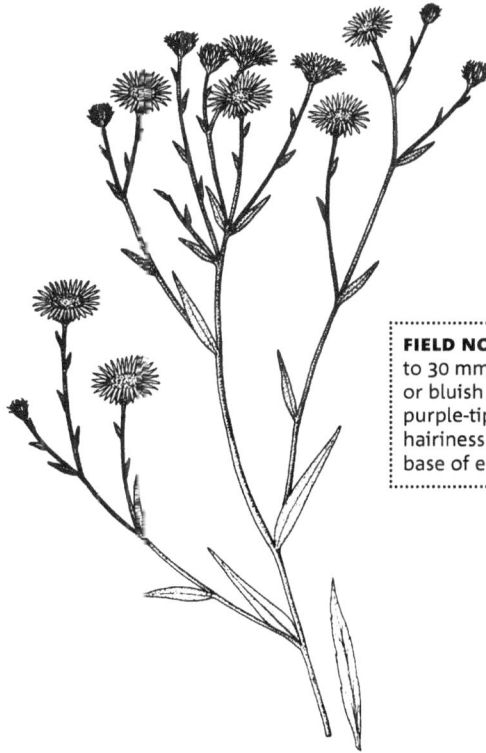

> **FIELD NOTES** Flower heads to 30 mm across, with white or bluish purple rays; bracts purple-tipped; stems with hairiness in lines below the base of each leaf.

Symphyotrichum lanceolatum (Willd.) Nesom
SISKIYOU ASTER

FAMILY Asteraceae (Aster)

FLOWERING August–October

HABITAT Along streams, in wet meadows.

HABIT Native perennial herb with creeping rhizomes.

STEMS Upright, usually branched, to 2 m tall, smooth except for lines of hairs below the base of each leaf.

LEAVES Alternate, simple, linear-lanceolate, to 15 cm long, to 20 mm wide, pointed at the tip, tapering to the base, the uppermost usually without teeth, the lowermost usually toothed, rough to the touch, ciliate along the edges.

FLOWERS Many united into heads to 30 mm across, each head with 20–35 white or bluish purple ray flowers and several short yellow tubular flowers forming a disk: bracts subtending each head linear, ciliate. **Sepals** absent. **Petals** 5, some of them united to form rays, others united to form tubes comprising the central disk. **Stamens** 5. **Pistils:** Ovary inferior, slightly hairy.

FRUIT Achenes elongated, slightly hairy, to 1 mm long, with a tuft of white or tawny hairs at the tip.

NOTE The achenes may be eaten by birds.

SYNONYMS *Aster hesperius* Gray

WETLAND STATUS AW OBL | GP FACW | WMV OBL

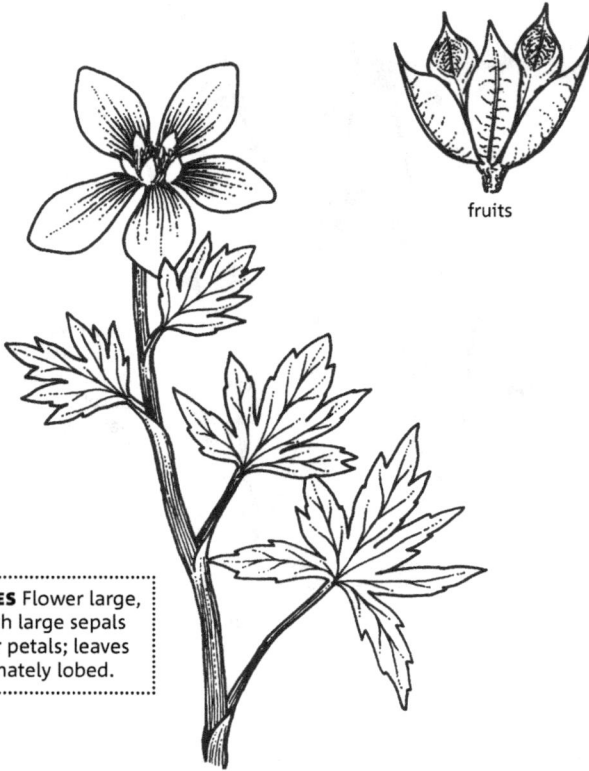

fruits

> **FIELD NOTES** Flower large, solitary, with large sepals and smaller petals; leaves deeply palmately lobed.

Trollius laxus Salisb.
AMERICAN GLOBEFLOWER

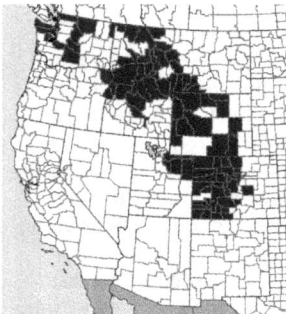

FAMILY Ranunculaceae (Buttercup)

FLOWERING June–August

HABITAT Swamps, wet woods, along streams, wet meadows, particularly in the mountains.

HABIT Native perennial herb with thick, fibrous roots.

STEMS Upright, slender, to 45 cm tall, smooth.

LEAVES Alternate, palmately 5-lobed, to 20 cm long and broad, the lobes coarsely toothed or partly divided again, smooth, the basal leaves on long stalks, the 1 or 2 cauline leaves on short stalks or sessile.

FLOWERS Solitary, to 6 cm across, showy, usually whitish or greenish yellow. **Sepals** usually 5, free from each other, petal-like, white or greenish yellow, to 2.5 cm long. **Petals** usually 5, free from each other, to 6 mm long, each with a basal gland. **Stamens** numerous, usually longer than the petals. **Pistils** numerous, each with a superior ovary.

FRUIT Follicles several, to 13 mm long, containing many seeds.

NOTE This handsome species varies considerably in flower color, and are sometimes grown as an ornamental.

SYNONYMS *Trollius albiflorus* (Gray) Rydb.

WETLAND STATUS AW OBL | GP OBL | WMV OBL

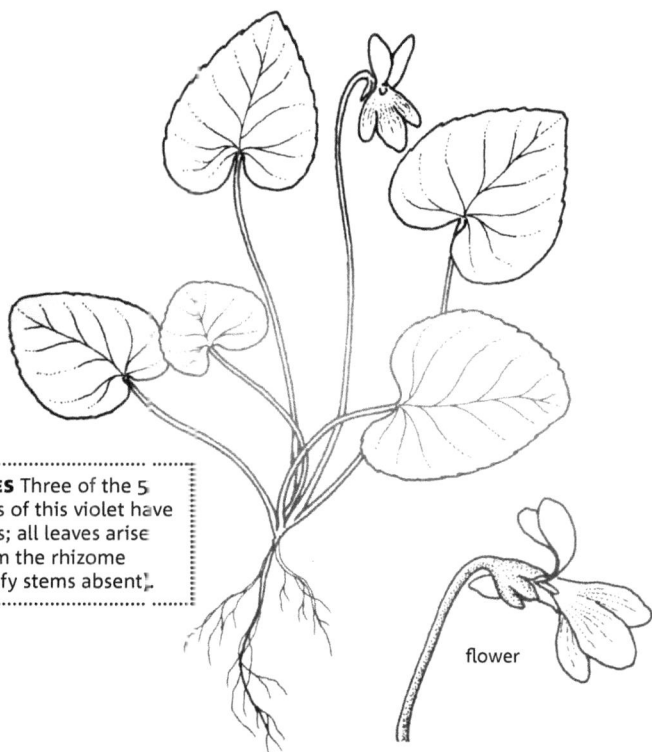

> **FIELD NOTES** Three of the 5 white petals of this violet have purple veins; all leaves arise directly from the rhizome (upright leafy stems absent).

flower

Viola macloskeyi Lloyd
SMALL WHITE VIOLET

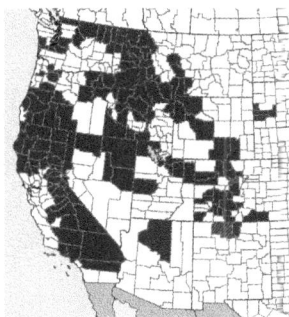

FAMILY Violaceae (Violet)

FLOWERING May–August

HABITAT Wet meadows, along streams, bogs.

HABIT Native perennial herb with creeping rhizomes and stolons.

STEMS Rhizomes below ground; horizontal stolons produced late in year; no upright leafy stems present.

LEAVES Simple, ovate to nearly round, rounded or more or less pointed at the tip, heart-shaped at the base, to 4 cm long, smooth or sparsely hairy, on smooth or hairy stalks to 10 cm long; stipules ovate, without teeth.

FLOWERS Solitary on stalks arising directly from the rhizomes, on stalks to 15 cm long, longer than the leaves. **Sepals** 5, green, free from each other, ovate to lanceolate, to 4 mm long, smooth. **Petals** 5, free from each other, white, the lower 3 also with purple stripes, 6 to nearly 13 mm long; one of the petals spurred, the spur to 3 mm long. **Stamens** 5. **Pistils:** Ovary superior, smooth.

FRUIT Capsules ovoid, green, to 6 mm long, smooth; seeds dark brown, about 1 mm in diameter.

NOTE The stolons are formed late in the year and often give rise to large mats of this species.

WETLAND STATUS AW OBL | GP FACW | WMV OBL

> **FIELD NOTES** Stemless violet whose leaf stalks and flower stalks arise directly from underground rhizomes. Flowers purple; stolons that trail across the soil absent.

flower

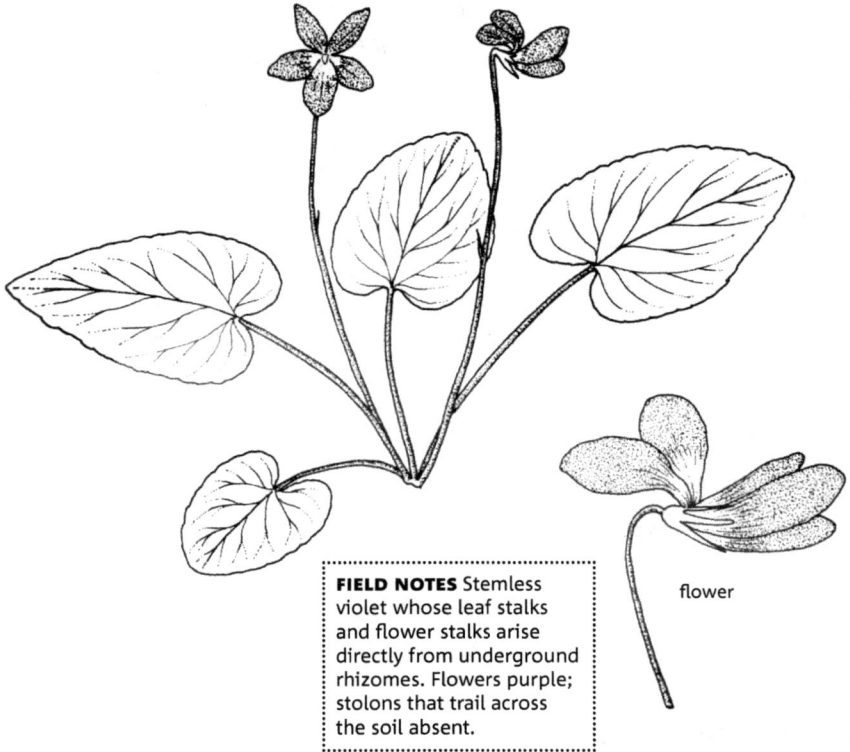

Viola nephrophylla Greene
NORTHERN BOG VIOLET

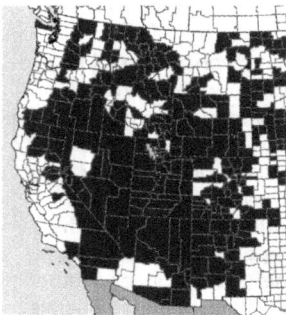

FAMILY Violaceae (Violet)

FLOWERING April–June

HABITAT Around springs, in bogs.

HABIT Native perennial herb with a thickened rhizome and long, fibrous roots; stolons not present.

STEMS Underground as a thickened rhizome; flowering stems without leaves, smooth, to 25 cm long.

LEAVES All basal, heart-shaped, to 6 cm long and wide, round-toothed, smooth, on long, smooth stalks to 15 cm long.

FLOWERS Solitary on long stalks that arise directly from the rhizomes, the flower stalks often longer than the leaves. **Sepals** 5, green, ovate-lanceolate, free from each other, smooth. **Petals** 5, purple, free from each other, 13–20 mm long, 3 of them with a dense tuft of hairs, 1 of them spurred. **Stamens** 5. **Pistils:** Ovary superior.

FRUIT Capsules oblongoid to ovoid, 6–13 mm long, smooth; seeds pale, shiny, to 2 mm long.

WETLAND STATUS AW FACW | GP FACW | WMV FACW

WESTERN WETLAND FLORA

female "cores"

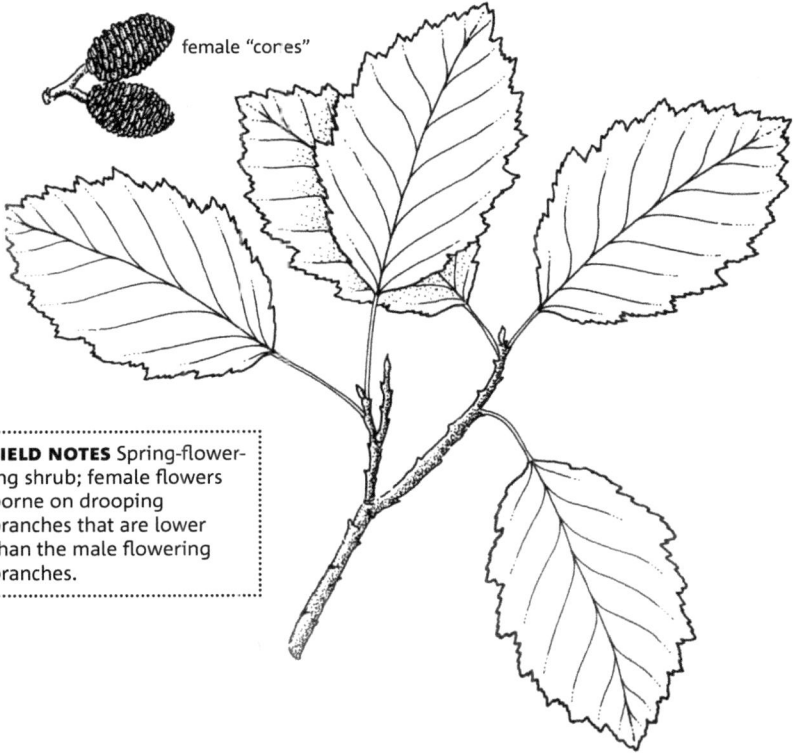

> **FIELD NOTES** Spring-flower-
> ing shrub; female flowers
> borne on drooping
> branches that are lower
> than the male flowering
> branches.

Alnus incana (L.) Moench
SPECKLED ALDER

FAMILY Betulaceae (Birch)

FLOWERING March–June

HABITAT Along streams, in swamps.

HABIT Native sprawling or upright shrub to 5 m tall.

STEMS Trunk and branchlets brown to blackish and marked with elongated, white lenticels.

LEAVES Alternate, simple, ovate to oval to broadly ellip-tic, rounded to pointed at the tip, rounded or somewhat heart-shaped at the base, usually doubly toothed, more or less smooth.

FLOWERS Male and female flowers borne separately but on the same plant, each kind arranged in clusters; male flowers usually 3 in a cluster, subtended by 4–5 small bracts; female flowers in spikes, each pair of flowers subtended by fleshy bracts. **Sepals** 3–5, united, greenish in the male flowers, absent in the female flowers. **Petals** absent. **Stamens** 3–5. **Pistils:** Ovary 1, apparently inferior.

FRUIT Woody "cones" to 13 mm long, with narrowly winged nutlets.

NOTE The fleshy bracts subtending the female flowers form the woody scales of the "cone" in fruit. The nutlets are eaten by birds and mammals. Plants of western North America are *Alnus incana* ssp. *tenuifolia* (Nutt.) Breitung.

WETLAND STATUS AW FACW | GP FACW | WMV FACW

> **FIELD NOTES** Leaves linear-lanceolate, toothed; flowers yellowish; flower heads of disk flowers only, in terminal panicles.

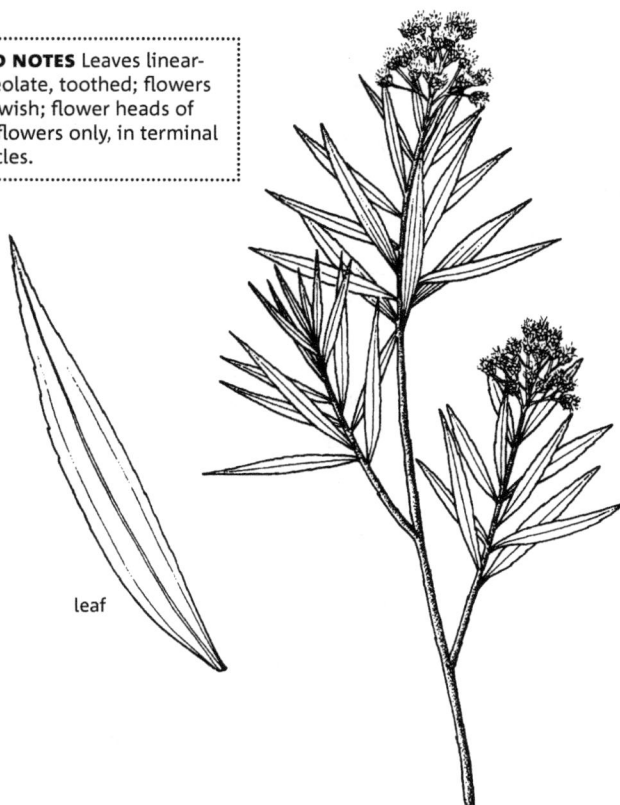

leaf

Baccharis glutinosa Pers.
STICKY FALSE-WILLOW

FAMILY Asteraceae (Aster)

FLOWERING April–October

HABITAT Along streams and other waterways, particularly in the desert.

HABIT Native shrub, woody at least at the base, to 3 m tall.

STEMS Upright, smooth, unbranched below, branched above.

LEAVES Alternate, simple, linear-lanceolate, to 15 cm long, to 2.5 cm wide, pointed at the tip, tapering to the base, without hairs but sticky, usually toothed; leaf stalks to 8 mm long.

FLOWERS Many crowded into heads, the male heads separate from the female heads and on separate plants, both types forming terminal panicles; heads containing only disk flowers, subtended by ovate to lance-ovate bracts to 4 mm long. **Sepals** absent. **Petals** 5, united to form yellowish tubular flowers that comprise the disk. **Stamens** 5. **Pistils:** Ovary inferior, smooth.

FRUIT Achenes narrowly ellipsoid, greenish, about 1 mm long, smooth.

WETLAND STATUS AW FACW | WMV OBL

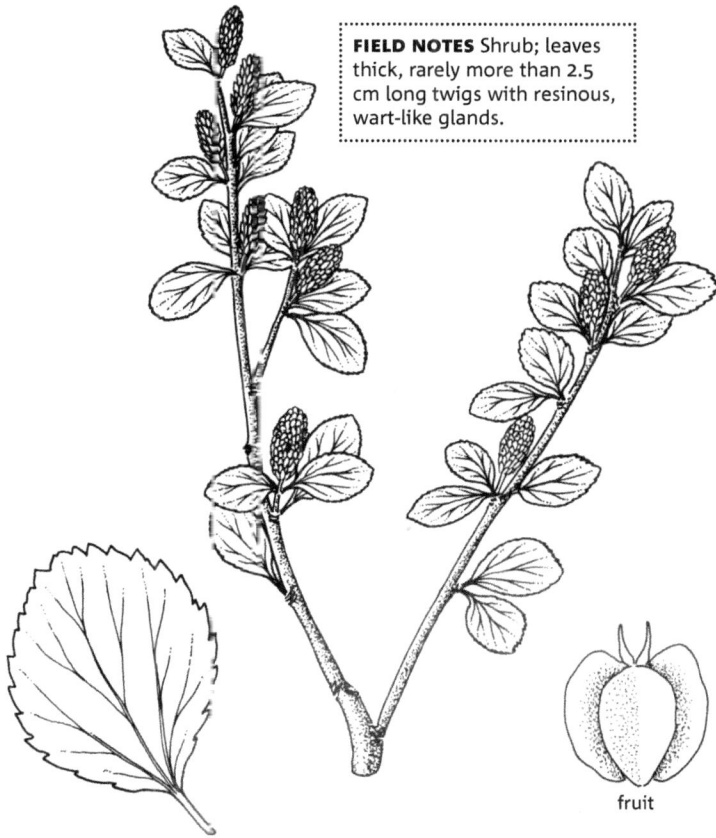

FIELD NOTES Shrub; leaves thick, rarely more than 2.5 cm long twigs with resinous, wart-like glands.

fruit

Betula glandulosa Michx.
TUNDRA DWARF BIRCH

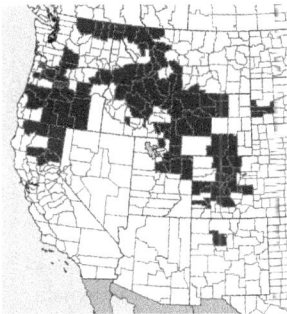

FAMILY Betulaceae (Birch)

FLOWERING April–June

HABITAT Bogs, fens, along streams, sometimes in shallow, standing water.

HABIT Native upright or sometimes prostrate shrub.

TWIGS Upright or prostrate, smooth except for wart-like, resin-producing glands.

BARK Brown, not peeling.

LEAVES Alternate, simple, orbicular to obovate, rarely more than 2.5 cm long, to 20 mm wide, rounded at the tip, rounded or tapering at the base, thick, toothed, smooth and shiny on the upper surface, glandular-dotted on the lower surface.

FLOWERS Crowded together into spikes, the male spikes borne separately from the female but on the same plant; male spikes to 13 mm long; female spikes longer and thicker. **Sepals** absent. **Petals** absent. **Stamens** 2. **Pistils:** Ovary superior.

FRUIT Nutlets with a pair of obscure wings.

WETLAND STATUS AW OBL | GP OBL | WMV OBL

> **FIELD NOTES** Small tree or large shrub; bark shiny, bronze-colored, that does not peel off; leaves thin, ovate, pointed.

fruit

Betula occidentalis Hook.
WATER BIRCH

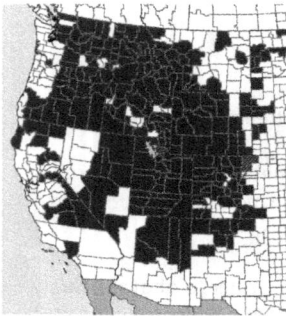

FAMILY Betulaceae (Birch)
FLOWERING May–June
HABITAT Along streams, in ravines, bogs.
HABIT Native small tree or large shrub.
BARK Smooth, shiny, bronze, not peeling.
STEMS Usually slightly glandular-sticky, with conspicuous pale lenticels.
LEAVES Alternate, simple, ovate, to 5 cm long, not quite as wide, usually pointed at the tip, rounded or tapering to the base, doubly toothed, smooth, thin.
FLOWERS Male and female flowers borne in separate spikes, the male spikes slender, pendulous, to 20 mm long, with flowers in groups of 3, each group subtended by a small bract, the female spikes thicker and shorter, erect, with flowers in groups of 3, each group subtended by a small, 3-parted bract. **Sepals** absent. **Petals** absent. **Stamens** 2. **Pistils:** Ovary inferior.
FRUIT "Cone"-like spikes consisting of 3-parted bracts that subtend tiny nutlets with narrow wings.
NOTE The immature male spikes are present during the winter.
SYNONYMS *Betula fontinalis* Sarg.
WETLAND STATUS AW FACW | GP FACW | WMV FACW

> **FIELD NOTES** Low shrub; leaves opposite, evergreen, oblong or oval, less than twice as long as broad.

flower

Kalmia microphylla (Hook.) Heller
ALPINE BOG LAUREL

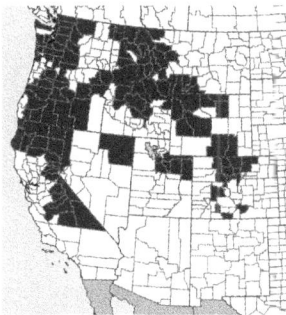

FAMILY Ericaceae (Heath)

FLOWERING June–August

HABITAT Moist ground.

HABIT Native low-growing, spreading shrub.

STEMS Much branched, to 20 cm long, smooth or hairy.

LEAVES Opposite, simple, evergreen, oblong or oval, rounded or barely pointed at the tip. tapering to the nearly sessile base, dark green above, paler beneath, to 2.5 cm long, more than 13 mm wide, smooth or slightly hairy, without teeth, flat.

FLOWERS Few in terminal clusters, to 20 mm across, on smooth, slender stalks to 5 cm long. **Sepals** 5, green, united below, 6–8 mm long. smooth, the lobes oblong to ovate. **Petals** 5, united to form a saucer, rose-purple, to 20 mm across. **Stamens** 10, not exserted beyond the petals. **Pistils**: Ovary superior, smooth; **style** slender.

FRUIT Capsules spherical, to 8 mm in diameter, smooth, subtended by the persistent stipules.

WETLAND STATUS AW OBL   WMV OBL

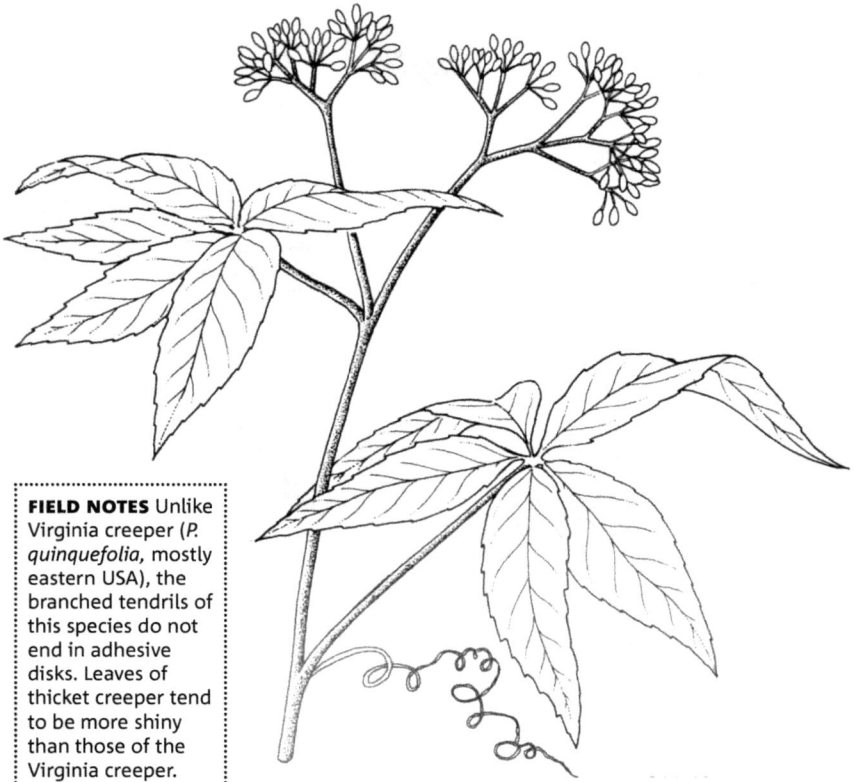

FIELD NOTES Unlike Virginia creeper (*P. quinquefolia*, mostly eastern USA), the branched tendrils of this species do not end in adhesive disks. Leaves of thicket creeper tend to be more shiny than those of the Virginia creeper.

Parthenocissus inserta (Kerner) Fritsch
THICKET CREEPER

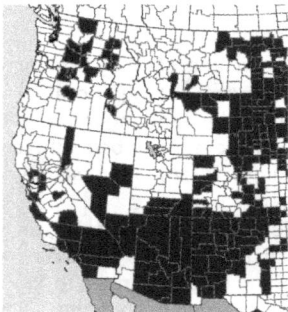

FAMILY Vitaceae (Grape)
FLOWERING May–July
HABITAT Thickets, open woods.
HABIT Native climbing or scrambling vine with tendrils.
STEMS Climbing or scrambling, to 1 m long; tendrils branched, not bearing adhesive disks at the tips.
LEAVES Alternate, palmately compound, with 5 leaflets; leaflets elliptic to obovate. to 15 cm long, pointed at the tip, tapering to the base, coarsely toothed, shiny on the upper surface, sparsely hairy on the lower.
FLOWERS Several in compound cymes. **Sepals** 5, green, tooth-like, united below, about 1 mm long. **Petals** 5, yellow-green, free from each other. 1–2 mm long. **Stamens** 5. **Pistils**: Ovary superior.
FRUIT Berries brown, to 13 mm in diameter, more or less veiny, containing 1–4 seeds.
SYNONYMS *Parthenocissus vitacea* (Knerr) A. S. Hitchc.
WETLAND STATUS AW FACU | GP FAC | WMV FACU

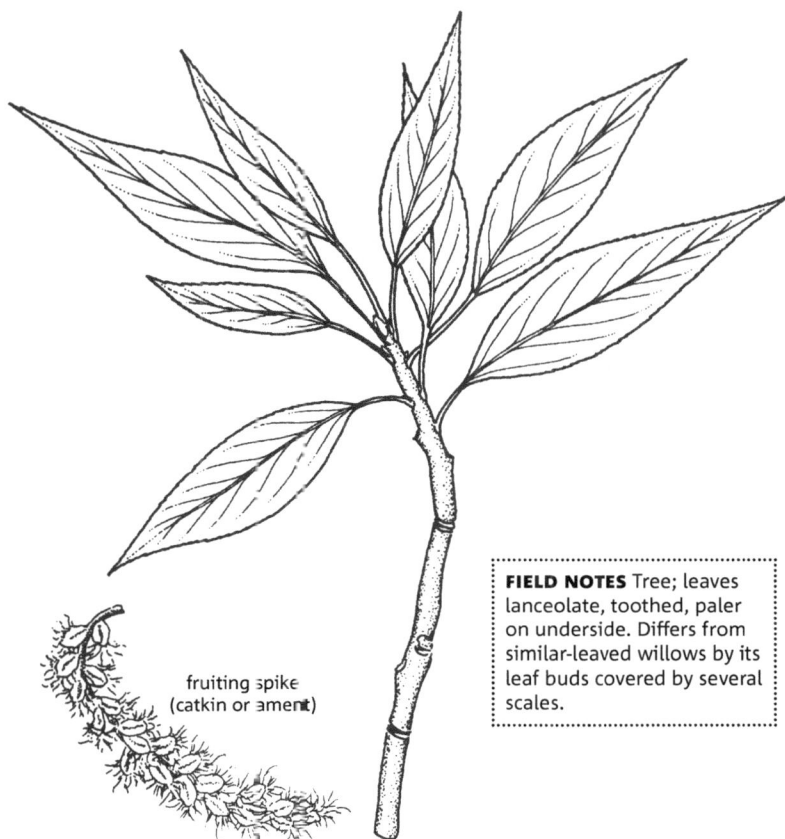

FIELD NOTES Tree; leaves lanceolate, toothed, paler on underside. Differs from similar-leaved willows by its leaf buds covered by several scales.

fruiting spike
(catkin or ament)

Populus angustifolia James
NARROW-LEAF COTTONWOOD

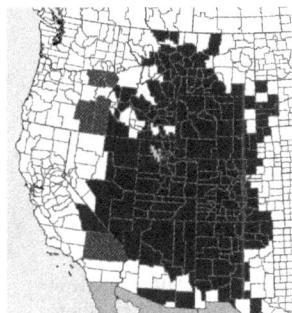

FAMILY Salicaceae (Willow)

FLOWERING February–April

HABITAT Along streams, in moist woods.

HABIT Native tree to 20 m tall; trunk light brown, shallowly furrowed.

STEMS Twigs orange-brown at first, becoming tan. not hairy.

LEAVES Alternate, simple, lanceolate, to 10 cm long, to 5 cm wide, long-pointed at the tip, rounded or tapering at the base, paler green on the lower surface, smooth, toothed.

FLOWERS Many crowded into elongated spikes to 10 cm long, the male spikes borne on separate trees from the female flowers each flower subtended by 1 bract. **Sepals** absent. **Petals** absent. **Stamens** 10–20. **Pistils:** Ovary superior.

FRUIT Capsules nearly spherical but with a short point at the tip. to 6 mm long: seeds with tufts of hairs.

NOTE The flowers bloom before the leaves unfold.

WETLAND STATUS AW FACW | GP FACW | WMV FACW

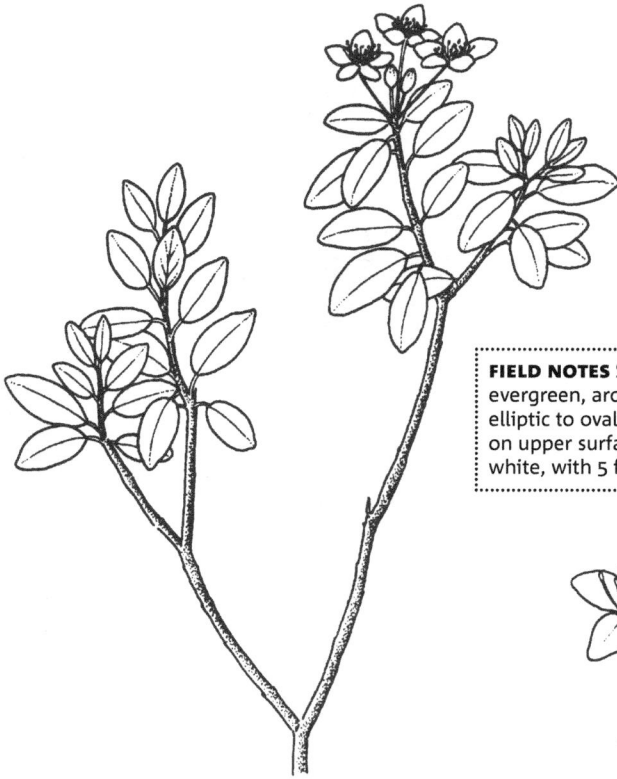

> **FIELD NOTES** Shrub; leaves evergreen, aromatic, are elliptic to oval and wrinkled on upper surface. Flowers white, with 5 free petals.

flower

Rhododendron columbianum (Piper) Harmaja
GLANDULAR LABRADOR-TEA

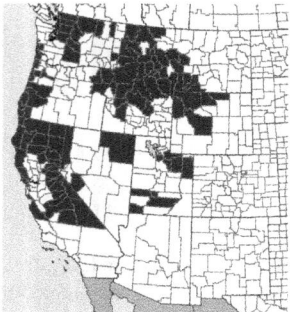

FAMILY Ericaceae (Heath)

FLOWERING May–August

HABITAT Wet meadows, bogs, swamps.

HABIT Native evergreen shrub with aromatic leaves.

STEMS Upright, usually branched, to 2 m tall; twigs glandular-hairy; bark at maturity peeling in shreds.

LEAVES Alternate, simple, evergreen, leathery, elliptic to oblong, usually rounded at the tip, rounded or tapering at the base, wrinkled-looking on the upper surface, glandular and paler on the lower surface, to 6 cm long, flat or rolled under along the edges; leaf stalks 6–13 mm long.

FLOWERS Several to many in terminal clusters; stalks usually glandular-hairy, yellow-green, to 2.5 cm long. **Sepals** 5, united below, green, to 2 mm long, the lobes ciliate. **Petals** 5, free from each other, oblong, white, 6–8 mm long. **Stamens** 10. **Pistils:** Ovary superior.

FRUIT Capsules nearly round to ovoid, to 6 mm long, with many elongated, winged seeds.

SYNONYMS *Ledum glandulosum* Nutt.

WETLAND STATUS AW FACW | WMV OBL

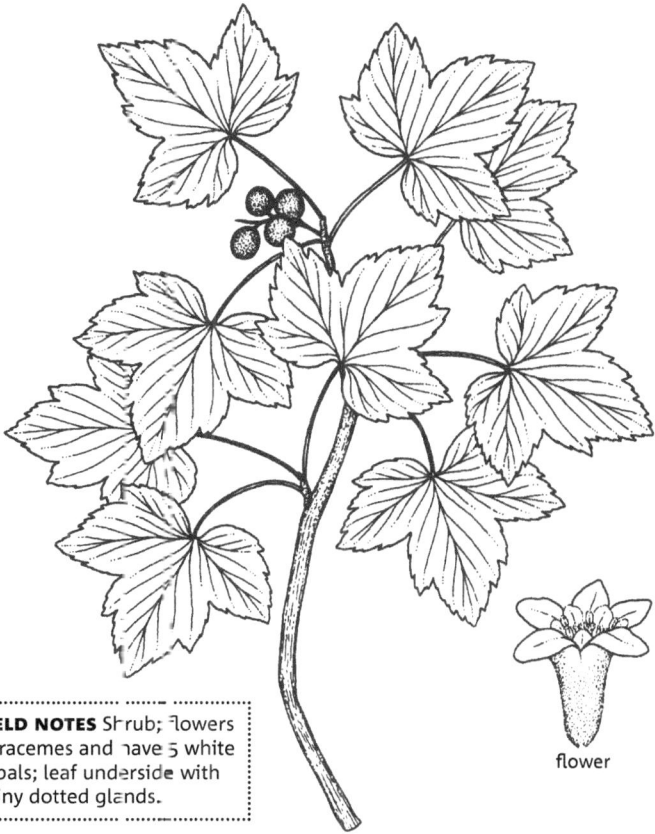

FIELD NOTES Shrub; flowers in racemes and have 5 white sepals; leaf underside with shiny dotted glands.

flower

Ribes hudsonianum Richards.
HUDSON BAY CURRANT

FAMILY Grossulariaceae (Currant)

FLOWERING May–June

HABITAT Along streams, wet woods, wet meadows, swamps.

HABIT Native upright shrub to 2 m tall.

STEMS Ascending to upright, without thorns.

LEAVES Alternate, simple, palmately 3- to 5-lobed, to 8 cm across, heart-shaped at the base, toothed, smooth or sparsely hairy, with shiny, glandular dots on the lower surface.

FLOWERS Several in ascending to upright racemes to 15 cm long; each flower on a stalk to 8 mm long. **Sepals** 5, united below to form a cup, white, to 6 mm long. **Petals** 5, white, free from each other, about 2 mm long. **Stamens** 5. **Pistils**: Ovary inferior, without hairs but with a few glands.

FRUIT Berries black, to 8 mm in diameter.

NOTE The berries are eaten by animals.

WETLAND STATUS AW FACW | CP OBL | WMV FACW

> **FIELD NOTES** Branches with few or no bristles or spines; leaves smooth.

flower

Ribes inerme Rydb.
WHITE-STEM GOOSEBERRY

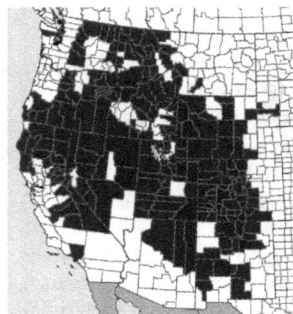

FAMILY Grossulariaceae (Currant)

FLOWERING May–June

HABITAT Along streams, shaded woods, particularly in the mountains.

HABIT Native much-branched shrub to 2.5 m tall. Twigs: Few or no spines at the nodes, and few or no bristles between the nodes.

LEAVES Alternate, simple, to 8 cm long, to 8 cm wide, palmately 3- or 5-lobed, each lobe with a few rounded teeth, smooth, rounded or heart-shaped at the base.

FLOWERS 1–4 in the axils of the leaves, on stalks shorter than the leaf stalks. **Sepals** 5, united to form a cup, green or purplish-tinged, smooth. **Petals** 5, free from each other, white, 1–3 mm long. **Stamens** 5. **Pistils:** Ovary inferior.

FRUIT Berries spherical, smooth, wine-colored, 6–13 mm in diameter.

NOTE The berries are eaten by birds and bears.

WETLAND STATUS AW FAC | GP FACW | WMV FAC

FIELD NOTES Tree; leaf underside paler than upper srface; leaves usually about 1/3 as broad as long; leaf stalks without glands; heart-shaped stipules often present.

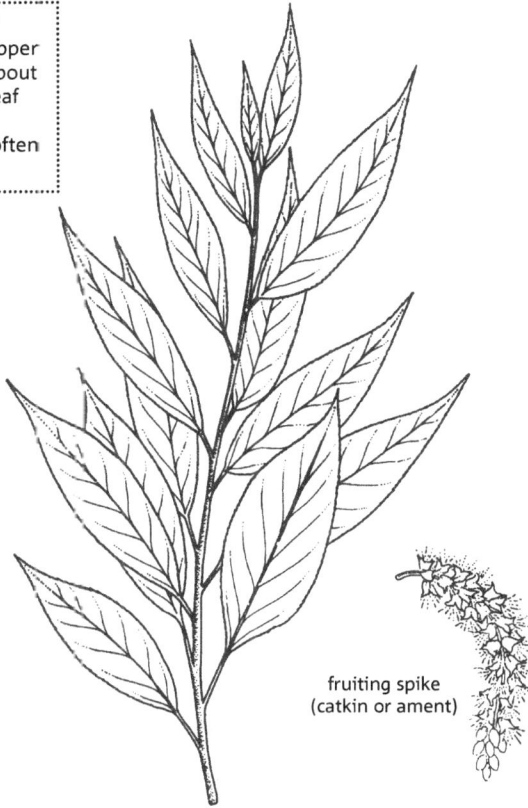

fruiting spike
(catkin or ament)

Salix amygdaloides Anderss.
PEACH-LEAF WILLOW

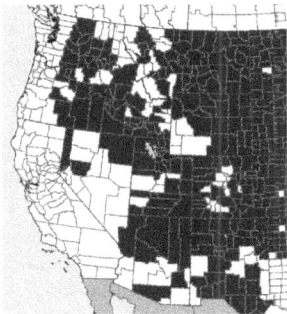

FAMILY Salicaceae (Willow)

FLOWERING April–May

HABITAT Along streams, around lakes and ponds, flood-p.ain woods, wet ditches.

HABIT Native tree to 15 m tall.

BARK Grayish brown, shallowly furrowed, becoming somewhat scaly.

STEMS Twigs shiny, gray to yellowish, smooth.

LEAVES Alternate, simple, lanceolate to broadly lanceolate, long-pointed at the tip, tapering or rounded at the base, finely toothed, to 8 cm long, to 4 cm wide, pale on the lower surface, smooth, without glands on the stalk; stipules often present frequently as much as 13 mm wide, heart-shaped.

FLOWERS Male and female borne in dense spikes on the same tree, opening as the leaves begin to open; male spikes very slender, to 10 cm long; female spikes not as slender, to 8 cm long. **Sepals** absent. **Petals** absent. **Stamens** 4–7. **Pistils**: Ovary smooth.

FRUIT Capsules ovoid, to 6 mm long, smooth, not crowded in the spike, each capsule on a very short stalk.

WETLAND STATUS AW FACW | CP FACW | WMV FACW

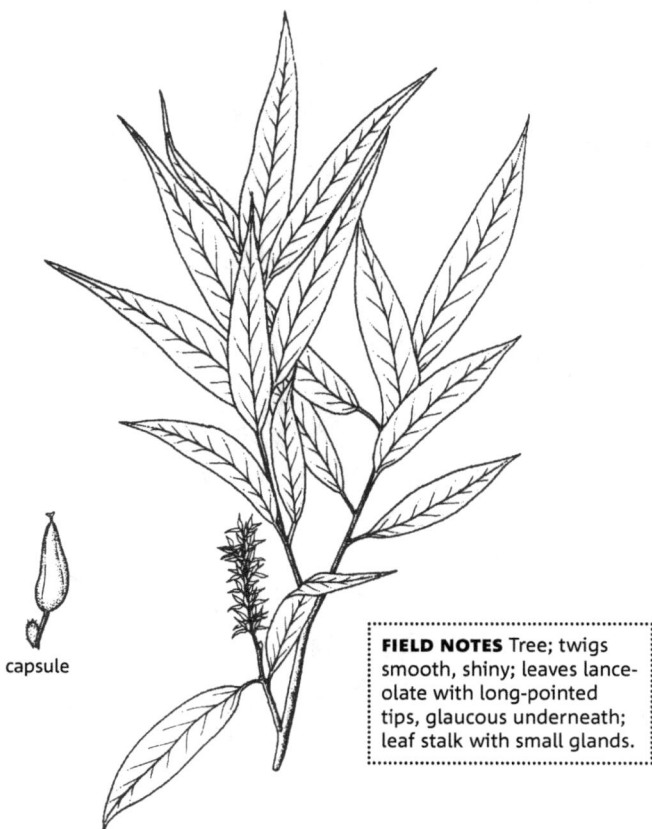

capsule

> **FIELD NOTES** Tree; twigs smooth, shiny; leaves lance-olate with long-pointed tips, glaucous underneath; leaf stalk with small glands.

Salix lasiandra Benth.
PACIFIC WILLOW

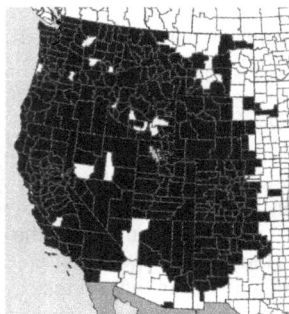

FAMILY Salicaceae (Willow)
FLOWERING March–May
HABITAT Along streams.
HABIT Native tree to 15 m tall.
BARK Bark rough, brown.
STEMS Twigs smooth, shiny, reddish.
LEAVES Alternate, simple, lanceolate, to 11 cm long, to 5 cm wide, long-pointed at the tip, tapering to the base, smooth, glaucous on the lower surface, with tiny glandular teeth along the edges; leaf stalk to 20 mm long, bearing small glands; stipules small.
FLOWERS Male and female flowers borne separately in separate spikes, the male spikes to 6 cm long, the female spikes to 11 cm long. **Sepals** absent. **Petals** absent. **Stamens** 4–5, the base of the filaments hairy. **Pistils:** Ovary superior, smooth.
FRUIT Capsules lanceoloid, to 8 mm long, smooth.
NOTE Deer and elk may browse the young shoots of this plant.
WETLAND STATUS AW FACW | GP FACW | WMV FACW

> **FIELD NOTES** Shrub or small tree; leaves seldom more than 3 times longer than wide; margins usually without teeth.

capsule

Salix lasiolepis Benth.
ARROYO WILLOW

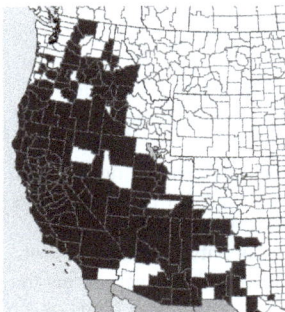

FAMILY Salicaceae (Willow)
FLOWERING February–April
HABITAT Along rocky streams and in arroyos.
HABIT Native shrub or small tree to 10 m tall.
STEMS Twigs yellow to dark brown, hairy.
LEAVES Alternate, simple, oblanceolate, to 10 cm long, to 5 cm wide, rounded or pointed at the tip, tapering to the base, smooth or slightly hairy, usually glaucous on the lower surface, usually without teeth, the edges sometimes rolled under.
FLOWERS Male and female flowers borne separately in separate spikes, to 8 cm long, appearing before the leaves begin to unfold. **Sepals** absent. **Petals** absent. **Stamens** 2, with smooth filaments. **Pistils**: Ovary superior, smooth.
FRUIT Capsules oblanceoloid, to 6 mm long, smooth, on stalks about 1 mm long.
NOTE Deer and elk may browse the young shoots of this plant.
WETLAND STATUS AW FACW | GP FACW | WMV FACW

> **FIELD NOTES** Shrub; leaves not more than four times longer than wide, usually smooth at maturity; margins toothed or toothless.

leaf

Salix lemmonii Bebb
LEMMON'S WILLOW

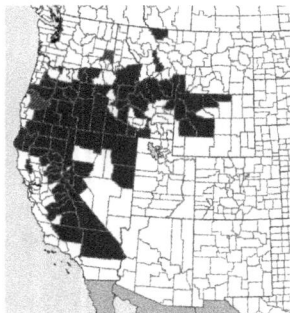

FAMILY Willow (Salicaceae)

FLOWERING May–June

HABITAT Wet areas in the mountains.

HABIT Native shrub to 5 m tall.

STEMS Slender, smooth or sparsely hairy, sometimes glaucous, yellow when young, becoming brownish black and shiny with age.

LEAVES Alternate, simple, lanceolate to oblanceolate, to 10 cm long, to 2.5 cm wide, pointed at the tip, tapering to the base, with or without teeth, silky-hairy when young, becoming smooth at maturity; stipules present.

FLOWERS Male and female flowers borne separately in separate spikes, the male spike to 4 cm long, the female to 6 cm long, both types of spikes appearing as the leaves are unfolding. **Sepals** absent. **Petals** absent. **Stamens** 2, the filaments hairy at the base. **Pistils:** Ovary superior, hairy.

FRUIT Capsules lanceoloid, silky-hairy, to 8 mm long, on stalks to 2 mm long.

NOTE Deer and elk may browse the young shoots of this plant.

WETLAND STATUS AW FACW | WMV FACW

FIELD NOTES Shrub; leaves smooth, to 20 mm wide; margins without teeth.

female catkin (ament)

Salix planifolia Pursh
DIAMOND-LEAF WILLOW

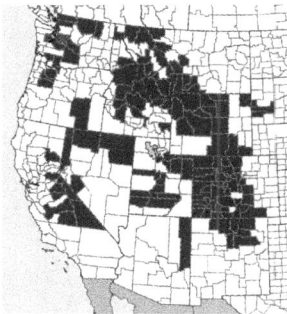

FAMILY Willow (Salicaceae)

FLOWERING May–June

HABITAT Along streams, moist meadows, fens.

HABIT Native shrub with numerous trunks, to 3 m tall. Twigs Reddish brown, smooth, shiny.

LEAVES Alternate, simple, elliptic to oblanceolate, to 6 cm long, to 20 mm wide, pointed or rounded at the tip, rounded or tapering to the base, paler on the lower surface, smooth at maturity, without teeth: leaf stalks to 8 mm long.

FLOWERS Crowded into slender spikes to 4 cm long, appearing as the leaves have begun to unfold; male spikes and female spikes borne separately; bracts black. **Sepals** absent. **Petals** absent. **Stamens** 2. **Pistils**: Ovary hairy.

FRUIT Capsules ovoid, with a long neck, hairy, to 4 mm long.

NOTE The fruits may be found between June and August.

WETLAND STATUS AW OBL | GP OBL | WMV OBL

> **FIELD NOTES** Inflorescence flat-topped; berries blue, covered by a whitish waxy coat. Leaves divided into 5–9 leaflets.

fruit

Sambucus nigra L.
BLACK ELDER

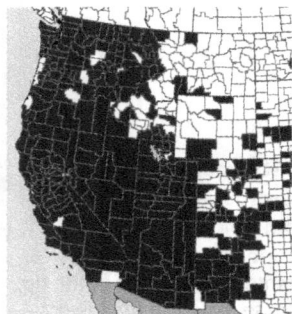

FAMILY Adoxaceae (Muskroot)

FLOWERING May–September

HABITAT Usually moist, open areas.

HABIT Native, many-stemmed shrub to 8 m tall, occasionally becoming tree-like.

STEMS Upright, soft and pithy, smooth, glaucous.

LEAVES Opposite, pinnately divided into 5–9 leaflets; leaflets lanceolate to narrowly ovate, to 15 cm long, to 5 cm wide, long-pointed at the tip, rounded at the asymmetrical base, toothed, smooth or slightly hairy.

FLOWERS Many in a flat-topped, compound umbel to 20 cm across; flower stalks short, slender, smooth. **Sepals** 5, green, united, very small. **Petals** 5, white or cream-colored, united, to 8 mm across, the lobes longer than the tube. **Stamens** 5, attached to the petals. **Pistils:** Ovary inferior, smooth.

FRUIT Berries spherical, to 6 mm in diameter, blue but covered with a whitish wax; nutlets wrinkled.

NOTE The berries can be eaten by humans, as well as by a variety of birds. Formerly placed in Caprifoliaceae.

SYNONYMS *Sambucus cerulea* Raf.

WETLAND STATUS AW FACU | GP FAC | WMV FAC

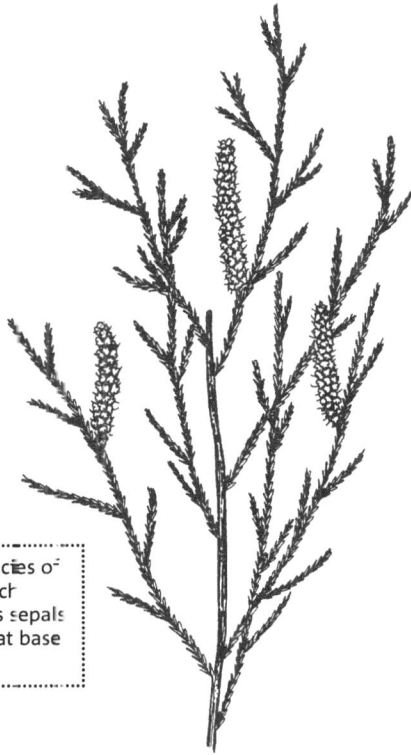

> **FIELD NOTES** All species of
> *Tamarix* are very much
> alike. *T. chinensis* has sepals
> more or less united at base
> and not toothed.

Tamarix chinensis Lour.
CHINESE TAMARISK

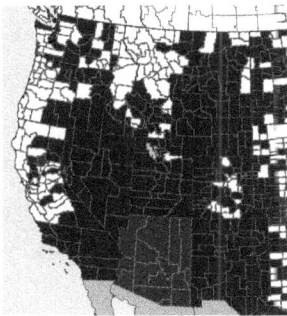

FAMILY Tamaricaceae (Tamarisk)
FLOWERING May–September
HABITAT Moist areas in the desert.
HABIT Introduced shrub or small tree to 6 m tall; bark reddish brown.
STEMS Erect, spreading, smooth.
LEAVES Alternate, scale-like, lanceolate to ovate-lanceolate, blue-green, sessile, to 4 mm long, smooth.
FLOWERS Many borne in racemes that are arranged in open panicles, each raceme to 5 cm long, each flower on a very short stalk; bracts about as long as or longer than the flower stalks. **Sepals** 5, green, free from each other, to 2 mm long, toothed along the edges. **Petals** 5, pink, free from each other, obovate, 1–3 mm long. **Stamens** 5, attached beneath a disk in the flower. **Pistils**: Ovary superior; **styles** 3.
FRUIT Capsule narrowed into a beak, longer than the petals, with tuft of soft bristles at the tip.
NOTE Native of Europe and Asia; formerly grown as an ornamental but very invasive along desert waterways.
SYNONYMS *Tamarix pentandra* Pallas.
WETLAND STATUS AW FAC | GP FACW | WMV FAC

GLOSSARY

Achene. A one-seeded, dry, indehiscent fruit with the seed coat not attached to the mature ovary wall.

Annual. Living only for one year.

Anther. The pollen-producing part of a stamen.

Arcuate. Curved.

auct. non Denotes a common misapplication or misinterpretation of a species name, i.e., a taxon that was identified erroneously as the named species.

Auriculate. Bearing ear-shaped lobes.

Awn. A bristle-like process.

Bearded. With a tuft of hairs.

Berry. A fruit with the seeds surrounded only by fleshy material.

Biennial. Living for two years

Bract. An accessory structure at the base of some flowers, usually appearing leaf-like.

Bracteole. A secondary bract.

Bractlet. A small bract.

Bristle. A stiff hair.

Calyx. All the sepals of a flower.

Capsule. A dry, dehiscent fruit splitting into 3 or more parts.

Cilia. Marginal hairs.

Ciliate. Bearing marginal hairs.

Clasping. Said of leaves that partially encircle the stem at the base.

Compressed. Flattened.

Connate. Union of like parts

Cordate. Heart-shaped.

Corm. An underground, tuber-like stem that stores food.

Corolla. All the petals of a flower.

Corymb. A type of flat-topped, branched inflores-cence.

Crest. A small ridge.

Cyme. A type of inflorescence in which the central flowers open first.

Cymose. In the form of a cyme.

Deciduous. Falling off.

Dehiscent. Splitting at maturity.

Diaphragmed. Divided by partitions.

Disk. The central group of flowers in the head of the Aster Family; a fleshy

growth that sometimes surrounds the ovary.

Drupe. A fruit with the seed surrounded by a hard, dry covering which, in turn, is surrounded by fleshy material.

Drupelet. A small drupe.

Ellipsoid. Referring to a solid object that is broadest at the middle, gradually tapering to both ends.

Elliptic. Broadest at the middle, gradually tapering to both ends.

Exfoliating. Stripping off.

Exserted. Projecting.

Fibrous. Referring to a cluster of slender roots, all with the same diameter.

Filament. The stalk of a stamen.

Fissured. Grooved.

Follicle. A dry, dehiscent fruit that splits along one side at maturity.

Furrowed. Grooved.

Glabrous. Smooth.

Glaucous. Having a bluish appearance.

Globose. Round.

Glume. A sterile scale found in grasses.

Hastate. Arrowhead-shaped, except that the basal lobes spread outward.

Hemispherical. Half-round.

Indehiscent. Not splitting open at maturity.

Inferior. Referring to the position of the ovary when it is below the point of attachment of the sepals and petals.

Inflorescence. A cluster of flowers.

Involute. Rolled up lengthwise.

Lanceolate. Lance-shaped; broadest near the base, gradually tapering to the narrower apex.

Lanceoloid. Referring to a solid object that is broadest near the base, gradually tapering to the narrower apex.

Latex. Milky sap.

Lemma. A fertile scale found in grasses.

Lenticel. A small opening on a stem.

Ligule. A structure on the inside at the junction of the leaf blade and leaf sheath.

Linear. Elongated and uniform in width throughout.

Mucronate. With a short point sticking out the tip.

Node. That place on a stem where leaves and buds arise.

Nutlet. A small nut.

Obconic. Reverse cone-shaped.

Oblanceolate. Reverse lance-shaped; broadest at the apex, gradually tapering to the narrower base.

Oblong. Broadest at the middle, and tapering to both ends, but broader than elliptic.

Oblongoid. Referring to a solid object that, in side view, is nearly the same width throughout.

Obovate. Broadly rounded at the apex, becoming narrowed below.

Obovoid. Referring to a solid object that is broadly rounded at the apex, becoming narrowed below.

Orbicular. Round.

Ovary. That part of the pistil that contains the ovules.

Ovate. Broadly rounded at the base, becoming narrowed above; broader than lanceolate.

Ovoid. Referring to a solid object that is broadly rounded at the base, becoming narrowed above.

Ovule. Immature seed.

Palmate. Divided radiately, like the fingers of a hand.

Panicle. An arrangement of flowers consisting of several racemes.

Pappus. Tufts of hairs attached to achenes of the aster family.

Peduncle. The stalk of an inflorescence.

Peltate. Attached at the middle.

Pendulous. Drooping.

Perennial. Living for 3 or more years.

Perforation. A circular opening.

Perianth. All the sepals and petals of a flower.

Perigynium. A sac-like structure enclosing the pistil and later the seed in *Carex*.

Petiole. Leafstalk.

Pinnate. Divided once along an elongated axis into distinct segments.

Pinnatifid. Divided nearly to the axis.

Pistil. The ovule-producing part of the flower.

Pith. The central, often soft, part of a stem.

Plumose. Feathery.

Prostrate. Lying flat on the ground.

Raceme. A grouping of flowers along an elongated axis where each flower has its own stalk.

Rachis. The axis of a flowering branch.

Ray. A flattened flower part in the aster family that is actually several petals fused together.

Receptacle. That part of the flower to which the sepals, petals, stamens, and pistils are usually attached.

Recurved. Curving downward.

Reflexed. Turned downward.

Reniform. Kidney-shaped.

Reticulate. Like a net.

Revolute. Turned under along the edges.

Rhizome. An underground, horizontal stem.

Ribbed. Bearing ridges.

Rosette. A cluster of leaves around the base of the plant.

Samara. An indehiscent winged fruit.

Scale. A tiny, leaf-like structure; the structure that subtends each flower in a grass or sedge.

Septate. With cross-walls.

Sessile. Without a stalk.

Setose. Bearing bristles.

Sheath. The base of a leaf that encircles the stem.

Sori. Structures found in ferns that contain the spore-producing sporangia.

Spadix. A fleshy axis in which flowers are embedded.

Spathe. A large bract subtending or sometimes enclosing a cluster of flowers.

Spatulate. Shaped like a spatula, broadest at the tip and tapering to the base.

Spike. A grouping of flowers along an elongated axis where each flower lacks a stalk.

Spikelet. A small spike.

Spinulose. Bearing small spines.

Sporangia. Structures that bear spores.

Spore. A microscopic reproductive body in ferns.

Spur. A slender, backward-pointing part of some flowers.

Stamen. The pollen-producing organ of a flower.

Stigma. The terminal part of a pistil.

Stipule. A green, often leaf-like, structure found at the base of some leaves.

Stolon. A horizontal stem lying on the surface of the soil.

Style. That part of the pistil between the ovary and the stigma.

Subglobose. Nearly round.

Succulent. Fleshy.

Superior. Referring to the position of the ovary when it is above the point of attachment of the sepals, petals, stamens, and pistils.

Sutures. Seams: areas where splitting occurs.

Tendril. A device, usually coiled, that enables some vines to climb.

Ternate. Divided into 3's.

Truncate. Cut straight across.

Tuber. A thickened, underground stem that stores food.

Tubercle. A wart-like process.

Umbel. A cluster of flowers in which the flower stalks arise from the same level.

Undulating. Wavy.

Valvate. Placed edge to edge.

Valve. The wing of the fruit in *Rumex*.

Whorl. An arrangement of 3 or more structures at a point on the stem.

A - Fruits

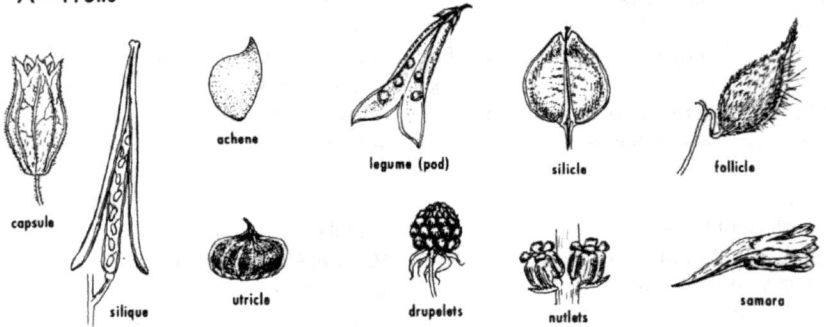

capsule

silique

achene

utricle

legume (pod)

drupelets

silicle

nutlets

follicle

samara

B - Roots and Stems

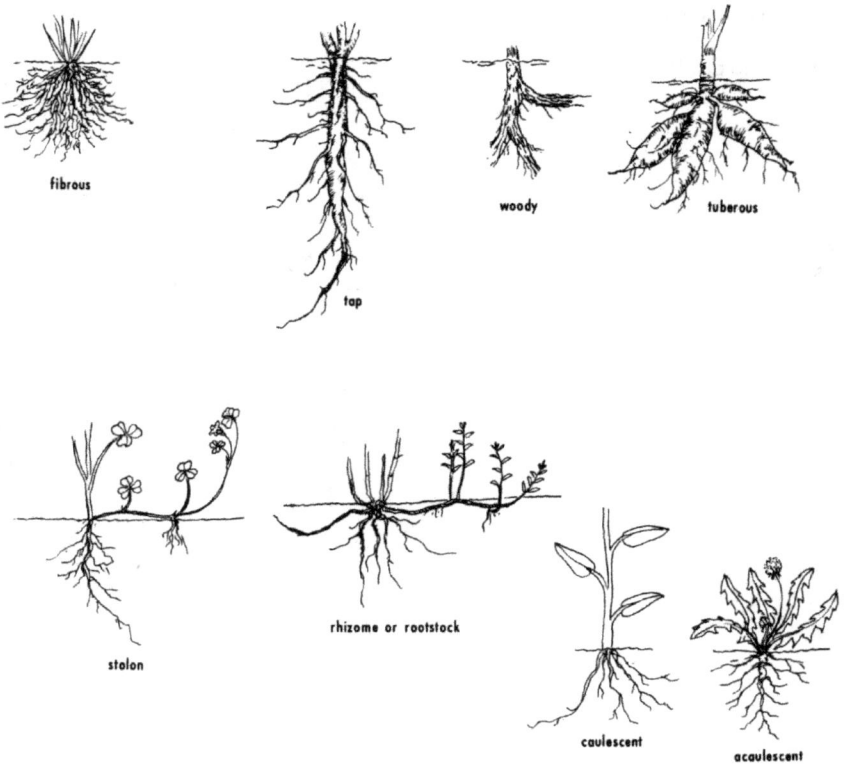

fibrous

tap

woody

tuberous

stolon

rhizome or rootstock

caulescent

acaulescent

Types of fruits, roots, and stems.

A - Simple and Compound Leaves

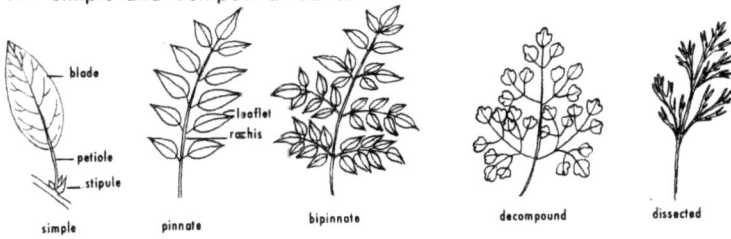

simple   pinnate   bipinnate   decompound   dissected

blade, petiole, stipule, leaflet, rachis

trifoliate   palmate

B - Margins

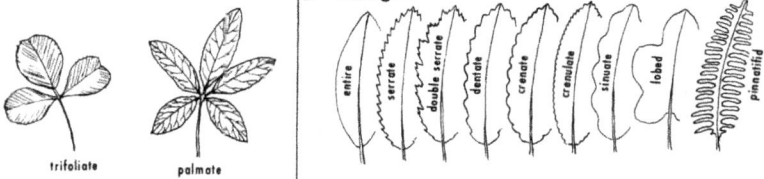

entire, serrate, double serrate, dentate, crenate, crenulate, sinuate, lobed, pinnatifid

C - Shapes

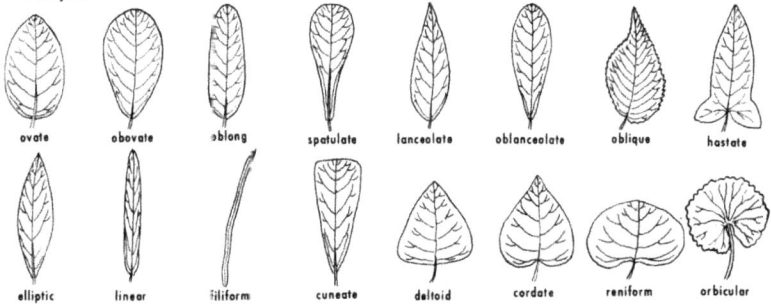

ovate   obovate   oblong   spatulate   lanceolate   oblanceolate   oblique   hastate

elliptic   linear   filiform   cuneate   deltoid   cordate   reniform   orbicular

D - Apices

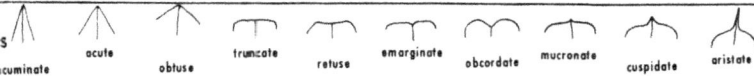

acuminate   acute   obtuse   truncate   retuse   emarginate   obcordate   mucronate   cuspidate   aristate

E - Bases

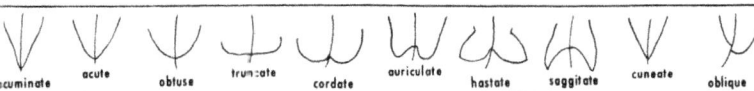

acuminate   acute   obtuse   truncate   cordate   auriculate   hastate   saggitate   cuneate   oblique

F - Attachments

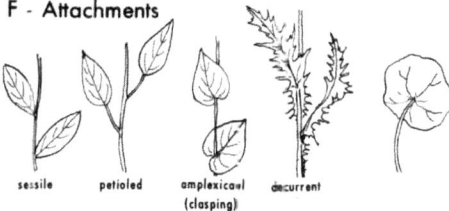

sessile   petioled   amplexicaul (clasping)   decurrent

G - Arrangements

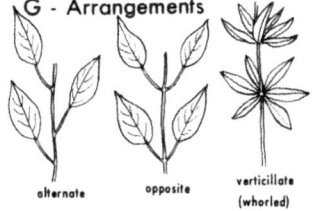

alternate   opposite   verticillate (whorled)

Leaf characters.

A - Flowers

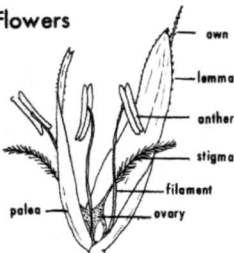

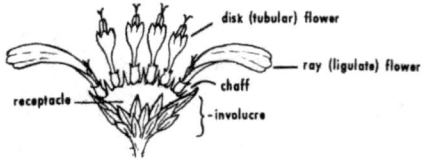

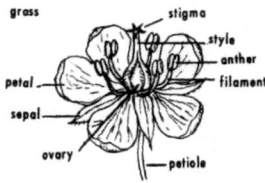

B - Inflorescences

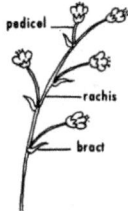

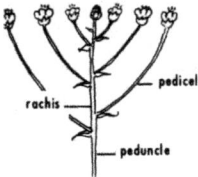

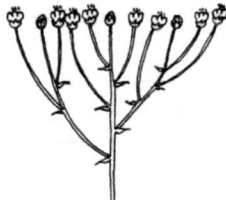

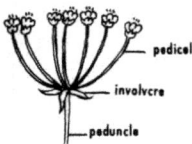

Types of flowers and inflorescences.

FERNS AND FERN-ALLIES

ATHYRIACEAE (LADY FERN FAMILY)

Athyrium cyclosorum Rupr. (Western Lady Fern) 54

BLECHNACEAE (CHAIN FERN FAMILY)

Woodwardia fimbriata J.E. Smith (Giant Chainfern) 61

ISOETACEAE (QUILLWORT FAMILY)

Isoetes echinospora Durieu (Spiny-spore Quillwort) 58

MARSILEACEAE (WATER-CLOVER FAMILY)

Marsilea vestita Hook. & Grev. (Hairy Water Fern) 59

OPHIOGLOSSACEAE (ADDER'S-TONGUE FAMILY)

Botrychium lanceolatum (S.G. Gmel.) Rupr. (Triangle Moonwort) 56
Botrychium lunaria (L.) Swartz (Moonwort) 57

PTERIDACEAE (MAIDENHAIR FERN FAMILY)

Adiantum capillus-veneris L. (Southern Maidenhair Fern) 53

SALVINIACEAE (WATER FERN FAMILY)

Azolla filiculoides Lam. (Large Mosquito Fern) 55
Salvinia molesta Mitchell (Karba-weed) 60

DICOTS

ADOXACEAE (MUSKROOT FAMILY)

Sambucus nigra L. (Black Elder) 322

AMARANTHACEAE (AMARANTH FAMILY)

Alternanthera philoxeroides (Mart.) Griseb. (Alligator-weed) 197
Amaranthus californicus (Moc.) S. Wats. (California Amaranth) 237
Bassia hyssopifolia (Pallas) Kuntz (Five-horn Smotherweed) 240
Bassia scoparia (L.) A. J. Scott (Mexican Summer-cypress) 241
Chenopodium rubrum L. (Coast-blite Goosefoot) 248
Dysphania ambrosioides (L.) Mosyakin & Clemants (American Wormseed) 257
Monolepis nuttalliana (J. A. Schultes) Greene (Nuttall's Poverty-weed) 269
Nitrophila occidentalis (Moq.) S. Wats. (Western Boraxweed) 223
Salicornia rubra A. Nels. (Red Saltwort) 225
Suaeda calceoliformis (Hook.) Moq. (Pursh Seepweed) 300
Suaeda occidentalis (S. Wats.) S. Wats. (Western Seepweed) 301

APIACEAE (CARROT FAMILY)

Angelica arguta Nutt. (Lyall's Angelica) 165
Berula erecta (Huds.) Coville (Cut-leaf Water Parsnip) 169
Cicuta douglasii (DC.) Coult. & Rose (Western Water-hemlock) 172
Conium maculatum L.(Poison-hemlock) 174

CAMPANULACEAE (BELLFLOWER FAMILY)

CARYOPHYLLACEAE (PINK FAMILY)

CERATOPHYLLACEAE (HORNWORT)

CONVOLVULACEAE (MORNING-GLORY)

ELATINACEAE (WATERWORT FAMILY)

ERICACEAE (HEATH FAMILY)

FABACEAE (PEA FAMILY)

FUMARIACEAE (FUMITORY FAMILY)

GENTIANACEAE (GENTIAN FAMILY)

GERANIACEAE (GERANIUM FAMILY)

GROSSULARIACEAE (CURRANT FAMILY)

Ranunculus eschscholtzii Schlecht. (Eschscholtz Buttercup) 290
Ranunculus flabellaris Raf. Yellow Water Buttercup) 43
Ranunculus flammula L. (Creeping Spearwort) 291
Ranunculus glaberrimus Hook. (Sagebrush Buttercup) 292
Ranunculus gmelinii DC. (Small Yellow Water Buttercup) 188
Ranunculus orthorhynchus Hook. (Straight-beak Buttercup) 189
Ranunculus pensylvanicus L. f. (Pennsylvania Buttercup) 293
Ranunculus repens L. (Creeping Buttercup) 190
Ranunculus subrigidus W. B. Drew (Pond Buttercup) 44
Trollius laxus Salisb. (American Globeflower) 304

ROSACEAE (ROSE FAMILY)

Comarum palustre L. (Marsh Cinquefoil) 173
Drymocallis glandulosa (Lindl.) Rydb. (Gland Cinquefoil) 176
Geum macrophyllum Willd. (Large-leaf Avens) 177
Potentilla anserina L. (Silverweed) 184
Potentilla glaucophylla Lehm. (Mountain-meadow Cinquefoil) 185
Potentilla gracilis Dougl. ex Hook. (Northwest Cinquefoil) 186
Potentilla plattensis Nutt. (Platte Cinquefoil) 187

SALICACEAE (WILLOW FAMILY)

Populus angustifolia James (Narrow-leaf Cottonwood) 313
Salix amygdaloides Anderss. (Peach-leaf Willow) 317
Salix lasiandra Benth. (Pacific Willow) 318
Salix lasiolepis Benth. (Arroyo Willow) 319
Salix lemmonii Bebb (Lemmon's Willow) 320
Salix planifolia Pursh (Diamond-leaf Willow) 321

SAURURACEAE (LIZARD'S-TAIL FAMILY)

Anemopsis californica (Nutt.) Hook. & Arn. (Yerba Mansa) 238

SAXIFRAGACEAE (SAXIFRAGE FAMILY)

Micranthes odontoloma (Piper) Heller (Brook Saxifrage) 266
Mitella pentandra Hook. (Five-point Bishop's-cap) 268

TAMARICACEAE (TAMARISK FAMILY)

Tamarix chinensis Lour. (Chinese Tamarisk) 323

VERBENACEAE (VERBENA FAMILY)

Phyla cuneifolia (Torr.) Greene (Wedge-leaf Frogfruit) 281

VIOLACEAE (VIOLET FAMILY)

Viola macloskeyi Lloyd (Small White Violet) 305
Viola nephrophylla Greene (Northern Bog Violet) 306 Violaceae (Violet)

VITACEAE (GRAPE FAMILY)

Parthenocissus inserta (Kerner) Fritsch (Thicket Creeper) 312

MONOCOTS

AGAVACEAE (AGAVE FAMILY)

Camassia quamash (Pursh) Greene (Common Camassia) 136

ALISMATACEAE (WATER-PLANTAIN FAMILY)

Alisma gramineum Lej. (Narrow-leaf Water-plantain) 134
Damasonium californicum Torr. (Small Fringed Water-plantain) 138
Sagittaria cuneata Sheldon (Northern Arrow-head) 155
Sagittaria latifolia Willd. (Broad-leaf Arrow-head) 156

ALLIACEAE (ONION FAMILY)

Allium validum S. Wats. (Tall Swamp Onion) 135

ARACEAE (ARUM FAMILY)

Lemna minor L. (Common duckweed) 24
Lemna trisulca L. (Star Duckweed) 25
Lemna valdiviana Philippi (Pale Duckweed) 26
Spirodela polyrhiza (L.) Schleic. (Greater Duckweed) 46

CYPERACEAE (SEDGE FAMILY)

Carex athrostachya Olney (Slender-beak Sedge) 101
Carex aurea Nutt. (Golden-fruit Sedge) 102
Carex bella Bailey (Showy Sedge) 103
Carex buxbaumii Wahlenb. (Brown Bog Sedge) 104
Carex canescens L. (Hoary Sedge) 105
Carex diandra Schrank (Lesser Panicled Sedge) 106
Carex douglasii Boott (Douglas' Sedge) 107
Carex kelloggii W. Boott (Kellogg's Sedge) 108
Carex lenticularis Michx. (Shore Sedge) 109
Carex leptalea Wahlenb. (Bristly-stalk Sedge) 110
Carex limosa L. (Mud Sedge) 111
Carex luzulina Olney (Wood-rush Sedge) 112
Carex microptera Mackenzie (Small-wing Sedge) 113
Carex nebrascensis Dewey (Nebraska Sedge) 114
Carex praegracilis W. Boott (Clustered Field Sedge) 115
Carex raynoldsii Dewey (Raynolds' Sedge) 116
Carex saxatilis L. (Russet Sedge) 117
Carex scopulorum Holm (Holm's Rocky Mountain Sedge) 118
Carex simulata Mackenzie (Short-beak Sedge) 119
Carex utriculata Boott (Beaked Sedge) 120
Carex vesicaria L. (Inflated Sedge) 121
Carex viridula Michx. (Little Green Sedge) 122
Cyperus squarrosus L. (Awned Flatsedge) 123
Eleocharis palustris (L.) Roemer & J. A. Schultes (Creeping Spikerush) 124
Eleocharis quinqueflora (F. X. Hartmann) Schwarz (Few-flower Spikerush) 125
Eleocharis rostellata (Torr.) Torr. (Beaked Spikerush) 126
Eriophorum gracile W. D. J. Koch (Slender Cotton-grass) 127
Eriophorum scheuchzeri Hoppe (Scheuchzer's Cotton-grass) 128

PONTEDERIACEAE (PICKERELWEED FAMILY)

POTAMOGETONACEAE (PONDWEED FAMILY)

RUPPIACEAE (DITCHGRASS FAMILY)

Ruppia maritima L. (Widgeon-grass) 45

RUSCACEAE (BUTCHER'S-BROOM FAMILY)

Maianthemum stellatum (L.) Link (Starry False-Solomon's-seal) 152

TYPHACEAE (CAT-TAIL FAMILY)

Sparganium natans L. (Small Burreed) 158
Typha angustifolia L. (Narrow-leaf Cat-tail) 162 Typhaceae (Cat-tail)
Typha domingensis Pers. (Southern Cat-tail) 162 Typhaceae (Cat-tail)
Typha latifolia L. (Broad-leaf Cat-tail) 163 Typhaceae (Cat-tail)
Valeriana occidentalis Heller (Western Valerian)195 Valerianaceae (Valerian)

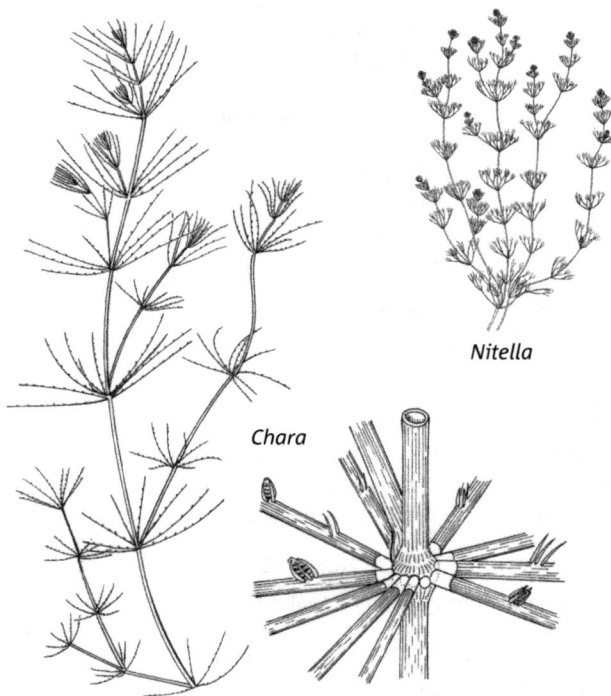

Nitella

Chara

ABOVE Muskgrass or Stonewort (*Chara* spp., Characeae) is a macro-algae easily confused with vascular aquatic plants. Plants of *Chara* are gray-green, and have a gritty texture and strong musky odor when rubbed. When dried, *Chara* will turn whitish due to calcium deposits. Although not having true roots, *Chara* will loosely attach itself to the bottom of lakes and ponds via rhizoids (thread-like structures), and will sometimes form extensive colonies in shallow water, especially where calciumrich. More than 30 species of *Chara* have been identified in the United States. Another stonewort, ***Nitella***, is similar but plants have no skunky odor, and stems and branches are typically bright green and smooth to the touch.

NOTE Synonyms listed in *italics.*

Aconitum columbianum 235
Adiantum capillus-veneris 53
Agoseris aurantiaca 236
Agropyron trachycaulum 79
Agrostis
  exarata 63
  idahoensis 64
Alisma gramineum 134
Alkali Bluegrass 92
Alkali Buttercup 259
Alkali Cordgrass 96
Alkali Muhly 87
Alkali Sacaton 97
Alligator-weed 197
Allium validum 135
Alnus incana 307
Alopecurus
  aequalis 65
  alpinus 66
  magellanicus 66
Alpine Bog Laurel 311
Alternanthera philoxeroides 197
Amaranthus californicus 237
American Alpine Speedwell 231
American Bistort 242
American Globeflower 304
American Licorice 178
American Primrose 285
American Purple Vetch 196
American Sloughgrass 69
American Speedwell 229
American White Water-lily 32
American Winter-cress 168
American Wormseed 257
Anemopsis californica 238
Angelica arguta 165
Annual Hairgrass 75
Antennaria corymbosa 239
Anthoxanthum hirtum 67
Arabis drummondii 244
Arctic Pearlwort 224
Arizona Centaury 232
Arnica
  amplexicaulis 199
  chamissonis 198
  xdiversifolia 203
  lanceolata 199
  latifolia 200
  longifolia 201
  mollis 202
  ovata 203

Arrow-leaf Groundsel 297
Arrow-leaf Sweet Coltsfoot 280
Arrow-weed 284
Arroyo Willow 319
Arundo donax 68
Aster
  chilensis 302
  hesperius 303
Astragalus
  agrestis 166
  canadensis 167
Athyrium
  cyclosorum 54
  filix-femina 54
Autumnal Water-starwort 17
Awned Flatsedge 123
Azolla filiculoides 55

Baccharis glutinosa 308
Barbarea orthoceras 168
Basin Wild-rye 85
Bassia
  hyssopifolia 240
  scoparia 241
Beaked Sedge 120
Beaked Spikerush 126
Bearded Sprangletop 77
Beckmannia syzigachne 69
Bergia texana 204
Berula
  erecta 169
  pusilla 169
Betula
  fontinalis 310
  glandulosa 309
  occidentalis 310
Bistorta
  bistortoides 242
  vivipara 243
Black Elder 322
Blitum nuttallianum 269
Boechera stricta 244
Bog St. John's-wort 211
Botrychium
  lanceolatum 56
  lunaria 57
Brasenia schreberi 16
Brassbuttons 251
Brazilian-waterweed 20
Brewer's Bitter-cress 170
Bristly Mousetail 270
Bristly-stalk Sedge 110
Broad-leaf Arrow-head 156

www.ingramcontent.com/pod-product-compliance
Lightning Source LLC
Chambersburg PA
CBHW051711020426
42333CB00014B/935